Indian Shield

Indian Shield

Precambrian Evolution and Phanerozoic Reconstitution

A.B. Roy

Ritesh Purohit

ELSEVIER

Elsevier
Radarweg 29, PO Box 211, 1000 AE Amsterdam, Netherlands
The Boulevard, Langford Lane, Kidlington, Oxford OX5 1GB, United Kingdom
50 Hampshire Street, 5th Floor, Cambridge, MA 02139, United States

Notices

Knowledge and best practice in this field are constantly changing. As new research and experience broaden our understanding, changes in research methods, professional practices, or medical treatment may become necessary.

Practitioners and researchers must always rely on their own experience and knowledge in evaluating and using any information, methods, compounds, or experiments described herein. In using such information or methods they should be mindful of their own safety and the safety of others, including parties for whom they have a professional responsibility.

To the fullest extent of the law, neither the Publisher nor the authors, contributors, or editors, assume any liability for any injury and/or damage to persons or property as a matter of products liability, negligence or otherwise, or from any use or operation of any methods, products, instructions, or ideas contained in the material herein.

Library of Congress Cataloging-in-Publication Data
A catalog record for this book is available from the Library of Congress

British Library Cataloguing-in-Publication Data
A catalogue record for this book is available from the British Library

ISBN: 978-0-12-809839-4

Publisher: Candice Janco
Acquisition Editor: Marisa LaFleur
Editorial Project Manager: Tasha Frank
Production Project Manager: Nilesh Kumar Shah
Cover Designer: Christian Bilbow

Typeset by SPi Global, India

Contents

PART 6 QUATERNARY GEOLOGY AND SEISMICITY

PART 7 EPILOGUE

Prologue

The book *Indian Shield: Precambrian Evolution and Phanerozoic Reconstitution* depicts precisely what is indicated in the title itself. Whereas in the first part we tried to provide a comprehensive geological account of the evolution of the Precambrian Indian Shield, the second part discusses the different post-Precambrian processes that caused reconstitution of the Indian Shield changing its pristine shape, size, and constitution into that of the present-day Indian Subcontinent, a geomorphic terrain comprising countries like Bangladesh, Bhutan, India, Nepal, and Pakistan. The complete geological history of the changeover is as spectacular as the temporal span of the two events. The evolutionary history of the Indian Shield covered a protracted span of about 3000 million years. By contrast, the period of reconstitution seems to have taken a little over 500 million years.

This book discusses the regional geology of the terrain in terms of the history of evolution of the Crust, describing how the Precambrian Shield evolved from a stable continental region to a tectonically unstable zone marked by frequent high-intensity earthquakes in a totally continental setting. It is a comprehensive and well-illustrated readable account of the history of growth and evolution of the Indian Subcontinent.

The strength of the book is the illustrations, both line drawings and photographs/images, used to supplement the text. Many illustrations mainly the photographs and images were drawn from different parts of the Indian Subcontinent in order to familiarize readers with the rocks and features of the terrain. This, we believed, would enhance the understanding of the subject we dealt with in the book.

An important point we would like to focus is that many terms and nomenclatures used in geology are derived from the names of common objects or features we are familiar with like 'crust', 'mantle', 'core', 'plate', 'shield', 'platform', and 'trap'. In order to avoid confusion in the mind of common readers whenever such common nouns are used for specific geological features, these are written with capitalized first letter. In addition, the text has been supplemented with 'boxes', providing additional information that is usually provided in footnotes.

Scripting a book like this is truly an arduous task. But we feel fortunate to have received help from innumerable friends, colleagues, acquaintances, and many others whom we requested for support. But before anyone else, we must thank Mukesh Suthar for drafting a large number of line drawings and figures, with utmost fineness and care, making them scientifically very useful documents. We also extend our gratitude to all those who have extended their helping hand in various ways in our book-writing project. The list is long that includes Manoranjan Mohanty, Sisir Mondal, H. N. Bhattacharya, Sarbani Patranabis-Deb, Dilip Saha, A. N. Sarkar, Sanjib Sarkar, Alokesh Chatterjee, Erfan Mondal, N. V. Chalapathi Rao, P. R. Golani, Surjaram Jakhar, Harsh Bhu, Indrani Roy, L. S. Chamyal, Om Bhargava, Rajneesh Bhutani, Debjani Roy, Indrajit Roy, J. Ganguly, V. S. Kale, Ikramul Hasan Sakil, Y. Sheedhar Murthy, Saibal Gupta, M. Jayananda, Adhir Basu, Raymond Duraiswami, Kamal Kant Sharma, N. K. Chauhan, S. K. Haldar, Arun Vyas, G. R. Ravindra Kumar, T. R. K. Chetty, C. Leelanandam, R. H. Sawkar, Jonali Medhi, Pranjali Kakoti, Bidyananda M., S. A. Sameeni, Sukanta Dey, Santosh Kumar, T. K. Biswal, Ashutosh Pandey, Mukund Sharma, Anupendu Chatterjee, Asima Saikia, Surya Prakash Singh, Vinod Singh, Sadhana M Chatterjee, B. P. Singh, and Rachit Parihar. Our sincere apologies to those whom we forgot to acknowledge for their help.

And finally, we individually acknowledge the personal helps and comfort provided by our dear family members during the process of scripting the book.

Ashit Baran Roy
Ritesh Purohit

PART

1

INDIAN SHIELD: CONCEPT AND PERSPECTIVE

CHAPTER

CONCEPT OF INDIAN SHIELD: EVOLUTION AND RECONSTITUTION

1

1.1 THE CONCEPT OF INDIAN SHIELD, DEFINITION AND EXTENT

A Shield is defined as a large area of exposed Precambrian terrane made dominantly of igneous and metamorphic rocks that remained tectonically stable since the youngest Precambrian. The oldest surviving rocks in the Shield areas of the world are generally as old as 3.5 billion years or even more. The youngest rocks on the other hand are those that evolved before the onset of Cambrian. Barring some minor cratonic deformations, the Shield area rocks have presumably remained free from any major tectonic/orogenic deformation during the latter 'post-Precambrian' Phanerozoic Eon.

BOX 1.1

The term Shield used here is the English translation of the original German word 'Shild' by H.B.C. Sollas (Suess, 1901). The Canadian Shield is considered a model example of Shield in the world today. The term 'Shield' itself is derived from the outline of the spatial extent of the Canadian Shield, which is quite similar to the shape of armour used by ancient warriors to protect their bodies.

Traditionally, the triangular-shaped Precambrian terrane of the southern Indian Peninsula is described as a Shield. The term 'Indian Shield' that receives wide reference in literatures does not truly fulfil the shape criteria implied in the definition of Shield. But even assuming that the shape criterion needs not be a necessary constraint, the available geological information, however, suggests that the Precambrian terrane of Indian Peninsula was once a part of much larger crustal block that evolved as a Shield like the Canadian or some other Shield areas of the world. The concept of 'Greater India' appeared to have emerged from such an understanding about nine decades ago (Argand, 1924).

BOX 1.2

The term 'terrane' is used following the definition given in Wikipedia, the free encyclopaedia, to simply describe a series of related rock formations or an area having a preponderance of a particular rock or rock groups. On the other hand, the term 'terrain' is used for a geographical ground or a piece of ground, especially with reference to its physical character.

The pristine size of this Precambrian crustal block that evolved as Indian Shield is difficult to ascertain because of the fact that a considerable part of it in the north has undergone extensive reconstitution during the 'Continent-Continent' collision that led to the growth of the Himalayas during the late Cenozoic time. The geological and geophysical data from the Himalayas also provide evidence that much of its edifice is made of components sliced off from the Indian Shield (Qureshy, 1969; Qureshy and Kumar, 1992; Warsi and Molnar, 1977). The concept is ingrained in the expression

Indian Shield. https://doi.org/10.1016/B978-0-12-809839-4.00001-1

'extrapeninsular rocks' used for all the ancient Precambrian rocks that constitute the youngest mountain belt, the Himalayas by the late 18th-century to early 19th-century geologists of the Geological Survey of India (Medlicott and Blandford, 1879–1881).

Several attempts have been made trying to reconstruct at least partially the true spatial extent (or the size) of the Indian Shield in its northern part. Though they differ in detail, the central strand in all these models is that the pristine Indian Precambrian crustal block constituting the Indian Shield had an extension varying between 500 and 950 km (Ali and Aitchison, 2008) in the north of the Himalayan Frontal Thrust (the southernmost base of the Himalayas, Valdiya, 1998). These estimates are compatible with some of the suggested geodynamic and geophysically derived models depicting the subducted Indian Lithosphere beneath Tibet and the estimates of the Himalayan shortening (DeCelles et al., 2002; Dewey et al., 1989; Le Fort, 1975; Molnar and Tapponnier, 1975; Searle et al., 1987; Virdi, 1987; Valdiya, 1984; Warsi and Molnar, 1977).

Apart from its reconstitution along the northern part, the slicing of the Indian Shield due to the separation of Antarctica in the southeast and Madagascar and Seychelles islands in the southwest during the late Phanerozoic has also added further complications in reconstructing the pristine size and shape of the Indian Shield prior to its decimation during the late Phanerozoic (Roy, 2004).

There are studies suggesting Indo-Antarctic connection based on the correlation of granulite belts of the two regions (Yoshida et al., 1992; Sen et al., 1995; Sengupta et al., 1999; Dasgupta and Sengupta, 2003; Bhadra et al., 2004; Gupta et al., 2005; Kelly et al., 2002). Such a correlation implies that the boundary of the Indian Shield does not end at the eastern margin of the Eastern Ghats Granulite Belt but extends far into East Antarctica. However, because of the lack of geological information, it is impossible even to guess what could even vaguely be the actual eastern boundary of the granulite belt in Antarctica (placing Antarctica against the present-day India).

Like the granulite belt of Eastern Ghats Granulite Belt and its continuity into the East Antarctica, the Southern Granulite Belt along with Sri Lanka and Madagascar formed a continuous Precambrian terrane in the south and southwest of Peninsular India (Harris et al., 1994; Jayananda and Peucat, 1996; Kröner et al., 1991; Radhakrishna et al., 1994, 1999; Storey et al., 1995; Torsvik et al., 2000; Veeraswamy and Raval, 2004; Yoshida et al., 1992). This suggests the extension of 'Greater Indian Shield' much beyond the boundary of the southern and southeastern peninsular India.

1.2 FROM INDIAN SHIELD TO INDIAN SUBCONTINENT: STORY OF PHANEROZOIC RECONSTITUTION

The history of geological evolution of the Indian Shield is quite long and complex and, broadly speaking, took place in two stages. The first stage covered the entire Precambrian, which was the period of its growth and eventual cratonization. In the second stage of its evolution during the 'post-Precambrian' Phanerozoic phase, the cratonized Indian Shield underwent repeated reconstitution, finally shaping into the present-day 'Indian continental block', known as Indian Subcontinent.

Several Phanerozoic (post-Precambrian) geological events not only facilitated in changing the shape and size of the pristine Indian Shield but also have grossly altered its geological, geomorphological, and geophysical characters (Roy, 2004). These are the following:

1. Pan-African magmatism during 550 ± 50 Ma
2. Lower Palaeozoic 'shelf-sea'-type basin development

3. Late Palaeozoic opening of the Gondwana basins
4. Jurassic break-up of Gondwanaland at ~165 Ma
5. Plume impingements during Cretaceous-Eocene
6. Himalayan collision and related orogeny, and
7. Postcollision tectonics and seismicity

The major Crust-building events that caused break-up and reconstitution of the pristine Precambrian Crust during the Phanerozoic are linked with three global bench-mark events (Roy, 2004): (i) Gondwana break-up at ~165 Ma ago, (ii) Plume impingements under the Indian Lithosphere during Cretaceous and Eocene (Fig. 1.1), and (iii) Himalayan-collision-related orogeny between ~55 and ~45 Ma (Fig. 1.2). The net result is the emergence of the landmass forming a distinctive geographic entity, the Indian Subcontinent. The almost entire Subcontinent's boundary in the north is bordered by the Himalayas and its associated branches in the northwest and in the southeast (Fig. 1.3).

Apart from the fragmentation of the pristine Indian Shield to the shape and size of Indian Subcontinent (Fig. 1.3), the different (post-Precambrian) Phanerozoic events have also grossly altered the geological, geomorphic, and geophysical characters of the terrain. The oldest reconstitution-related event was in the form of intrusion of alkali granite and syenite and syenodiorite having 'A'-type anorogenic character. This phase of magmatic activity centring around 500 ± 50 Ma broadly shows a near peripheral concentration virtually defining the boundaries of the Indian Subcontinent (partly including

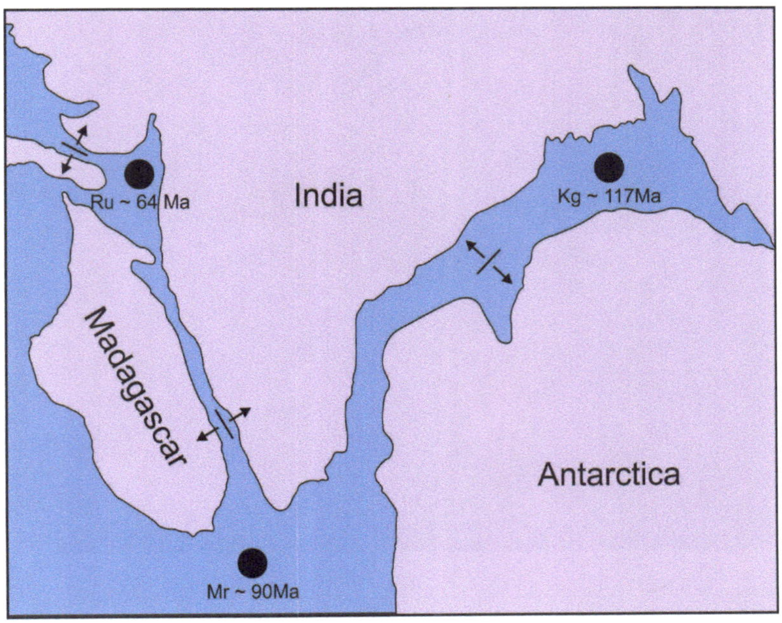

FIG. 1.1

Fragmentation of the Precambrian crustal block constituted of India, Antarctica, and Madagascar during Plume outbursts. Possible position of Plumes, indicating the age of outburst. Kg, Kerguelen; M, Marion; Ru, Reunion.

Reproduced from Raval and Veeraswamy (2003) with permission.

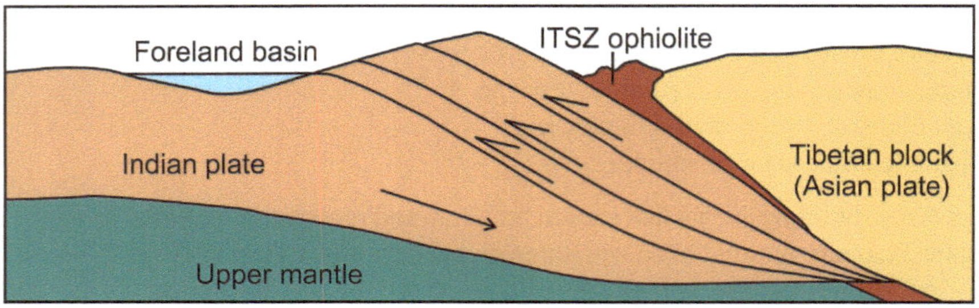

FIG. 1.2

Schematic illustration showing evolution of Himalaya through upthrusting of slices of the northern part of the Indian Shield during the process of 'continental collision'. ITSZ, Indus-Tsangpo Suture Zone.

FIG. 1.3

The satellite imagery indicating three physical divisions of the Indian Subcontinent.

Modified from Google Earth image.

Madagascar) is popularly known as Pan-African events in India (Roy, 1999a, 1999b, 2004; Valdiya, 1993). Some Shear Zones have reportedly developed in the Southern Granulite Belt in the Kerala region of southern Peninsula.

Reports of Cambrian fossil-bearing beds in the Salt Range region are indications of the earliest marine shelf deposits in the northwestern part of the Subcontinent. Scattered occurrence of lower Palaeozoic fossil-bearing marine sediments also occurs in the northwest of the Kashmir Valley, which continue for some distance into the Liddar Valley around Pahelgam east of Srinagar and in the Spiti Valley in the Tethys Himalayas. Continuity of these fossil-bearing lower Palaeozoic rocks is not traceable further east of this. Records of continuity of sedimentation in these basins during the Ordovician and Silurian are, however, quite equivocal, although both the periods are represented by fossils.

After a break in sedimentation of about 200 Ma between the Ordovician and the early Permian, the deposition of sediments in the Indian Subcontinent started with the formation of tillites and glacial boulder beds in close association with Permian marine beds. This was accompanied by the deposition of fluvial and fluviolacustrine sediments in linear intracontinental rift basins. These sediments, along with the intercalated plant remains that ultimately turned into coal seams, constitute the Gondwana Supergroup.

The Gondwana sedimentation, which began in the Permian, continued until the Lower Jurassic. The next major global event that had a major effect on the Indian continental block was the break-up of Gondwana at around 165 Ma. The initial separation resulted in marine incursions and deposition of sediments in the northwestern Rajasthan and in the Kachchh region of Gujarat along WNW-ESE trending rift basins. The deposition of continental sediments, which had earlier stopped in different Gondwana basins before the Lower Jurassic, was resumed at least in certain cases. The Gondwana break-up event was also responsible for the development of arrays of fracture systems in the Indian continental block. Geomorphologically expressed as Lineaments, these features developed either as a new set of ruptures or as a reactivation of old tectonic structures. The newly emerged fracture system helped in shaping the geomorphologic and geophysical character of the Indian Crust in a variety of ways.

As the Indian continental block along with attached parts of Madagascar, Seychelles, and Antarctica moved northwards following the dismemberment of the Gondwana Supercontinent, it was affected successively by the outbursts of four 'Plume heads' centred at Marion, Reunion, Crozet, and Kerguelen Islands. The manifestations of the Crozet Plume outbursts are virtually unknown. The Marion Plume outbursts resulted in the separation of Madagascar from the Indian continental block at around 88–90 Ma. Evidence for this comes from the occurrence of acid and mafic rocks of similar age in different parts of central and north Kerala, St. Mary's Island off the Karnataka coast, and also Madagascar. The Rajmahal Traps, the Sylhet Traps, and also those that underlie the Bengal Basin are the manifestations of Kerguelen Plume activities during 117 ± 1 Ma. The Plume that triggered the successive separation of Antarctica from the Indian continental block induced Cretaceous marine ingress, both in the southeastern part of the Peninsula and in the southeast of the Bengal Basin. Record of another Plume outburst around 80/82 Ma linked with the 85° East Ridge in Bay of Bengal has been correlated with opening of the Bengal Basin (Roy and Chatterjee, 2015). Plume outbursts have grossly affected the shaping of the Indian Shield as indicated in Fig. 1.1, apart from the opening of new basins.

Deposition of shelf facies sediments was resumed during the Carboniferous in zones of marine incursions along intracratonic rift basins that had developed in the Salt Range and in the Kashmir region. The deposition in some basins continued until the Triassic. There are records of contemporary volcanism in the Pir Panjal Range and in other places in the east. Both marine and continental sediments can be correlated with those of the Gondwana deposits which occur in parts of Nepal and Sikkim Himalayas. Continuous sedimentation from the Cambrian to the Eocene with a number of breaks is recorded in the Tethys belt. Fossil records from different parts of this belt indicate that the extent of these breaks was not of uniform duration everywhere. The closure of the Tethys Ocean by the end of Eocene caused a brief pause in sedimentation, which was resumed by around mid-Miocene in two different basins. One in the north opened as a major intermontane (Back Arc) basin in the Suture Zone, leading to the deposition of the Indus Group. The Siwalik Group was deposited in the southern foreland basin that developed in front of the rising Himalayas from around 18 Ma.

Most remarkable reconstitution of the Indian Shield is noted along the northern margin when the Himalayas emerged as grand collisional mountain incorporating considerable part of Indian Shield elements (Figs. 1.2 and 1.3).

The Himalayas represent a classic example of Continent-Continent collision. Palaeomagnetic data indicate that the initiation of the continental collision had started at equatorial latitudes, resulting in the progressive suturing from the Paleocene in the northwestern Himalayas to the Eocene in the eastern Himalayas. Continued convergence and indentation of the Indian continental block with southern Asia (or Tibet) up to the Early Miocene resulted in the doubling of the crustal thickness over a large region in the Himalayas, Pamir-Hindu Kush, and Tibet. The total area of thickened Crust may account for about 2000 km of crustal shortening in the entire orogen. As to the origin of the Himalayan Arc, palaeomagnetic observations seem to favour a steady-state model of formation of the arcuate bending of the mountain ranges due to late Cenozoic anticlockwise rotational underthrusting of the Indian continental block beneath the Tibetan Plateau after the latest Miocene.

A unique feature of the Himalayas is their crustal thickness, which rises from about 35 km in the Indo-Gangetic Alluvial Plain to between 65 and 80 km over the Higher Himalayas. The last Himalayan upheaval at around 1.7 Ma caused shifting of depocentres to the south, to build up the flood plains of the Indo-Gangetic Alluvial Plain formed over the linear zone of subsidence south of the Himalayan Front. While the closure of Tethys marked the end of sedimentation in the north, marine shelf sedimentation continued both along the eastern and western margins of the Indian continental block, in the Naga Hills and Arakan Yoma in the east and the Sulaiman and Kirthar Ranges in the west. Sedimentation in these basins, which began in the Eocene, continued at least until the Oligocene. The earliest Himalayan deformation coincided with the final closure of Tethys at around 50 Ma, affecting the rocks on either side of the Suture Zone. There was a distinct southward polarity of deformation across the Tethyan region to the Higher Himalayan Crystalline Complex. A series of south-directed recumbent folds and thrusts was produced in the Higher Himalayas, resulting in thickening of the Crust with the attendant Barrovian-type metamorphism, anatexis, and generation of leucogranites. The southward transmission of the thrust nappes by the Main Central Thrust continued until around 22 Ma (Atlos et al., 2004). This was also the time when the Barrovian metamorphic isograds underwent inversion. Almost simultaneously with the piling up of the fold-thrust nappes in the Higher Himalayas, the Indus Molasse Basin in the north and the Siwalik Molasse Basin in the south developed as rapidly subsiding troughs. As in a model of 'piggyback' thrusting, the southward transmission of the fold-thrust nappes, which were initially along the Main Central Thrust, was later shifted to the Main Boundary Thrust in the south. The Himalayan Frontal Thrust that overrides the recent sediments was the latest dislocation in the form of an 'upthrust', and evolved during the Himalayan tectonism (Valdiya, 1998).

The Quaternary geology, which began with the waning phase of the Siwalik sedimentation, came to an end with the most recent upheaval of the Himalayas. The period is known for unique sedimentation process, geomorphic changes marking attainment of great heights in some locales, active tectonics, and associated seismicity. The subsiding depocentres had by then shifted to the south of the Himalayan domain and ultimately evolved as the huge flat plain known as the Indo-Gangetic Alluvial Plain. Geographically, the Indo-Gangetic Alluvial Plain also includes the narrow basin of the Brahmaputra River in the east and in the Thar Desert (along with the North Gujarat Plain) in the west. The alluvial sediments over the entire Indo-Gangetic Alluvial Plain belt range in thickness between 400 and 800 m, with a maximum of about 6 km along the edge of the Himalayas. The belt is divided into a number of subbasins by several submerged ridges (basement highs) lying across it. Quaternary sediments outside the Indo-Gangetic Alluvial Plain occur in the Narmada and Tapti Basins in Peninsular India along the western coastlines, and in the Bengal Basin in the east (Roy and Chatterjee, 2015). Thick laterite formations (some of which contain rich bauxite deposits) were produced during this time in parts of Central India, Eastern Ghats, and Konkan coasts in the Western Ghats. The Thar Desert in the east of Indus Basin had a fluvial prehistory, and its formation is linked with the establishment of the monsoon system over the Subcontinent by the mid-Pleistocene, with the high rising Aravalli Mountains producing the rain shadow zone to its west. The saline lakes that occur throughout the entire desert region were formed by the segmentation and blocking of river channels due to Neotectonic movements (Roy, 1999b). The Quaternary tectonic movements caused spectacular geomorphic changes in the entire Subcontinent, primarily through the movements along the fault-bounded blocks. Movement along rigid blocks caused pronounced geomorphic changes like formation of horst-type mountain (Mt. Abu and Nilgiri Hills), rift grabens (Narmada Basin), drainage disorganization, migration and extinction of Vedic Saraswati (Roy and Jakhar, 2001) and the repeated changes in the course of the rivers in the Bengal Basin, and many other features attributed to neotectonism. Movements are continuing. The spectacular landform changes during this last period of geological history and describes archetypal examples of tectonic geomorphology *vis-à-vis* palaeoseismicity. On the other hand, the Rann of Kutch in Northern Gujarat has the records of repeated regional uplift and subsidence during historical times (Roy et al., 2013). The development of the Ganga-Brahmaputra-Meghna Delta Complex (also known as the Sundarban Delta) is a very important geological landform feature, which evolved in three stages of tectonically influenced deltaic sedimentation processes during the Late Pleistocene and the Holocene.

In summary, the Crust of the Indian Subcontinent bears records of the evolutionary history of the Precambrian Shield, which evolved (geophysical speaking) as a 'Stable Continental Region', and gradually turned into a tectonically unstable zone marked by frequent high-intensity earthquake.

REFERENCES

Ali, J.R., Aitchison, J.C., 2008. Gondwana to Asia: plate tectonics, palaeogeography and biological connectivity of the Indian sub-continent from the Middle Jurassic through latest Eocene (166–35 Ma). Earth-Sci. Rev. 88, 145–166.

Argand, E., 1924. La tectonique de l'Asie. Proc. Int. Geol. Cong. 7, 171–372.

Atlos, E.J., Dubey, C.S., Harrison, T.M., Edwards, M.A., 2004. Late Miocene movement within the Himalayan Main Central Thrust shear zone, Sikkim, north-east India. J. Metamorphic Geol. 22, 207–226. https://doi.org/10.1111/j.1525-1314.2004.00509.x.

Bhadra, S., Gupta, S., Banerjee, M., 2004. Structural evolution across the Eastern Ghats Mobile Belt–Bastar cratonboundary, India: hot over cold thrusting in an ancient collision zone. J. Struct. Geol. 26, 233–245.

Dasgupta, S., Sengupta, P., 2003. Indo-Antarctic correlation: a perspective from the Eastern Ghats granulite belt, India. Geol. Soc. Lond. Spec. Publ. 206, 131–143.

DeCelles, P.G., Robinson, D.M., Zandt, G., 2002. Implications of the shortening in the Himalayan fold-thrust belt for uplift of the Tibetan Plateau. Tectonics 21, 12-1–12-25. https://doi.org/10.1029/2001tc001322.

Dewey, J.F., Cande, S., Pitman III, E.C., 1989. Tectonic evolution of India/Eurasia collision zone. Eclogae Geol. Helv. 82, 717–734.

Gupta, S., Nanda, J., Mukherjee, S.K., Santra, M., 2005. Alkaline magmatism versus collision tectonics in the eastern ghats belt, India: constraints from structural studies in the koraput complex. Gondwana Res. 8, 1–17.

Harris, N.B.W., Santosh, M., Taylor, P., 1994. Crustal evolution in South India: constraints from Nd isotopes. J. Geol. 102, 139–150.

Jayananda, M., Peucat, J.J., 1996. Geochronological framework of Southern India. In: Santosh, Yoshida (Eds.), The Archaean and Proterozoic Terrains in Southern India Within East Gondwana. vol. 3. Mem. Gondwana Research Group, Nagpur, India, pp. 53–75.

Kelly, N.M., Clarke, G.L., Fanning, V.M., 2002. A two-stage evolution of the Neoproterozoic Rayner structural episode: new U-Pb sensitive high resolution ion microprobe constraints from the Oygarden Group, Kemp Land, east Antarctica. Precambrian Res. 116, 307–330.

Kröner, A., Cooray, P.G., Vitanage, 1991. In: Lithotectonic subdivision of the Precambrian basement in Sri Lanka., Pt.1. Summary of Research of the German-Sri Lankan Consortium. Geol. Surv. Dept. Sri Lanka, Professional Paper. vol. 5, pp. 5–25.

Le Fort, P., 1975. Himalayas: the collided range. Present knowledge of the continental arc. Am. J. Sci. 27, 1–44.

Medlicott, H.B., Blandford, W.T., 1879–1881. Manual Geology of India (Part I, Part II). Geol. Surv. of India, Calcutta.

Molnar, P., Tapponnier, P., 1975. Cenozoic tectonics of Asia: effects of a continental collision. Science 189, 419–426.

Qureshy, M.N., 1969. The thickening of the basalt layer as a possible cause of for the uplift of the Himalayas—a suggestion based on gravity data. Tectonophysics 7, 137–157.

Qureshy, M.N., Kumar, S., 1992. Isostasy and neotectonics of the northeast Himalaya and foredeep. Mem. Geol. Soc. India 23, 201–222.

Radhakrishna, T., Dallmeyer, R.D., Joseph, M., 1994. Palaeomagnetism and $^{36}Ar/^{40}Ar$ vs. $^{39}Ar/^{40}Ar$ isotope correlation of basic dyke swarms in central Kerala: tectonic implications. Earth Planet. Sci. Lett. 121, 213–226.

Radhakrishna, T., Maluski, H., Michell, J.G., Joseph, M., 1999. $^{40}Ar/^{39}Ar$ and K/Ar geochronology of the dykes from south Indian granulite terrain. Tectonophysics 304, 109–129.

Raval, U., Veerasswamy, K., 2003. India Madagascar separation: break-up along a pre-existing mobile belt and chipping of the Craton. Gondwana Res. 6, 467–485.

Roy, A.B., 1999a. Neoproterozoic crustal evolution and India-Gondwana linkage—an Epilogue. Gondwana Res. 2, 193–198.

Roy, A.B., 1999b. Evolution of saline lakes in Rajasthan. Curr. Sci. 76, 290–295.

Roy, A.B., 2004. Phanerozoic reconstitution of Indian Shield as the aftermath of break-up of the Gondwanaland. Gondwana Res. 7, 387–406.

Roy, A.B., Chatterjee, A., 2015. Tectonic framework and evolutionary history of the Bengal Basin in the Indian subcontinent. Curr. Sci. 109, 271–279.

Roy, A.B., Jakhar, SR., 2001. Late Quaternary drainage disorganization and migration and extinction of the Vedic Saraswati. Curr. Sci. 81, 1188–1195.

Roy, A.B., Chatterjee, A., Chauhan, N.K., 2013. Late quaternary deformation and palaeoseismicity: insight into geomorphotectonic evolution of Kachchh, Western Indian Subcontinent. J. Geol. Soc. India 81, 31–40.

Searle, B.P., Windley, B.F., Coward, M.P., Cooper, D.J.W., Rex, A.J., Tingdong, L., Xuchang, X., Jan, M.Q., Thakur, V.C., Kumar, S.M., 1987. The closing of the Tethys and the tectonics of the Himalaya. Bull. Geol. Soc. America 78, 678–701.

Sen, S., Bhattacharya, S., Acharya, S., 1995. A multi-staged pressure temperature record in the Chilka Lake granulites: the epitome of the metamorphic evolution of Eastern Ghats, India? J. Meta. Petro. 13, 287–298.

Sengupta, P., Sen, J., Dasgupta, S., Raith, M., Bhui, U.K., Ehl, J., 1999. Ultra-high temperature metamorphism of metapelitic granulites from Kondapalle, Eastern Ghats Belt: implications for the Indo-Antarctic correlation. J. Petrology 40, 1065–1087.

Storey, M., Mahoney, J.J., Saunders, A.D., Duncan, R.A., Kelly, S.K., Coffin, M.F., 1995. Timing of hot spot-related volcanism and break-up of Madagascar and India. Science 267, 852–855.

Suess, E., 1901. Face of Earth. Wikipedia, the free Encyclopedia.

Torsvik, T.H., Tucker, R.D., Ashwal, L.D., Carter, L.M., Jamtveit, B., Vidyadharan, K.T., Venkataraman, P., 2000. Late Cretaceous India-Madagascar fit and timing of break-up related magmatism. Terra Nova 12, 220–224.

Valdiya, K.S., 1984. In: Tectonics of the folded fringe of the Indian platform. Reports of Colloquium 05, 27th Int. Geol. Congress, Moscow. pp. 110–137.

Valdiya, K.S., 1993. Evidence of Pan-African-Cadomian tectonic upheavals in Himalaya. J.Palaeontological Soc. India 38, 51–62.

Valdiya, K.S., 1998. Dynamic Himalaya. Universities Press, Hyderabad. 178p.

Veeraswamy, K., Raval, U., 2004. Chipping and breakup along mobile belts of a supercontinent. Earth Planets Space 56, 491–500.

Virdi, N.S., 1987. Northern margin of Indian plate: some litho-tectonic constraints. Tectonophysics 134, 29–38.

Warsi, W.E.K., Molnar, P., 1977. Gravity anomalies and plate tectonics in the Himalaya. Colloques Int. du CNRS. In: Himalaya. Sciences de la Terra, pp. 463–478.

Yoshida, M., Fumaki, M., Vinitage, P.W., 1992. Proterozoic to Mesozoic East Gondwana: the juxtaposition of India-Sri Lanka-Antarctica. Tectonics 11, 381–391.

CHAPTER

INDIAN SUBCONTINENT: GEOMORPHIC AND GEOPHYSICAL TRAITS

2

2.1 GEOMORPHIC CHARACTERISTICS OF INDIAN SUBCONTINENT

It is the common practice to discuss the physiography of a terrain before describing the geology of the region in detail. This is done merely to provide a first-hand account of the terrain that exposes different types of lithology or more precisely the lithological formations that formed over a long period of geological time. Traditionally, the term physiography is used to imply the broad, all-comprehensive aspects of physical geology including the structure (implying landform types), relief, and drainage pattern. Geomorphology, on the other hand, emphasizes on geomorphologic processes and landform patterns and their origin and expressions of relief.

The Indian Subcontinent constitutes a distinctive geographic entity that is virtually cut off from the rest of Asia by lofty mountain chains; the countries included are Bangladesh, India, Nepal, and Pakistan (Roy, 2014). Almost the entire Subcontinent's boundary in the north is bordered by the Himalayas and its associated branches, the Sulaiman and Kirthar Ranges passing on to the Hindu Kush in the northwest and the Naga Hills and the Arakan Yoma constituting the Indo-Myanmar Ranges in the southeast (Fig. 2.1).

In the following description of the physical geology of the Indian Subcontinent (the remnant of the pristine Indian Shield), the emphasis will be on aspects of geomorphology highlighting the landform pattern and the land-sculpting processes, and the typology of landforms especially of the ground surface rather than that of the subsurface. The present landform pattern of the Indian Subcontinent has been described as 'palimpsest' (Sen and Prasad, 2002) assuming that it has evolved through repeated superimposition of geotectonic history since the early Precambrian.

On the basis of common geologic and geomorphic attributes, the Indian Subcontinent can be divided into three geomorphic provinces:

(a) The Indian Peninsula
(b) The Indo-Gangetic Alluvial Plain, and
(c) The Himalayas

In addition to these major (macro-)geomorphic features that characterize the Indian Subcontinent, an important physiographic entity is the 5700 km-long coastline that borders the south-pointing Peninsula from the Sundarban Delta in the Bengal region in the east to the Gujarat Plain in the west.

Indian Shield. https://doi.org/10.1016/B978-0-12-809839-4.00002-3

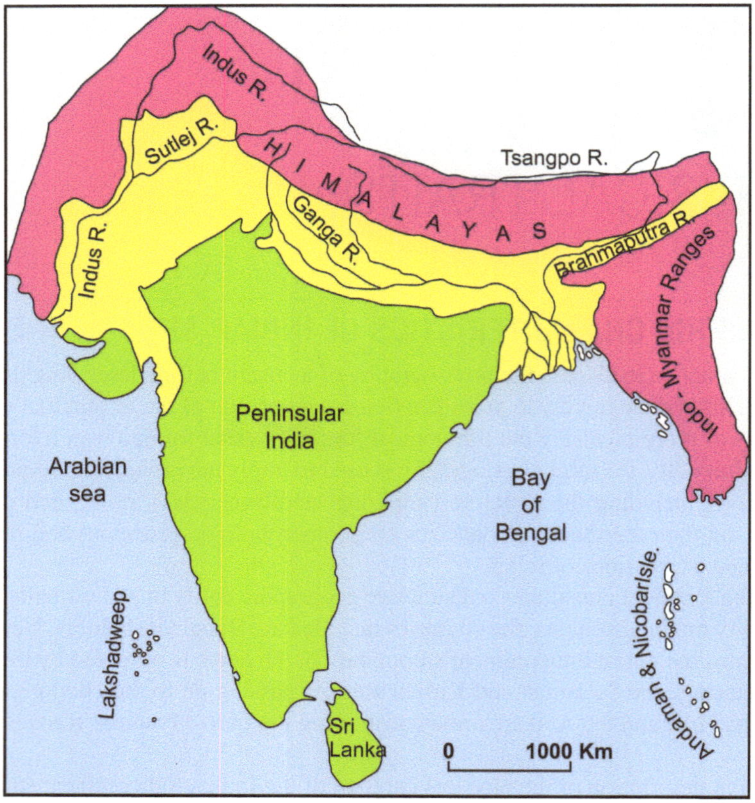

FIG. 2.1

Physiographic map of Indian Subcontinent, showing distribution of major physiographic constituents. The Indo-Gangetic Alluvial Plain lies in between the Himalayas in the north and the Peninsular India in the south.

2.1.1 THE INDIAN PENINSULA

The Indian Peninsula, lying south of Indo-Gangetic Alluvial Plain (also known as the Indus-Ganga-Brahmaputra Plains) (Fig. 2.2) is a unique macrogeomorphic terrain representing a huge triangular-shaped landscape made of ancient landmass with records of a prolonged post-Precambrian history of erosion, denudation, and resurgent tectonic activities. This geomorphic entity is also considered the geologically most exposed part of the pre-existing Gondwanaland (Kale, 2014). The average elevation of this easterly tilted undulating tableland is about 300 m above the mean sea level and is dissected by several narrow river valleys, chiefly the Mahanadi, Godavari, Krishna, Kaveri (Cauvery), Narmada, and Tapti. On an average, these river valleys lie below 150 m above the mean sea level.

Constituting the largest geomorphic component of the Indian Subcontinent, the Indian Peninsula shows extreme diversity in relief because of the presence of high mountains and several isolated plateaus characteristically having steep scarp faces. The Deccan Plateau that makes up most of the southern part of the Subcontinent covering about 422,000 km^2 area is the principal subprovince of the Indian Peninsula. In the south, the Plateau is mostly over 1000 m above the mean sea level, while in the north, it is between 300 and 500 m above the mean sea level. The Peninsular Plateau forms a familiar

FIG. 2.2

The Principal geomorphic provinces of the Indian Peninsula superimposed on the physical map of Indian Subcontinent.

southward-pointing triangle of the Indian Peninsula almost converging at the southern tip and is nested between two mountain ranges, the Western Ghats and the Eastern Ghats, each of which rises from the respective nearby coastal plains.

Both the Western and Eastern Ghats form elongate ranges along the west and the east coast, respectively. The attribute 'Ghats' is used to the name of these geomorphic features because of the fact that these mountain ranges are characterized by steplike succession of elevation from east to west. The Western Ghats with steep west-facing escarpments almost all along the western edge of the Deccan Plateau (Fig. 2.3) span for about 1500 km in the north-south direction. The southward continuity of the Western Ghats is through the Nilgiri Hills in Tamil Nadu and beyond that the Anaimalai and Cardamom Hills in Kerala. There are at least 24 peaks above 2000 m in the southern part of the Western Ghats. The highest ones are Doddabetta (2637 m in the Nilgiri Hills), Anamudi (2695 m) (Fig. 2.4), Chembra (2100 m), and Banasura (2073 m) in the southern most ranges.

FIG. 2.3

High scarp face of the Western Ghats Mountains, viewing easterly from Mahabaleshwar near Mumbai.

FIG. 2.4

The towering Anamudi Mountain (2695 m) in the southern part of Western Ghats Mountains, Kerala.

FIG. 2.5

View of the Eastern Ghats Mountains at Kollimalai (Kolli Hills), Tamil Nadu.

Image from en.wikipedia.org.

The Nilgiri Mountains is in northwestern Tamil Nadu and the Biligirirangana Hills southeast of Mysore in Karnataka, which meet the Shevaroys (Servarayan Range) and the Tirumala Ranges farther east, linking the Western Ghats with the Eastern Ghats. The Eastern Ghats constitute a series of discontinuous, much denuded mountain ranges running northeast-southwest parallel to the coastline of the Bay of Bengal (Fig. 2.5). It is breached by a number of easterly flowing rivers. The largest single sector of the Eastern Ghats, the remnant of an ancient mountain range that eroded and subsequently rejuvenated, is found in the Dandakaranya region between the Mahanadi and Godavari Rivers. The high mountains occur in the northern part of the mountain range in between the Mahanadi and Krishna Rivers with an average elevation of 1100 m. The highest peak of the Eastern Ghats is Jindhagada (1690 m). Other important peaks are the Arma Konda (1680 m), Deomali (1672 m), Gali Konda (1643 m), and Sinkram Gutta (1620 m).

The northern margin of the Deccan Plateau is a wide zone of upland comprising a number of narrow mountain ranges, discontinuous ridges, and several plateaus with ill-defined boundaries. In the west occurs the prominent mountain range, the Aravalli Mountains, which separate the Peninsular India from the Thar Desert (Fig. 2.2) and the western plains of Indus Valley. The Aravalli Mountains conventionally thought as the most ancient mountain range in Indian Peninsula, which stretch for about 650 km between Delhi in the north and the southern end at Palanpur north of Ahmedabad in Gujarat. The highest point in the Aravalli Mountains is Guru Shikhar (about 1722 m above the mean sea level) in Mount Abu in the southwestern part of the mountains. Like the Western Ghats, the Aravalli Mountains show easterly tilt controlling the drainage pattern in the region. A scarp-like western face of the Aravalli Mountains represents a spectacular topographic feature against the backdrop of the flat-lying arid plain of western Rajasthan.

Between the Aravalli Mountains in the west and the Bengal Basin in the east, there are several elevated topographic features like high plateaus, narrow linear ridges, and hills having precipitous heights. The most prominent of the plateaus are the Malwa Plateau in the western part, the Chhota Nagpur Plateau in the east, and Bastar Plateau in the southeast (Fig. 2.2). Geologically, the Malwa Plateau refers to the volcanic upland north of the Vindhya Range. The average elevation of the Malwa Plateau is 500 m above the mean sea level, and the landscape generally slopes towards the north. The Chhota Nagpur Plateau covers a large area in the northeastern part of the Indian Peninsula rising abruptly above the general plain lands of the terrain. The Plateau shows stepwise increase in the topographic height from 910 to 1070m in the western part, which drops down to around 300m in the eastern part. The highest point is the Parasnath Hills, which rise about 1370m above the mean sea level. The topography is undulating with prominent dome-like granite hills.

The Bastar Plateau covers about 92,200 km^2 of land that includes the Abujhmar Hills in the west and the Eastern Ghats in the east. The region is characterized by undulating plateau-like topography with well-marked elevations and depressions. To the northwest of the region lies the Kanke basin (450m above the mean sea level), which is a southward extension of the Chhattisgarh Plain in the north. In the southwestern part occurs the Godavari-Sabari Plain which itself is an elevated land known as Malkangiri Plateau. A highly undulating region marked by steep ghats, valleys, and plateaus.

Apart from the plateau and other tablelands, the two most important mountain ranges in the northern part of the Indian Peninsula are the Vindhya and the Satpura Ranges, which run subparallel to each other and separated by the steep valley of the Narmada and Son Rivers mainly in the western part. The Vindhya Range refers to a complex, discontinuous chain of hill ranges and highlands with elevation ranging between 450 and 1100m (above mean sea level) forming the southern escarpment of the central upland of India. The term 'Vindhyas' is defined by convention, and therefore, the exact definition of the Vindhya Range has varied at different times in geological history. The hills of the Satpura Range (Fig. 2.6), with peaks >1200m high (above the mean sea level), are considered a part of the Deccan Plateau, which stretches for about 900km across the widest part of the Indian Peninsula through Maharashtra and Madhya Pradesh. The range forms the watershed between the Narmada and Tapti Rivers. The Satpura Range includes the Mahadeo Hills to the north, the Maikala Range to the east, and the Rajpipla Hills to the west.

A number of isolated plateaus separated by deep river valleys occur in the northeastern part of the Indian Subcontinent, of which the Shillong Plateau (also called Meghalaya Plateau) is the largest.

FIG. 2.6

Steep scarp faces and deep valleys along the eastern plateau region of the Baghelkhand Plateau lying between the Maikal Ranges and Chhattisgarh Plain area with an elevation of 1033m above mean sea level.

Image from en.wikipedia.org.

These isolated tablelands occurring south of the east-west-running Brahmaputra River are now detached from the Indian Peninsula due to subsidence in the Bengal Basin that occurs in between. The topographic break is commonly known as the Garo-Rajmahal Gap with boundary faults at both ends (Roy and Chatterjee, 2015). These northeastern highlands show polycyclic topography with erosion surfaces at different elevations, between 1880 and 1410 m (above the mean sea level), and evidence of multiple tectonic activities at different times.

Summarizing, though the gross physiographic character of the Indian Peninsula is considered a single 'plateau-like' geomorphic entity, there are significant diversities not only in its geological make up but also in the physiographic types and character. There are several mountains criss-crossing the entire belt (Fig. 2.2), which did not evolve as fold mountains. The topography of individual geomorphic entities is generally undulating with prominent granite hills having dome-like outlines. Another interesting feature is the retention of the pristine horizontality of bedding or other depositional features even in those which had undergone considerable vertical uplifts (Fig. 2.3). Added to these are the invariable presence of scarp faces and the development of water falls especially when such faces cut across the river valleys and stream channels (Fig. 2.7). All these features provide unmistakable evidence that the major landforms (valleys, high-level surfaces, and lateritic and duricrust landforms) could not have been formed merely as the legacy of long history of prolonged weathering and erosion. On the other hand, much of the mega- and the microscale geomorphic traits can be attributed to the more recent block uplift-type tectonic activities that affected the entire Indian Peninsula. The evidence of ancient gently rolling, almost featureless 'peneplain' surfaces marking the top of upland areas like plateaus and mountains provides proofs of a prolonged period of denudation reaching the base level of erosion much before its elevation to the mountainous height (Fig. 2.8). Heron (1953) while discussing the physiography of the Aravalli Mountain region mentioned about

FIG. 2.7

Spectacular water falls developed along the vertical scarp face of Satpura Range at Pachmarhi, Madhya Pradesh.

FIG. 2.8

The landscape around Munnar Hills characteristically showing rolling ground at the top of the Munnar Plateau at around 450 m above the mean sea level.

the uplifted 'Jurassic' erosion surfaces. Presumably, the combination of two features resulted because of vertical uplift of much denuded peneplained surfaces. The landscape around Munnar Hills (1450 m above the mean sea level) lying on the western foot hill of the Anamudi Mountain (Fig. 2.4) might have been fashioned in that way. We have, therefore, reasons to consider the present physiographic traits of the Indian Peninsula as examples of tectonic geomorphology.

BOX 2.1

A mountain is a large landform that stretches above the surrounding land in a limited area, usually in the form of a peak. These can form due to (i) volcanic eruption (e.g. Mt. Kilimanjaro), (ii) tilting of a large block of Earth's Crust (e.g. the Sierra Nevada), or (ii) horizontal squeezing of rocks along some belts (e.g. the Himalayas or Alps). The last mentioned types are popularly described as fold mountain.

2.1.2 THE INDO-GANGETIC ALLUVIAL PLAIN

The Indo-Gangetic Alluvial Plain (also known as Indus-Ganga-Brahmaputra Plain) constitutes the vast plain land encompassing an area of about 1.17 million km² between the Himalayas in the north and the tableland of the Peninsular India in the south. Extending from the Indus Plains in the western part of the Subcontinent to the Bengal Delta in the east, this low-relief alluvial plain is irrigated by three important rivers, Indus (Sindhu), Ganga (Ganges), and Brahmaputra. The narrow plain of the Brahmaputra Basin merges with the Ganga Basin in the northeastern part of the long arc-shaped basin.

The Indus River rises beyond the Himalayas, and its major tributaries are the Jhelum, Chenab, Ravi, Sutlej, and Beas. The Punjab Plains are benefited by the Indus river system. The literal meaning of the term 'Punjab' is the land of five rivers. Sind is situated at the lower valley of the Indus. The River Ganga rises in the Himalayas and flows south and then towards the east. The river Yamuna flows almost parallel to the Ganges before joining it near Allahabad. The area between these two rivers is called 'doab', meaning the land between two rivers. The important tributaries of the Ganga are the Gomati, Sarayu, Ghagra, Kosi, and Gandak. In the eastern India, the Ganga Plains merge with the plains of Brahmaputra. The river Brahmaputra rises beyond the Himalayas, flows across Tibet, and then after turning southward continues to flow through the plains of northeast India. In the plains, it is a vast but a slow-moving river forming several islands.

Geologically, this is the youngest geomorphic unit evolved as the foreland basin in the frontal region of the rising Himalayas and its associated mountain ranges both in the west and east. The Indus-Ganga belt is the world's most extensive expanse of uninterrupted alluvium formed by the deposition of silt by numerous rivers. The plains are the world's most intensely farmed areas and rank amongst the world's most densely populated areas. The average height of the plain is about 200 m above the mean sea level, which increases to about 300 m in the Punjab region. The lowest parts lie in parts of the Bengal Delta having a height nearer to that of the mean sea level. The apparent monotony of the topographic simplicity is, however, broken by some diversities of microrelief features brought about by the sedimentation pattern and the nature of fluvial deposits.

The most distinctive relief and geomorphology are observed in the piedmont zone of the Himalayas, which are known locally as Bhabar and Tarai. The Bhabar is generally narrow about 8–16 km-wide belt adjacent to the foothills of the Himalayas, lying between the Indus in the west and the Teesta in the east. It is a zone built up of unsorted debris, boulders, and pebbles, mixed with coarse-textured sands brought down from the Himalayas by a number of rivers. As the porosity of this belt is very high, the streams flow underground and emerge onto the surface in the belt known as Tarai. The Tarai belt is between 15 and 30 km wide lying south of the Bhabar region It is generally a marshy and forested track in the Himalayan foothills, and is composed of newer alluvium consisting of finer silts. The high water table with improper drainage pattern in the belt causes waterlogging leading to the formation of swamps and marshes.

The two types of alluvial deposits generally recognized in the vast stretch of the Indo-Gangetic Alluvial Plain are the Bhangar (older alluvium) and Khadar (newer alluvium). The former generally occupys the higher interfluves of the Indus-Ganga system while the later constitues the lower flood plains. The older alluvium of the Bhangar plains is generally infertile in character composed of coarser materials and is more overleached and quite widespread in the upper Ganga plains, which at places occur above 275 m in the interfluves like the Ganga-Yamuna Doab and some other plains in western Uttar Pradesh rising as high terraces, >20–30 m from the flood plains. Locally, these are invaded by mass-wasting processes like gulley and sheet erosion giving rise to the formation of badlands (Sen and Prasad, 2002). Wherever the highlands are created by stones and sand, it is locally called as 'Bhur'. For example, Bhurs are found in the upper parts of Ganga-Yamuna Doab. The Khadar belt, lying in the lowland areas of Bhangar, is made up of fresh newer alluvium, which is deposited by the rivers flowing down the plain. In this region, floods bring new alluvium every year. Khadar region is mainly found along the river banks and contains fine particles or clays making this a fertile region.

BOX 2.2

Badlands are a type of dry terrain where softer sedimentary rocks and clay-rich soils have been extensively eroded by wind and water. They are characterized by steep slopes, minimal vegetation, lack of a substantial regolith, and high drainage density. Chambal valleys are especially characterized as badland area.

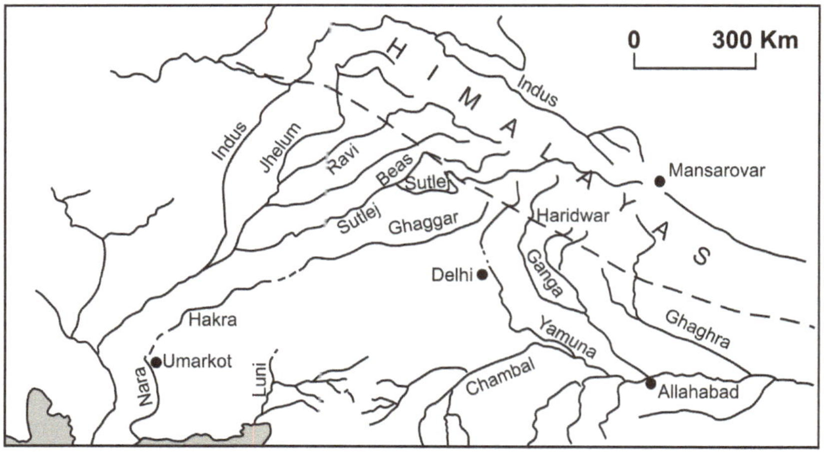

FIG. 2.9

Spectacular drainage divide between the Indus and the Ganga river systems. The eastern rivers flow towards east to drain into Bay of Bengal, while the western rivers flow in the southwesterly direction to the Arabian Sea.

Reproduced from Roy and Jakhar (2002) with permission.

A very significant feature of the Indo-Gangetic Alluvial Plain is the spectacular division of the drainage basins along a narrow zone of virtually featureless country providing a the perception that the land between the Indus and Ganga Plains is continuous between the two drainage basins (Fig. 2.9). As an explanation, it is suggested that a topographic high did exist in the form of a mountain, which had acted as the drainage divide between the two different river systems. The geophysical studies in the region help in detecting a subsurface feature in the region recorded as the Delhi-Haridwar Ridge. The subsurface feature is interpreted as the pre-existing mountain or the topographic high that was responsible for the observed drainage divide between the Indus and the Ganga River systems. The feature has now been subsided below the ground level during the process of deepening of the frontal basin simultaneously with the rise of the Himalayas. The location of the 'submerged' Delhi-Haridwar Ridge suggests that it existed as the northern continuity of the Aravalli Mountains in the south (Roy and Jakhar, 2002).

2.1.3 THE HIMALAYAS

The Himalayas constitute an imposing crescent-shaped mountain range extending for over 2500 km from south of the Indus Valley beyond Nanga Parbat (height, 8114 m) in the west to Namcha Barwa (height, 7755 m) in the east (Fig. 2.10). With a prominent southward convexity, the majestic mountain chain stands like a wall bordering the entire northern margin of the Indian Subcontinent. Topographically, the Himalayas are bent sharply at the western end to join with the Sulaiman and Kirthar Ranges, south of the Pamir. There is a similar sharp bending at the eastern end, where the mountain range joins the north-south trending Indo-Myanmar Range, represented by the Naga Hills and Arakan Yoma. One of the most striking aspects of the Himalayan orogen is the lateral continuity of its major tectonic elements.

A unique feature of the Himalayas is the crustal thickness, which rises from about 35 km in the Indus-Ganga-Brahmaputra Plains to between 65 and 80 km over the Higher Himalayas. The increasing crustal thickness is reflected in the dip of the MOHO, which is estimated to be 7–8°N under the Sub-Himalayas but over 15° further north.

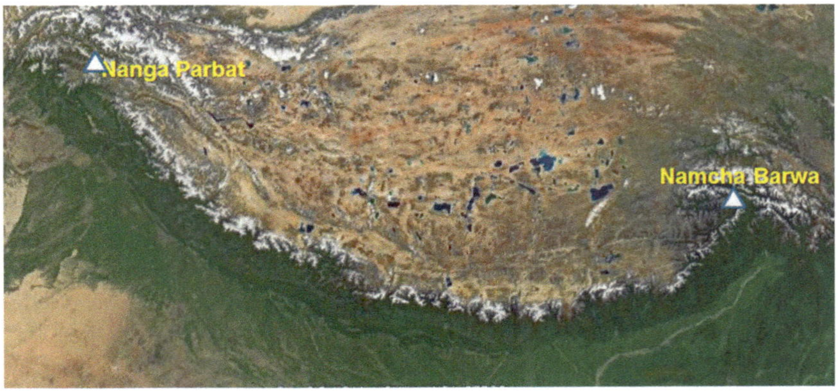

FIG. 2.10

Satellite imagery of the Himalayas showing the spectacular southward curving snow-capped mountain ranges between Namcha Barwa in the east and Nanga Parbat in the west.

Image from Google Earth.

The Himalayan geomorphology significantly portrays the different aspects of structural-geomorphic features which evolved in response to the orogenic forces during the relatively recent erosional history. The mountain range is divided axially into the following six units, each showing distinct lithotectonic and geomorphic character and evolutionary history (Fig. 2.11):

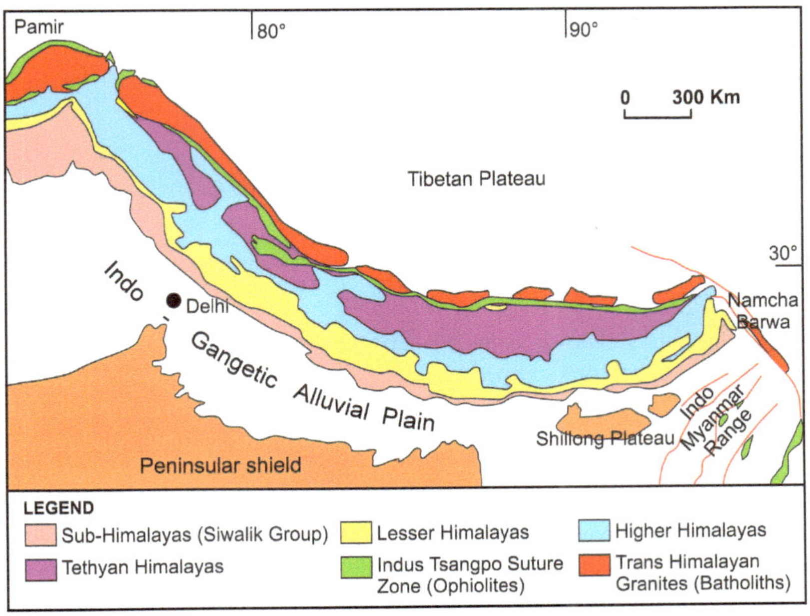

FIG. 2.11

Geological map of the Himalayas showing important tectonic elements.

Reproduced from Roy (2014) with permission.

1. The Sub-Himalayas

It forms a 10–50 km-wide belt constituting the foothills of the Himalayan Range and is essentially made of Neogene (Miocene to Pleistocene)-type sediments derived from the erosion of the Himalaya. The deposits are known as the Murree and Siwalik Formations, which occur in the narrow piedmont plain at the base of the mountains. These are internally folded and imbricated, and bounded between the Main Boundary Thrust in the north and the Himalayan Frontal Thrust (Valdiya, 2010) in the south. The latter fault/thrust belt is represented by a series of en échelon faults, which are parallel to the strike of the hills between Punjab in the west and Assam in the east. The width of the Sub-Himalayas gradually narrows down from about 50 km in the west until it nearly disappears in Bhutan and eastern India. A characteristic feature of the Sub-Himalayas is the large number of long, flat-bottomed valleys known as Duns or Doons, which are usually spindle-shaped areas filled with gravelly alluvium. South of the foothills lies the Tarai and Dooars plains (Fig. 2.12). The southern part of the Tarai and Dooars plains is heavily farmed.

FIG. 2.12

Dooars in the Eastern Himalayas is a classic example of the Sub-Himalayan landscape.

2. The Lesser Himalayas (also known as the Mahabharata range in Nepal)

This is a 60–80 km-wide belt predominantly comprising Proterozoic low-grade metamorphic rocks overlain by sheets of granites and metamorphic rocks, which are thrust over the Sub-Himalayas along the Main Boundary Thrust. The Lesser Himalayas often appears as tectonic windows within the High Himalayan Crystalline Sequences. The general elevation of this geomorphotectonic unit is between 3000 and 3500 m above the mean sea level, with the greatest elevations being up to 6000 m. The mountains are dissected by narrow, steep-walled valleys. Traces of ancient glaciation have been preserved in many ranges. The snow line is at an elevation of 4500 m in the east and 5000 m in the west. There are small glaciers on the northern slope of the Pir Panjal Range. The Lesser Himalayas have a dense river network (primarily tributaries of the Ganges). The rivers have a high-water period in the summer, caused by monsoon rains and the melting of mountain snow.

3. The Higher (or Great) Himalayas

It is a 10–15 km-thick belt of dominantly Precambrian metamorphic rocks and is also a zone of highest uplift. The Higher Himalayas form the backbone of the Himalayan Orogen and encompass the areas with the highest topographic relief (Fig. 2.13). Although most of the metasediments forming the Higher Himalayas are of late Proterozoic to early Cambrian age, much younger metasediments have also been reported to occur in some parts. It is generally believed that the metasediments of the Higher Himalayan Crystalline Complexes represent the metamorphic equivalents of the sedimentary series forming the base of the overlying Tethys Himalayas. The Higher Himalayan Crystalline Complexes form a major nappe, which is thrust over the Lesser Himalayas along the Main Central Thrust.

The Higher Himalayas are also known as the Central Crystalline Zone, composed of deformed metamorphic rocks, and marks the axis of orogenic uplift. Mica schist, quartzite, paragneiss, migmatite, and leucogranite bodies characterize this uppermost Himalayan zone. They represent multiphase metamorphic events, the first being Barrovian type under normal geothermal gradient. There was then a shift to Buchan-type metamorphism under low pressure and high temperature conditions, with temperatures greatly exceeding normal gradient temperatures (Sorkhabi and Arita, 1997). Local retrograde events have also been noted. Analyses show that the peak orogenic temperatures and pressures were 475–825°C and 500–800 MPa, respectively. Corresponding mineral assemblages are dominated by biotite to sillimanite, representing greenschist to amphibolite facies metamorphism. Deformation seems to have occurred in a north-to-south direction and is associated with the Main Central Thrust Fault, which brings the Higher Himalayas on top of the Lower Himalayas (Sorkhabi and Arita, 1997). Initially, it was thought that approximately 350 km of shortening had occurred in the Greater Himalayan sequence of rocks. However, based on the studies by DeCelles et al. (1998), a shortening between 600 and 650 km has been estimated. Some questions have been raised on the provenance of

FIG. 2.13

The majestic Mt. Everest, crowning glory of the Higher Himalayan Crystalline Complexes, represent the highest topographic point in the world.

Image courtesy: Ikramul Hasan Shakil.

the Great Himalayan rocks. Previous work suggested that lower Indian Crust comprised this area. New interpretations indicated the possibility that the Higher Himalayas are actually made of supracrustal rock. The new idea emphasizes that upper crustal material of India had accreted northward onto the Asian continent and that the crustal material was originally an appendage of India that was itself accreted to India during Paleozoic. This might imply that the Indian Subcontinent probably had significantly more Continental Crust than previously thought, implying much more crustal shortening during the formation of the Greater Himalayas.

4. The Tethys Himalaya

This is a 100km-wide strongly folded and imbricated belt comprising weakly metamorphosed sedimentary rocks. A virtually complete stratigraphic record ranging from the Upper Proterozoic to the Eocene is preserved within the sediments of the Tethys Himalayas. Stratigraphic studies of these sediments yield important indications on the geological history of the northern continental margin of the Indian Subcontinent. Several nappes, termed North Himalayan Nappes, have also been described within this unit. The transition between the generally low-grade sediments of the Tethys Himalayas and the underlying low- to high-grade rocks of the High Himalayan Crystalline Sequence is usually progressive.

5. Indus-Tsangpo Suture Zone of ophiolites

The belt defines the zone of collision between the Indian Plate in the south and the Trans-Himalayan belt in the north. The belt comprises the following:

1. The ophiolite mélanges which are composed of an intercalation of flysch and ophiolites from the Neotethys Oceanic Crust
2. The Dras Volcanics, which are relicts of a Cretaceous-Jurassic volcanic Island Arc consisting of basalt, dacite, volcano-clastics, and some radiolarian cherts
3. The Indus Molasse, which is a continental clastic sequence (with rare interbeds of marine saltwater sediments) comprising alluvial fan, braided stream, and fluviolacustrine sediments derived not only from the Ladakh Batholith but also from the Suture Zone itself and the Tethyan Himalayas. These molasses are post collisional and thus Eocene to post-Eocene in age.
4. The Indus Suture Zone represents the northern limit of the Himalayas. Further to the north is the so-called Trans-Himalayas or Ladakh Batholith in the northwest, which corresponds essentially to an active margin of the Andean type. Widespread volcanism in this volcanic arc was caused by the melting of the Mantle at the base of the Tibetan Block, triggered by the dehydration of the subducting Oceanic Crust.

The tectonic architecture of the Himalayas is built on three prominent intracrustal thrusts. From south to north, these thrusts are:

1. The Himalayan Frontal Thrust, which marks the tectonic and physiographic boundary between the Siwalik and the Indo-Gangetic Alluvial Plain. It is a 2500km-long fault system and is one of the great fault systems of the world with many attributes similar to those which occur in the Subduction Zone. The Himalayan Frontal Thrust marks a sharp physiographic and tectonic boundary between the Himalayan foothills and the recent alluvial plain (Fig. 2.1).
2. The Main Boundary Thrust is a major dislocation zone that regionally separates the Lesser Himalayas from the Sub-Himalayas. The Main Boundary Thrust is defined as the southernmost thrust that places metasedimentary rocks of the Lesser Himalayas over unmetamorphosed clastic rocks of the Himalayan foredeep. Because it folds the Main Central

Thrust and cuts the Quaternary sedimentary rocks preserved in the Himalayan foreland, the Main Boundary Thrust is thought to have formed sometime between the present and 25–20 Ma (the age of formation of the Main Central Thrust), although an older middle Paleocene age has been suggested by some authors. The oldest well-dated evidence (4–5 Ma) for the displacement of Main Central Thrust comes from the western region in Pakistan and northwestern India. Subsidence histories across the western Himalayan foreland, the provenance of a newly dated conglomerate, and new cooling ages from the leading edge of the Main Boundary Thrust are herein interpreted to define its formation at 0.10 Ma along at least 1000 km of the Himalayan front.

3. The Main Central Thrust separates the crystalline rocks of the Higher Himalayas from the low-grade metamorphic rocks of the Lesser Himalayas. The Main Central Thrust is a major dislocation that continues along 2200 km of the Himalayan mountain belt. The fault slopes down to the north and is exposed on the surface in the NW-SE direction. It is a thrust fault, marked by ductile shear zone along which the High-grade Great Himalayan Crystalline complex occurs. The Himalayan Main Central Thrust, with its zones of inverted metamorphic isograds from sillimanite grade down to biotite grade, is one of the largest ductile shear zones known from any collision-related mountain belt in the world. It dips north, and transported the high-grade metamorphic rocks of the Greater Himalayas to the south over the unmetamorphosed rocks of the Lesser Himalayas.

2.2 GEOPHYSICAL CHARACTERIZATION OF INDIAN SHIELD (INDIAN SUBCONTINENT)

The most significant information about the geophysical character of the Indian Shield comes from the study of Bouguer gravity anomaly map of the Indian Subcontinent. The colour image prepared by Murthy (1999) (Fig. 2.14) using the Bouguer gravity anomaly map of National Geophysical Research Institute, Hyderabad, comes in handy providing excellent visual impression of the gravity anomaly pattern over the region. This is in spite the lack of information from very sensitive regions that fall outside the political boundary of India.

Based on the image pattern, it is possible to divide the gravity pattern into three distinctive domains. The first is Domain 1 representing much of the Peninsular India, which, barring patches of the coastal region (significantly more in the east coast rather than in the west) shows a low negative Bouguer anomaly value (between ~ −60 and ~ −120 mGal). The gravity value coupled with low heat flow data, and the evidence of between 38 and 40 km MOHO depth over the major part of the terrain are in conformity with the stable cratonic condition (Mahadevan, 1994), which compares well with the gravity pictures of the well-known Precambrian Shield areas of the world (Windley, 1973). The Deccan Traps that cover a considerable part in the north and locally attaining a height of about 2000 m, did not affect the Precambrian geophysical character of the domain. In strong contrast to this, the gravity image pattern, a distinct low-gravity zone (varying from −5 to −200 mGal) is observed over much of the Indo-Gangetic Alluvial Plain and in the Himalayas. The development of this prominent zone of low gravity finds explanation in the presence of thick low-density sediments over a sialic basement ducking under the Himalayas. Increase in the presence of low-density sediments accounts for the corresponding decrease in the gravity values.

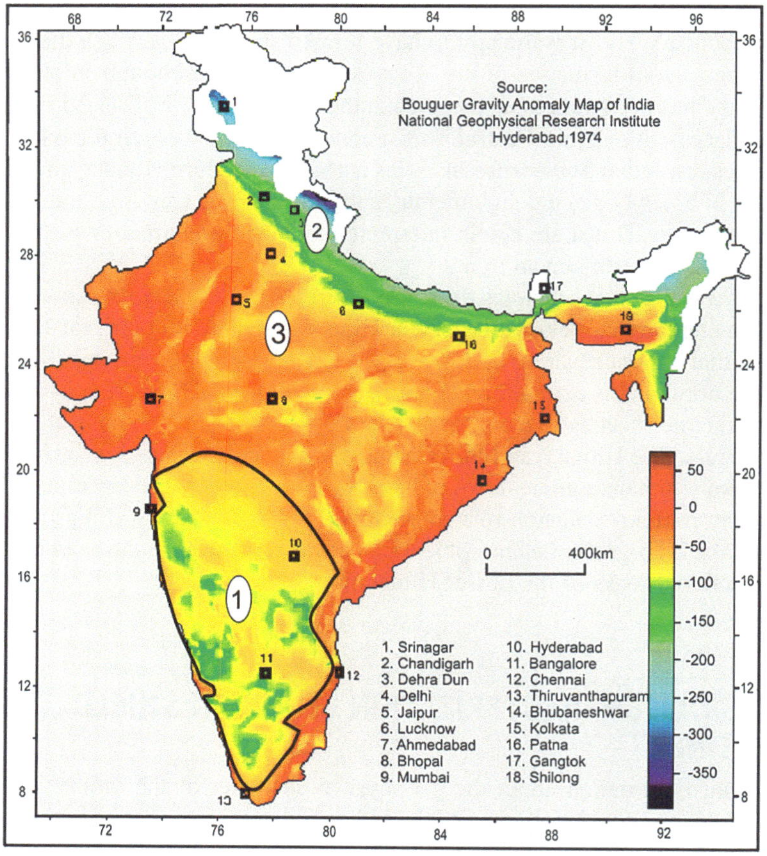

FIG. 2.14

Bouguer gravity colour image of Indian Subcontinent.

Reproduced with modification from Murthy (1999) with permission.

Between these two situations of pristine Precambrian character of the Peninsular India (including the area occupied by the Deccan Traps) and the totally reconstituted conditions in the Indo-Gangetic Alluvial Plain occur the domain of high gravity ranging from very low negative to over +50 mGal at places. Some broad generalizations about the Bouguer gravity anomaly pattern are possible if *a priori* assumption is made that high gravity is a reflection of the presence of denser rocks (like gabbro or dunite/peridotite) in the Crust. This implies the possible magma underplating channelized along ancient orogenic belts and lineaments. The gravity patches are interpreted as zones affected by magma plutons or pouches. There are enough data to confirm effects of repeated Plume impingements in changing the gravity anomaly pattern of the once cratonized Shield region (Roy, 2004). Much of the high-gravity zones are marked by belts of high heat flow.

The Peninsular Indian Shield is traditionally considered a stable continental block remaining rigid even during the post-Cambrian period, the last ~550 million years of Earth's history. The concept of stability receives support from the fact that there is virtually no record of any orogeny-related post-Precambrian tectonothermal reconstitution of the Indian Crust except along the belt of the Himalayas in the north. The idea of stability of the Indian Crust is so deeply entrenched in our thoughts that we tend to characterize the incidence of earthquakes in the region as 'Stable Continental Region' seismicity. The following facts on the other hand tell us a different story:

1. Not only the Indian Shield is situated close to the active Himalayan belt, a considerable part of it now forms a part of the Himalayan edifice. The huge Indo-Gangetic Alluvium Plain that developed on a depression in front of the rising mountain has resulted from the bending of the Shield under the mountain.
2. There are historical records of high-intensity earthquakes (> 8 in the Ritchter scale) rocking the Indian Shield areas, which usually occur only in collision-type Suture Zones. The two most sensitive areas in this respect are (i) Kachchh in western India (falling in seismic zone V, Bureau of India Standard, IS 1893: 2002) and (ii) Kolkata region, which has the 'dubious' record of witnessing the world's third most devastating earthquake in human history, killing over 300,000 people on 11 October 1737 (USGS Earthquake Hazards Programme: Earthquake Report).
3. The localized, unusually high heat flow values (reaching up to 107 $\mathrm{mWm^{-2}}$) over parts of the Peninsular Indian Shield speak of an abnormally hot Crust, unlike the heat flow pattern in different Shield areas of the world (Gupta, 1993; Quereshy, 1964).
4. Bouguer gravity anomaly maps published by the National Geophysical Research Institute, Hyderabad show high positive values (over 40 mGal) in several parts of India, which is considered a reflection of 'mobility' rather than stability.

All the above features indicate that the Indian Shield is tectonically quite unstable, unlike the Canadian and other similar shield areas of the world. An understanding of the unique character of the Indian Shield in geophysical terms needs a critical analysis of the different Phanerozoic (post-Cambrian) geological events which significantly altered its pristine geological character of the Indian Shield.

REFERENCES

DeCelles, P.G., Gehrels, G.E., Quade, J., Ojha, T.P., Kapp, P.A., Upreti, B.N., 1998. Neogene foreland basin deposits, erosional unroofing, and the kinematic history of the Himalayan fold-thrust belt, western Nepal. Geol. Soc. Am. Bull. 110, 2–21.

Gupta, M.L., 1993. Is Indian Shield hotter than other Gondwana shields. Earth Planet. Sci. Lett. 115, 275–285.

Heron, A.M., 1953. Geology of Central Rajputana. Memoir 79. Geol. Surv. India, Calcutta, 339p.

Kale, V.S., 2014. Geomorphic history and landscapes of India. In: Kale, V.S. (Ed.), World Geomorphological Landscapes. Springer Science+BuisnessbMedia, Dordrecht, https://doi.org/10.1007/978-94-107-8029-2_3.

Mahadevan, T.M., 1994. Deep Continental structures of India: a review. Mem. Geol. Soc. India 28, 569p.

Murthy, Y.S., 1999. Images of gravity field of India: their salient features. J. Geol. Soc. India 54, 221–235.

Quereshy, M.N., 1964. In: Gravity anomalies as related to regional tectonics of Peninsular India. In Proc. 22nd Int. Geol. Cong., New Delhi, Pt III. pp. 490–506.

Roy, A.B., 2004. The Phanerozoic reconstitution of Indian Shield as the aftermath of the break-up of the Gondwanaland. Gondwana Res. 7, 387–406.

Roy, A.B., 2014. Indian Subcontinent, Reference Module in Earth Systems and Environmental Sciences. 2014. Elsevier, https://doi.org/10.1016/B978-0-12-409548-9.09372-6.

Roy, A.B., Chatterjee, A., 2015. Tectonic framework and evolutionary history of the Bengal Basin in the Indian subcontinent. Curr. Sci. 109, 271–279.

Roy, A.B., Jakhar, SR., 2002. Drainage disorganization, migration and extinction of the Vedic Saraswati. Curr. Sci. 81, 1188–1195.

Sen, P.K., Prasad, N., 2002. An Introduction to the Geomorphology of India. Allied Publishers Pvt, Mumbai. 378p.

Sorkhabi, R.B., Arita, K., 1997. Towards a solution for the Himalayan puzzle: mechanism of inverted metamorphism constrained by the Siwalik sedimentary record. Curr. Sci. 72, 862–867.

Valdiya, K.S., 2010. The Making of India: Dynamic Evolution. Macmillan Publish India Ltd., New Delhi. 816p.

Windley, B., 1973. Crustal development in the Precambrian. Philos. Trans. R. Soc. Lond. A 273, 321–341.

CHAPTER

FRAMEWORK OF INDIAN SHIELD: TECTONIC SUBDIVISIONS

3

3.1 FRAMEWORK OF INDIAN SHIELD

Understanding of the tectonostratigraphic framework of the Indian Shield is beset with several constraints mainly because of the reconstitution and destruction of a considerable part of the 'Greater Indian Shield' about which we discussed in the earlier chapter. In view of this fact, there is hardly any option but to depend on the geological information that we get from the Indian Subcontinent for understanding the growth and evolutionary history of the Indian Shield. However, as far as the Precambrian history is concerned, the only source of information is the Peninsular India, and that too in spite of the fact that a considerable part of the terrain is under the cover of Phanerozoic rocks like Deccan Traps, Gondwana formations, and other younger sediments (Fig. 3.1).

Several authors attempted to study different aspects of the Precambrian geology of the Indian Peninsula. Krishnan (1956) was amongst the earliest geologists who proposed subdivision of the Precambrian terrane of Peninsular India (henceforth described as Indian Shield) by noting divergent patterns of structural trend lines in different parts. The concept was challenged by Naha (1964), who questioned the usefulness of structural trend lines for Precambrian crustal blocks as a basis of subdivision especially in regions of complex superimposed folding. However, in spite of the criticism, Krishnan's suggestion merits consideration assuming that the different crustal blocks, which evolved under divergent stress systems, are likely to have distinctive trend lines or 'structural grains'. According to Naqvi et al. (1974), the Indian Shield is made up of a collage of 'Protocontinents' (*smaller fragments of Precambrian Crust forming continental nuclei*) separated by rift valleys. Rogers (1986), however, preferred to follow the Canadian model, and recognized five different 'Provinces' in the Indian Shield separated by four major 'Joins'.

In the present analysis, Rogers' (1986) nondescript term 'Join' has been used as the separation plane (trace of which appears as a line on the surface) between the individual Protocontinent that formed nucleus of initially growing Crust. In the context of the Indian Shield, all these Joins are the sites of Gondwana Rift basins generally overlying 'unclassified Precambrian granitoids', except along the Pranhita-Godavari Valley where cratonic, platformal Proterozoic sediments form the basement for the Gondwana rift basins. Implication of using a nondescript term is that it speaks of the contact zone between two separate crustal blocks without specifying any tectonic model of clustering of these terranes, which is truly unknown.

Ramakrishnan and Vaidyanadhan (2008), while proposing subdivisions of the Indian Shield, advocated use of terminologies like 'Cratons' and 'Mobile belts' stressing that these are widely adopted in Precambrian literature (see also Sharma, 2010). They chose to accept these terms because of the

Indian Shield. https://doi.org/10.1016/B978-0-12-809839-4.00003-5

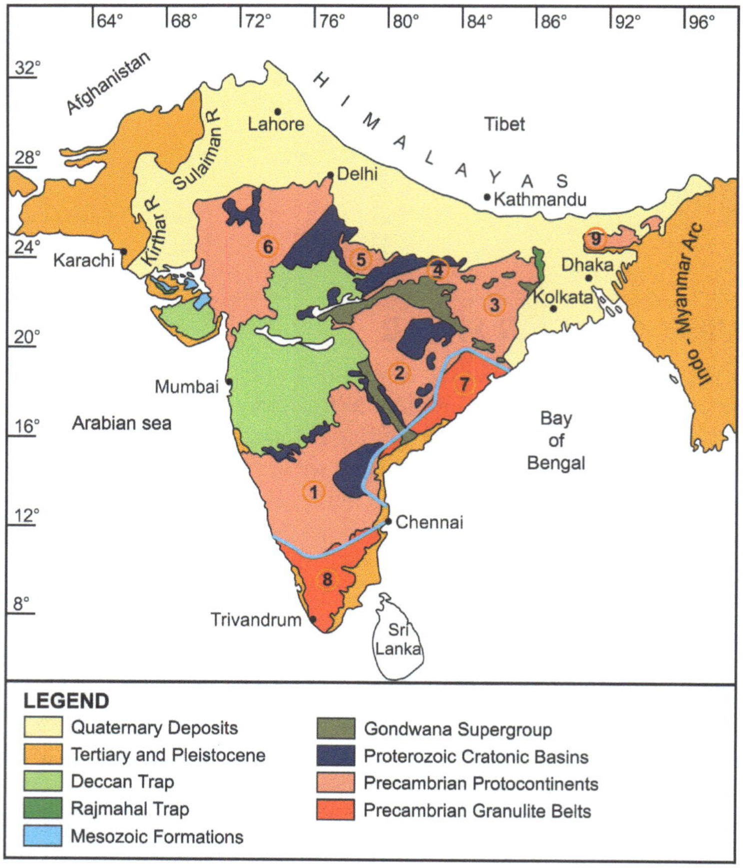

FIG. 3.1

Geological map of the Indian Subcontinent showing extent of the Phanerozoic covers over the Precambrian basement (based on the Geological Survey of India Map, 1993). The inset numbers represent (1) Dharwar, (2) Bastar, (3) Singhbhum, (4) Rajmahal, (5) Bundelkhand, (6) Aravalli, (7) Eastern Ghats Granulite Belt, (8) Southern Granulite Terrain, and (9) Shillong Plateau.

genetic flavour they carry in preference to the *sterile* nomenclatures like 'Provinces' and 'Blocks'. It may sound a bit of a semantic exercise, but a critical review of nomenclatures and terminologies that are being used to describe components of the Indian Shield may appear instructive especially in view of the divergent and at times contradictory opinions expressed by different authors.

It is true that terms 'Shield' and 'Craton' are widely used in geological literature dealing with Precambrian geology in different parts of the world, but their usage generally carries a different connotation. Shield in most usage is considered synonymous with Craton. On the other hand, some authors define Craton 'as large, stable Precambrian terrane that include both Shield and Platform' (Hatcher, 1978). A completely different meaning is intended in the usage of terms by Ramakrishnan and Vaidyanadhan (2008) while proposing division of the Peninsular Indian Shield into Cratons and Mobile belts.

BOX 3.1

The term 'Craton' is rooted in German word 'Kraton', which is a modification of Greek 'kratos' meaning 'power' or 'strength' (Wikipedia, the free encyclopedia). The etymological meaning of the word is reflected in the description that 'it is an old and stable part of the Continental Crust that survived merging and splitting of Continents and Supercontinents since Precambrian' (Hatcher, 1978).

Ingrained in the definition is the subtle thought that Cratons are virtually permanent features on Earth. The idea, itself, may appear preposterous knowing well that the Earth has undergone a number of tectonic or Supercontinent-building cycles since the formation of the earliest Crust about 4.5 billion years ago. The notion of permanency (of Cratons) perhaps comes from the perception that the Canadian, Ukrainian, and some other Shield areas in the world remained tectonically undisturbed since Precambrian.

The geological history of the Indian Shield tells us a different story. This is amongst the very few Precambrian terranes in the world that was involved in continental collision, and losing its components while building up much of the Himalayan edifice. We may, therefore, conclude that a Craton, contrary to its etymological meaning, need not be a permanently stable feature on Earth that would not change with time. The evolution of the Himalaya tells us how a Craton or parts of it turned into a Mobile belt. That the reverse is also possible is implied in the expression like 'cratonization', which stands for enlargement of cratonic blocks through accretion (i.e. addition by attachment) of Mobile belts. Precisely speaking, Cratons and Mobile belts are time-dependent relative terms, and because of that, their usage in describing Precambrian geology of a region without its characterization is liable to add unnecessary confusion. Regarding other terms and nomenclatures used by different authors, the sheer multiplicity of propositions certainly calls for an objective view on the question of the use of terms. In the present discussion, the term 'Protocontinent' suggested by Naqvi et al. (1974) will be used for the Precambrian crustal blocks, which evolved as fundamental units during the early days of the evolutionary history of the Earth's Crust.

3.2 TECTONIC SUBDIVISIONS OF THE INDIAN SHIELD

The earliest proposal on the broad tectonic division of the Indian Shield (should read Peninsular Indian Shield) comes from Fermor (1936) who divided it into two broad tectonometamorphic domains: Charnockite and Noncharnockite belts.

Since Fermor's (1936) division of the Indian Shield into two exclusive terranes, a number of reports came up indicating occurrence of charnockite and granulite facies rocks from different parts of the Indian Shield outside the Fermor's Charnockite Line. Yet, Fermor's proposal deserves consideration as it provides an important clue to differentiate the Precambrian rocks of Indian Peninsula into two basic crustal types having distinctly different tectonic evolutionary history.

A feature of great significance is the cross-cutting relationship between the Charnockite and Noncharnockite terranes indicated in the 'Geological Map of India' published by Geological Survey of India (1993), especially in the case of the Eastern Ghats Granulite Belt, which lies astride the structural grains of the three Precambrian crustal blocks: Dharwar, Bastar, and Singhbhum, occurring on its western side (Fig. 3.1). The cross-cutting relationship is most obvious especially in the domain north of the Godavari Valley. According to Ramakrishnan et al. (1998), the Eastern Ghats Granulite Belt represents a zone of westerly directed thrust slices abutting against the Precambrian block of Bastar. Based on this, the western margin of the Eastern Ghats Granulite Belt is interpreted as an ancient Suture Zone resulting from continental collision (very much like the Himalayan collisional tectonics) in response to the 'compression from the "Enderby Land" (in Antarctica)'. Quoting Gupta et al. (2005) and Leelanandam et al. (2006), Ramakrishnan and Vaidyanadhan (2008) interpreted the occurrences of 'deformed alkaline rocks and carbonatites' along the western margin of the Eastern Ghats Granulite Belt as a marker of an ancient Suture Zone. The interpretation has a strong support from the multidisciplinary studies of Gupta et al. (2000) and Bhadra et al. (2004).

Further confirmation of collisional tectonic comes from the field and Satellite Imagery study of Chetty (2001, 2010), Biswal et al. (2000), and Biswal and Sinha (2003, 2004) who characterized the Eastern Ghats Granulite Belt as a collage of nappes thrusted over the western Indian crustal blocks.

Regarding the tectonic status of the western margin of the Eastern Ghats Granulite Belt south of Pranhita-Godavari Valley, a little elaboration is necessary in view of some studies made in recent years, especially on the tectonostratigraphic status of the Nellore-Kammam Schist Belt and associated ensembles (Ramam and Murthy, 1997; Babu, 1998; Ramakrishnan et al., 1998; Rickers et al., 2001; Dobmeir and Raith, 2003). According to Ramakrishnan and Vaidyanadhan (2008), the western margin of the Eastern Ghats Granulite Belt coincides with the 'Western Charnockite Zone', which disappears into Bay of Bengal near Ongole (about 100 km north of Nellore). The Nellore-Kammam Schist Belt, which is traditionally considered a component of the Dharwar Protocontinent (Krishnan, 1956; Valdiya, 2010), occurs immediately west of Eastern Ghats Granulite Belt. Ramakrishnan (2003), making a departure from the conventionally held view, described the status of the Nellore-Kammam Schist Belt as enigmatic. He made special mention of the parallelism of the belt with the tectonic trend of the Eastern Ghats Granulite Belt, describing a Mesoproterozoic age (>1600 Ma) for the Nellore-Khammam Schist Belt. The author also hinted at the possible correlation of the Nellore-Khammam Schist Belt with the evolution of the Nallamalai Fold Belt occurring along the tectonized eastern margin of the Cuddapah Basin. Based on such a correlation, one cannot rule out the possibility that Nallamalai folds are indentation structures resulting from the westerly directed compressive force.

Dobmeir and Raith (2003), based on critical evaluation of the existing geological and isotopic data, suggested inclusion of the granulites of the Ongole block as also the Nellore-Khammam Schist Belt into the domain of the Eastern Ghats Granulite Belt. The revision implies not only the enlargement of the concept of Fermor's 'Charnockite Line' but also the shifting of this 'Line' further west coinciding with the eastern sheared margin of the Cuddapah Basin. Apart from shifting of the terrane boundary further, the tectonostratigraphic reinterpretation of the ensembles occurring between the 'East Coast Granulite Belt' and the Cuddapah Basin implies that the Eastern Ghats Granulite Belt not only is made entirely of granulites but also includes middle to low-grade metamorphic components in parts.

The structural relationship shown in the schematic map (Fig. 3.2) is based on the tectonostratigraphic reinterpretation of the rock ensembles discussed above.

In the southern part of the (Peninsular) Indian Shield, the irregular, broadly east-west trending Charnockite Line of Fermor (1936) is said to demarcate contact between the N–S trending Dharwar granite-greenstone ensembles in the north and the dominantly charnockite-bearing granulite terrane in the south (Fig. 3.3). As in the case of the Eastern Ghats Granulite Belt, here also the Charnockite Line is considered a terrane boundary between two distinctly different crustal types having different tectonostratigraphic significance.

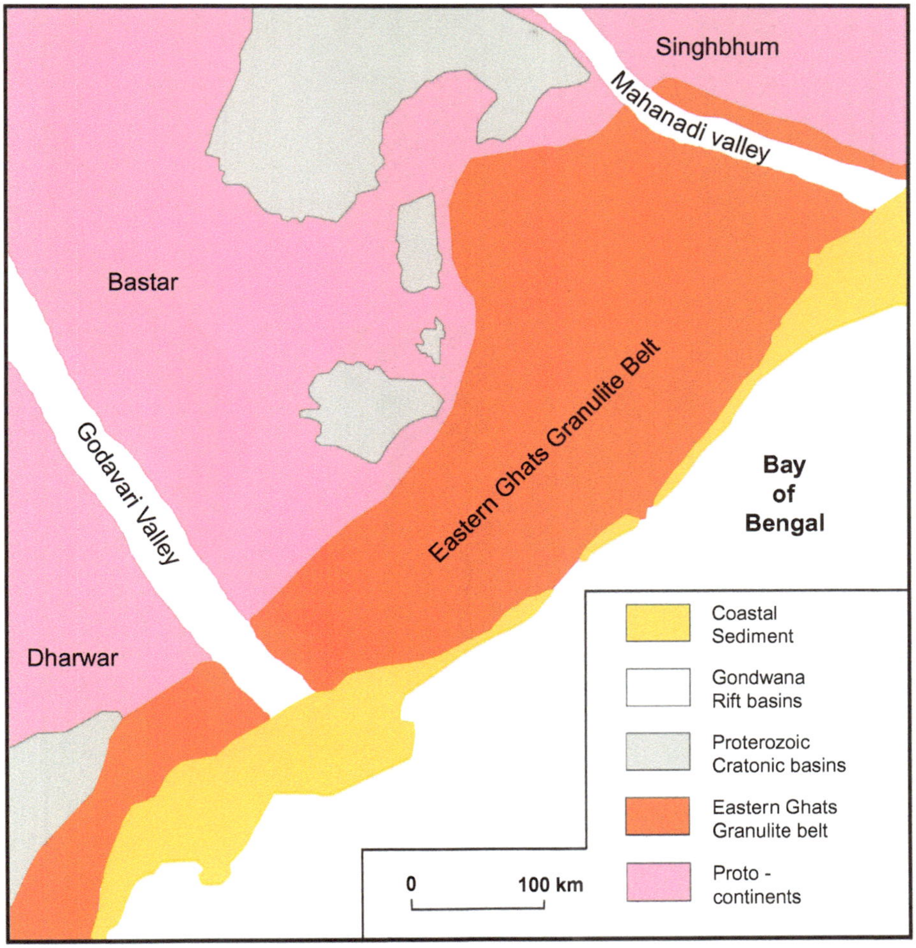

FIG. 3.2

Geological map showing extent of the Eastern Ghats Granulite Belt (with the redefined western boundary in the south), truncating the structural grains of the Dharwar, Bastar, and Singhbhum Protocontinents.

Reproduced from Roy (2014) with permission.

Contradicting Fermor's concept, some later workers interpreted the Charnockite Line as a metamorphic isograd boundary (Allen et al., 1983; Ramakrishnan, 2003) marking a regional southward progression of metamorphism from the greenstone to upper amphibolite grade in the north to the hypersthene-bearing granulite grade in the south (Janardhan, 1983). In other words, the concept implies a continuity of the Archaean Dharwar rocks further south of the Charnockite Line of Fermor (1936).

The proof of the presence of a single continuous belt of rocks also comes from the occurrences of the trails of the coevally metamorphosed Archaean granite-greenstone belt rocks showing a metamorphic state compatible with the enclosing granulite bodies along with the 'tonalite-trondhjemite' bearing Peninsular Gneiss, metabasaltic dykes of different generations, and their granulite equivalents well

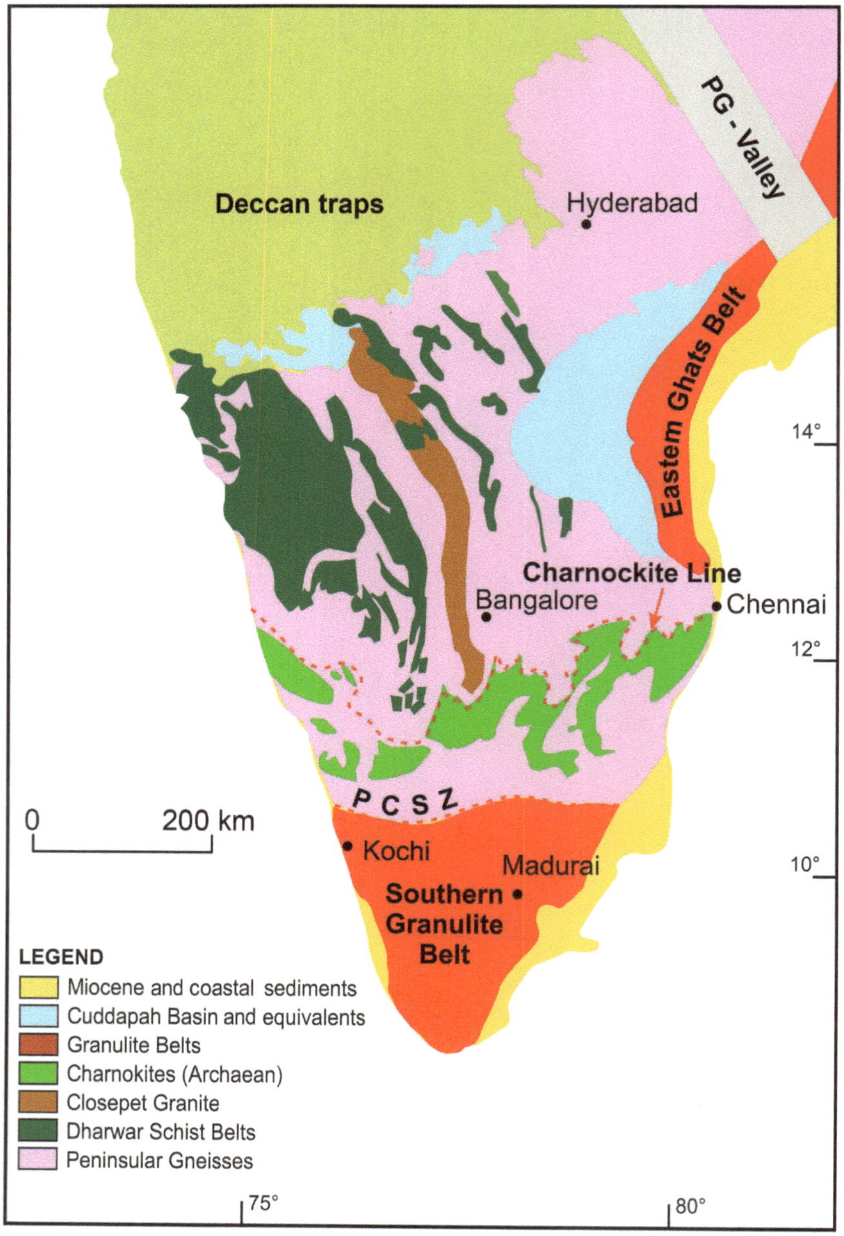

FIG. 3.3

Lithotectonic map of the southern part of the Peninsular Indian Shield showing major tectonic features. PCSZ, Palghat-Cauvery Shear Zone.

Reproduced from Roy (2014) with permission.

within the massifs of Biligirirangan, Coorg, and Nilgiri Hills in the south (Rama Rao, 1940; Janardhan, 1983). The reports of end-Archaean ages ranging between 2500 and 2900 Ma both from the charnockite granulites and also from the Peninsular Gneiss from the southernmost part of the Peninsular India (Venkatasubramanian, 1975; Friend, 1981; Friend and Nutman, 1992; Jayananda and Peucat, 1996) further confirm the existence of a single continuous Archaean granulite terrane, described as the High-Grade Terrane (Allen et al., 1983) south of the Charnockite Line.

Based on the study of the fault system and lineaments, Vemban et al. (1977) was the first to propose that the southernmost tip of the granulite mass occurring south of the Cauvery Fault (marking the southern boundary of the Palghat-Cauvery Shear Zone) represents a crustal block that is tectonically different from the indisputably Archaean granulite terrane in the north. The new tectonic model received weighty support of Drury and Holt (1980) and Drury et al. (1984). The geochronological support for the concept of the two diverse terranes comes from the reports of 'younger' Pan-African tectonothermal reconstitution ages centring around 500 ± 50 Ma from the granulites and associated rocks from the south of the Palghat-Cauvery Shear Zone (Harris et al., 1994; Chaudhary et al., 1992; Bartlett et al., 1995; Unnikrishnan-Warrier et al., 1995; Santosh et al., 1992, 2003; Shabeer et al., 2004; Collins et al., 2007). Both the Satellite Imagery and isotope data, therefore, provide undeniable proof for the existence of two juxtaposed terranes across the Palghat-Cauvery Shear Zone. The northern granulites are now considered an extended zone of the Dharwar Protocontinent, characteristically bearing signatures of Archaean evolutionary history, while the Southern Granulite Terrain, shows evidence of strong Pan-African tectonothermal reconstitution south of the Palghat-Cauvery Shear Zone (Srikantappa et al., 2003).

Naming the southern terrane, south of the Palghat-Cauvery Shear Zone as the Southern Granulite Belt may appear misleading because the same nomenclature is also in use for the entire granulite terrane south of the Charnockite Line of Fermor. Possibly because of that, Ramakrishnan (1988) introduced a new term 'Pandyan Mobile Belt' for the southernmost granulite terrane. However, this new nomenclature though appears quite rational did not find general acceptance in literatures (Sharma, 2010).

The Palghat-Cauvery Shear Zone, which acts as a tectonic divide between two distinctive terranes, is considered a Suture Zone marking accretion (or welding) of diverse crustal blocks (Radhakrishna and Naqvi, 1986; Gopalkrishnan, 1996). Some authors advocate that the accretion was caused during collision (of the Dharwar crustal block) with a 'Craton' in Africa or Antarctica (Ramakrishnan, 2003; Sharma, 2010).

The rocks of the redefined Southern Granulite Belt show many similarities in lithologic and metamorphic character with those of the Eastern Ghats Granulite Belt. Both the belts show Archaean-Palaeoproterozoic protolithic ages of the gneisses and granulites but have undergone significant reconstitution (under granulite facies condition) by the late-Proterozoic tectonothermal events. The combined terranes are known to have their counterparts outside the domain of the Indian Peninsula. Apart from the possibility of their being 'exotic' or 'suspect terranes' (Radhakrishna, 1989), the granulite and other high-grade metamorphic rocks comprising the Eastern Ghats Granulite Belt and the Southern Granulite Belt (even considering the revised boundaries) appear distinctly different from the components of the Noncharnockite belt in terms of their lithostratigraphic character, tectonometamorphic evolutionary history, geochronologic framework, and metallogenic traits (Ramakrishnan and Vaidyanadhan, 2008; Sharma, 2010; Valdiya, 2010). Characterization of the charnockite belts as exotic or suspect terranes implies that the Noncharnockite domain constituted the primary crustal block or 'Proto-India' (Dobmeir and Raith, 2003), the frontiers of which was enlarged because of the subsequent accretion of Charnockite terranes. The concept is ingrained in the thought of Mahadevan (1994) who conceived of a 'Central Core Region' as the primary crustal block constituting the Peninsular Indian Shield to which the granulite belts were later accreted. In the present description, however, the term Proto-India (Dobmeir and Raith, 2003) is preferred as it appears a little more evocative in describing the framework of the Indian Shield.

Following the suggestion of the early geologists of the late nineteenth century of the Geological Survey of India (Medlicott and Blandord, 1879–81), the Proto-India comprising components of Fermor's (1936) Noncharnockite terrane can be divided into a northern Aravalli-Bundelkhand domain and a southern Gondwana domain. The two domains are physiographically separated by the Narmada-Son River Valley, which today is better known as the Narmada-Son Lineament (Mahadevan, 1994). The Gondwana rift basins developed only in the southern Gondwana domain and partly along the southern fringe of the Aravalli-Bundelkhand domain.

The Aravalli-Bundelkhand domain can be further subdivided into an eastern 'Bundelkhand Mass' and a western Aravalli Mountains (including the western sandy plain of Thar Desert). In spite of many similarities, the emerging thought is that these two Precambrian crustal blocks have evolved more or less contemporaneously but under two different tectonic settings. The Archaean Crust of the Aravalli Mountains enlarged its dimension through stabilization (cratonization) of three successive orogenic belts during Palaeoproterozoic-early Neoproterozoic (Roy and Jakhar, 2002). The Bundelkhand Mass, on the other hand, portrays a different picture of late-Archaean cratonization with the deposition of Palaeo-Mesoproterozoic sedimentary-volcanic succession under stable platformal condition. The difference between the two is also reflected in some way in the lithologic character, the tectonic pattern, and the metallogenic behaviour (Roy and Jakhar, 2002). Considering all these, it may be rational to subdivide the northern domain into two fundamental nuclei, the Bundelkhand and Aravalli Protocontinents, separated by the Great Boundary Fault.

The southern Gondwana domain can be subdivided into a number of separate 'fundamental nuclei' or Protocontinents, which are separated by narrow belts of 'Gondwana rift basins'. For example, the Gondwana rift basin of the Pranhita-Godavari Valley divides the Dharwar from the Bastar Protocontinent. On the other hand, the Gondwana rift basin of the Mahanadi Valley separates the Bastar from the Singhbhum Protocontinent. Thus, looking into the faithful regularity with which the Gondwana rift basins occur between the Protocontinents, we may consider these as zones separating the Protocontinental nucleus (Fig. 3.4). In the present description, the nondescript term 'Join' (hiring the term from Rogers, 1986) is used as marker separating the primary crustal blocks. Very little is known about these zones along which Joins are traced except the antiquity of the rift zones over which the Gondwana Basins evolved (Naqvi et al., 1974). In the Pranhita-Godavari Valley, which marks the Join between the Dharwar and Bastar Protocontinents, the cratonic platformal Proterozoic sediments form the basement for the Gondwana Rift Basins. In other instances, the Gondwana Rift Basin (marking Joins) overlies the unclassified Precambrian granite and gneisses.

The proposed scheme of dividing the Gondwana domain by Joins marked by the Gondwana Rift Basins would imply delimitation of the Singhbhum Protocontinent by Gondwana Rift Basin of the Damodar Valley in the north. This leaves a small region between the Narmada-Son Lineament in the northwest and the Damodar Valley Gondwana Basins in the south. If we accept the poorly exposed, north-south trending coal occurrences of the Rajmahal area as the remnant of a Gondwana Basin, then the extent of this continental nucleus would be the small triangular area bounded by the Narmada-Son

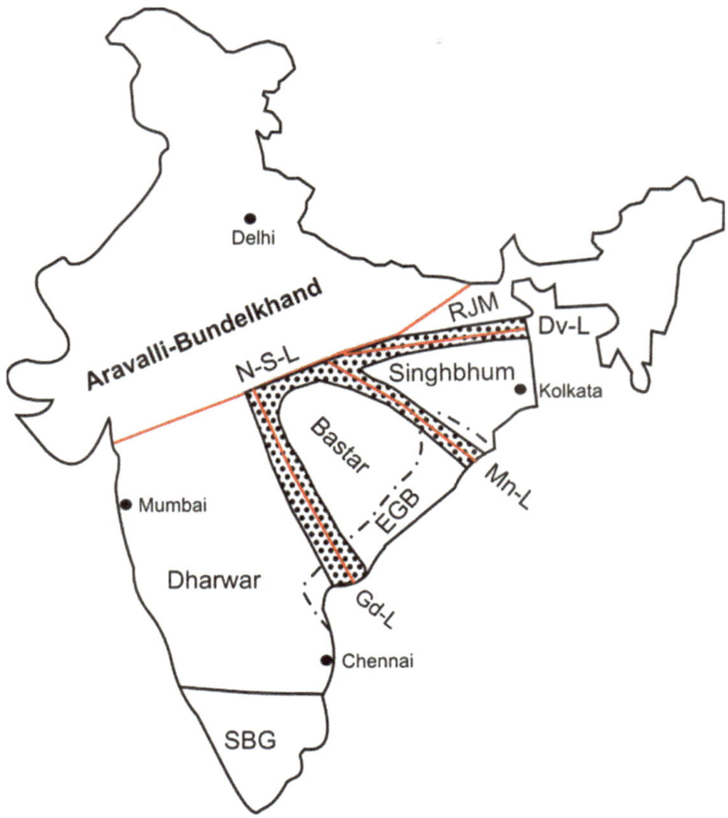

FIG. 3.4

Schematic map showing distribution of Joins separating the Protocontinents of the Gondwana domain of Proto-India. N-S-L, Narmada-Son Lineament; Dv-L, Damodar Valley Lineament; Gd-L, Pranhita-Godavari Valley Lineament; Mn-L, Mahanadi Valley Lineament; RJM, Rajmahal Protoconttinent.

Reproduced from Roy (2014) with permission.

Lineament in the northwest, the Damodar Valley Gondwana Basins in the south, and the Rajmahal coal occurrences in the east. On the other hand, the reported presence of Gondwana rift basins in the Bengal Basin under the thick cover of 'Cenozoic to Recent' sediments in the 'Garo-Rajmahal Gap' (in Bangladesh) across the north-south lying outcrops of the Rajmahal Traps strongly suggests the continuity of the Damodar Valley Basins further east (Khan et al., 1994). Thus, if we neglect the Rajmahal coal belts as misidentified *freak* occurrences, then there lies the possibility of extension of this northern Precambrian nuclei further east, and is now covered under the Bengal Basin. We may describe this ill-defined crustal block as the Rajmahal Protocontinent.

Some Gondwana-type coal basins are known from different parts of the eastern Himalayas. However, because of the highly dismembered and tectonized state of these Gondwana Basin bodies, it is hard even to guess their lithotectonic/stratigraphic significance.

REFERENCES

Allen, P., Condie, K.C., Narayana, B.L., 1983. The Archaean low- to high-grade transition in the Krishnagiri-Dharmapuri area. Tamilnadu, Southern India. In: Naqvi, S.M., Rogers, J.J.W. (Eds.), Precambrian of South India. Mem. Geol. Soc. India, Bangalore, vol. 4, pp. 450–461.

Babu, V.R.R.M., 1998. Nellore schist belt – an Archaean green stone belt. Andhra Pradesh. In: Rao, A.T., Divi, S.R., Yoshida, M. (Eds.), Precambrian Crustal Processes in East Coast Granulite-Greenstone Regions of India Within East Gondwana (IGCP). Gondwana Research Group Memoir, vol. 4, pp. 97–136.

Bartlett, J.M., Harris, N.B.W., Hawkesworth, C.J., Santosh, M., 1995. New isotope constraints on crustal evolution of South India and Pan-African Granulite metamorphism. Yoshida, M., Santosh, M. (Eds.), Indian and Antarctica During the Precambrian. Mem. Geol. Soc. India, Bangalore, vol. 34, pp. 391–397.

Bhadra, S., Gupta, S., Banerjee, M., 2004. Structural evolution across the Eastern Ghats Mobile Belt–Bastar craton boundary, India: hot over cold thrusting in an ancient collision zone. J. Struct. Geol. 26, 233–245.

Biswal, T.K., Sinha, S., 2003. Deformation history of the NW salient of the Eastern Ghats Mobile Belt, India. J. Asian Earth Sci. 22, 157–169.

Biswal, T.K., Sinha, S., 2004. Fold-thrust belt structure of the Proterozoic Eastern Ghats mobile belt: a proposed correlation between India and Antarctica in Gondwanaland. Gondwana Res. 7, 43–56.

Biswal, T.K., Jena, S.K., Datta, S., Das, R., Khan, K., 2000. Deformation of Terrane Boundary shear Zone (Lakha Shear Zone) between the Eastern Ghats Mobile belt and the Bastar Craton, in Bolangir and Kalahandi districts of Orissa. J. Geol. Soc. India 55, 367–380.

Chaudhary, A.K., Harris, N.B.W., Van Clasteren, P.C., Hawkesworth, C.J., 1992. Pan-African charnockite formation in Kerala, South India. Geol. Mag. 129, 257–264.

Chetty, T.R.K., 2001. The Eastern Ghats Mobile Belt, India: a collage of juxtaposed terranes (?). Gondwana Res. 4, 319–328.

Chetty, T.R.K., 2010. Structural architecture of the northern composite terrane, the Eastern Ghats Mobile Belt, India: implications for Gondwana tectonic. Gondwana Res. 18, 565–582.

Collins, A.S., Santosh, M., Braun, I., Clark, C., 2007. Age and sedimentary provenance of Southern Granulites of Southern India: U-Th-Pb SHRIMP secondary ion mass spectrometry. Precambrian Res. 155, 125–138.

Dobmeir, C., Raith, M.M., 2003. Crustal architecture and evolution of the Eastern Ghats Belt and adjacent regions of India. In: Yoshida, M., Windley, B.F., Dasgupta, S. (Eds.), Proterozoic East Gondwana Supercontinent Assembly and Breakup. vol. 206. Geol. Soc., London, Sp. Pub. pp. 145–168.

Drury, S.A., Holt, R.W., 1980. The tectonic framework of the south Indian craton: a reconnaissance involving Landsat imagery. Tectonophysics 65, 1–15.

Drury, S.A., Harris, N.B.H., Holt, R.W., Reeves-Smith, G.J., Wightman, R.T., 1984. Precambrian tectonics and crustal evolution in south India. J. Geol. 92, 3–20.

Fermor, L.L., 1936. An attempt at the correlation ancient schistose formation of Peninsular India. Mem. Geol. Surv. India 70 (1), 51.

Friend, C.R.L., 1981. The timing of charnockite and granite formation in relation to influx of CO_2 at Kabbaldurga, Karnataka, South India. Nature 294, 550–552.

Friend, C.R.L., Nutman, A.P., 1992. Response of U-Pb isotopes and whole rock geochemistry of CO_2 induced granulite facies metamorphism, Kabbaldurga, Karnataka, south India. Contrib. Mineral. Petrol. 111, 299–310.

Gopalkrishnan, K., 1996. An overview of Southern Granulite Terrain, India-Constraints in Reconstruction of Precambrian Assembly of Gondwanaland. In: Gondwana Nine. vol. 2. Oxford-IBH, New Delhi, pp. 1003–1026.

Gupta, S., Bhattacharya, A., Raith, M., Nanda, J.K., 2000. Contrasting pressure–temperature-deformation history across a vestigial craton mobile belt boundary: the western margin of the Eastern Ghats belt at Deobhog, India. J. Metamorph. Geol. 18, 683–697.

Gupta, S., Nanda, J., Mukherjee, S.K., Santra, M., 2005. Alkaline magmatism versus Collision Tectonics in the Eastern Ghats Belt, India: constraints from structural studies in the koraput complex. Gondwana Res. 8, 1–17.

Harris, N.B.W., Santosh, M., Tayler, P., 1994. Crustal evolution of south India: constraints from Nd isotopes. J. Geol. 102, 139–150.
Hatcher Jr., R.D., 1978. Enclyclopedia of Geological Sciences. McGraw-Hill. p. 816.
Janardhan, A.S., 1983. In: Naqvi, S.M., Rogers, J.J.W. (Eds.), Precambrian of South India. Mem. Geol. Soc. India, Bangalore, vol. 4, pp. 417–435.
Jayananda, M., Peucat, J.J., 1996. Geochronological framework of Southern India. Santosh, M., Yoshida, M. (Eds.), The Archaean-and Proterozoic Terrains in Southern India Within East Gondwana. Gondwana Research Group Memoir, vol. 3, pp. 53–75.
Khan, A.A., Sattar, G.S., Rahman, T., 1994. In: Tectogenesis of the Gondwana rifted basins of Bangladesh in the so-called Garo-Rajmahal Gap and their drift regional tectonic correlation. Proc. Ninth International Gondwana Symposium, Oxford-IBH, New Delhi, pp. 647–655.
Krishnan, M.S., 1956. Geology of India and Burma. Higginbothams (Private) Ltd., Madras. p. 555.
Leelanandam, C., Burke, K., Ashwal, L.D., Webb, S.J., 2006. Proterozoic Mountain Building in Peninsular India: analysis based primarily on alkaline rock distribution. Geol. Mag. 143, 195–212.
Mahadevan, T.M., 1994. Deep continental structure of India. Mem. Geol. Surv. India 28, p. 569.
Medlicott, H.B., Blandord, W.T., 1879–1881. Manual Geology of India (Part I, Part II). Geol. Surv. of India, Calcutta.
Naha, K., 1964. A critique of the orogenic trends in Archaean correlation in India. Tectonophysics 1, 431–438.
Naqvi, S.M., Divakara Rao, V., Narayan, H., 1974. Archaean protocontinental growth of the Indian Shield and the antiquity of the rift valleys. Precambrian Res. 1, 345–398.
Radhakrishna, B.P., 1989. Suspect tectono stratigraphic terrane elements in the Indian Subcontinent. J. Geol. Soc. India 34, 1–24.
Radhakrishna, B.P., Naqvi, S.M., 1986. Precambrian continental crust of India and its evolution. J. Geol. 94, 145–166.
Rama Rao, B., 1940. The archaean complex of Mysore. Mysore Geol. Dept. Bull. 17, 1–101.
Ramakrishnan, M., 1988. Tectonic evolution of the Archaean high-grade terrain of south India. J. Geol. Soc. India. 31, 118–119.
Ramakrishnan, M., 2003. Craton-Mobile Belt relations in Southern Granulite Terrain. Ramakrishna, M. (Ed.), Tectonic of the Southern Granulite Terrain: Kuppam-Palani Geotransect. Mem. Geol. Soc. India, Bangalore, vol. 50, pp. 1–24.
Ramakrishnan, M., Vaidyanadhan, R., 2008. Geology of India. vol. I. Geological Society of India, Bangalore. p. 556.
Ramakrishnan, M., Nanda, J.K., Augustine, P.F., 1998. Geological evolution of the Proterozoic Eastern Ghats mobile belt. Geol. Surv. India Spec. Publ. 44, 1–21.
Ramam, P.K., Murthy, V.N., 1997. Geology of Andhra Pradesh. Geol. Soc India, Bangalore. p. 245.
Rickers, K., Mezger, K., Raith, M.M., 2001. Evolution of continental crust in the Proterozoic Eastern Ghats Belt, India and new constraints for Rodinia reconstruction: implications from Sm-Nd, Rb-Sr and Pb-Pb isotopes. Precambrian Res. 112, 183–212.
Rogers, J.J., 1986. Dharwar craton and the assembly of Peninsular India. J. Geol. 94, 129–143.
Roy, A.B., 2014. Indian Subcontinent, Reference Module in Earth Systems and Environmental Sciences. Elsevier. 11-Aug-14. https://doi.org/10.1016/B978-0-12-409548-9.09372-6.
Roy, A.B., Jakhar, S.R., 2002. Geology of Rajasthan: Precambrian to Recent. Scientific Publishers (India), Jodhpur. p. 421.
Santosh, M., Kagami, H., Yoshida, M., Nada-Kumar, V., 1992. Pan-African charnockite formation in East Gondwana: geochronologic (Sm-Nd and Rb-Sr) and petrologic constraints. Bull. Indian Geol. Assoc. 25, 1–10.
Santosh, M., Yokoyama, K., Biju-Sekhar, S., Rogers, J.J.W., 2003. Multiple tectonothermal events in the granulite blocks of southern India revealed from EPMA dating; implication for theory of supercontinent. Gondwana Res. 6, 29–63.
Shabeer, K.P., Satish Kumar, M., Armstrong, M., Buick, I.S., 2004. Constraints on timing of Pan-African granulite facies metamorphism in the Kerala Khondalite Belt of south India: SHRIMP mineral ages and Nd isotope systematics. J. Geol. 113, 95–106.

Sharma, R.S., 2010. Cratons and Fold belts of India, Lecture notes in Earth Sciences. vol. 127. Spinger Verlag. p. 304.

Srikantappa, C., Srinivas, G., Basavarajappa, H.T., Prakash Narasimha, K.N., BasavaLingu, B., 2003. Metamorphic evolution and fluid regime in the deep continental crust along the N-S geotransect from Vellar to Dharapuram, southern India. In: Ramakrishnan M. (Ed.), Tectonics in Southern Granulite Terrane: KuppampPalani Geotransect. Mem. Geol. Soc. India, Bangalore, vol. 50, pp. 318–374.

Unnikrishnan-Warrier, C., Santosh, M., Yoshida, M., 1995. First report of Pan-African Sm-Nd and Rb-Sr mineral isochron ages from regional charnockites of southern India. Geol. Mag. 132, 253–260.

Valdiya, K.S., 2010. The Making of India. Macmillan Publishers India Ltd. p. 816.

Vemban, N.A., Subramanian, K.S., GopalKrishna, K., Venkata Rao, V., 1977. Major faults, dislocation lineaments of Tamilnadu. Geol. Surv. India Misc. Publ. 31, 53–56.

Venkatasubramanian, V.S., 1975. Studies in the geochronology of the Mysore Craton. Geophys. Res. Bull. NGR India. 13, 239–246.

PART

PRECAMBRIAN CRUSTAL BLOCKS OF INDIAN SHIELD

2

CHAPTER

DHARWAR PROTOCONTINENT 4

4.1 GEOLOGY OF THE ARCHAEAN CRUST: CONCEPTS AND PERSPECTIVES

The Dharwar Protocontinent that covers a large area of the Indian Peninsula is known as one of the best-studied Precambrian crustal block of the Indian Shield. However, in spite of the extensive studies that began as early as late nineteenth century (Bruce Foote, 1876), the lithotectonic relationship between different entities remained grossly misunderstood and baffling (Valdiya, 2010). Recent studies that rely heavily on geochronology and isotope data did not help much in sorting out the relationship between different lithotectonic components of the Dharwar Protocontinent. Added to this is the uncritical acceptance of some time-honoured thoughts that have never received rational scientific scrutiny. One of this is the model of continuous increase in the grade of metamorphism from the greenschist facies condition in the north through amphibolite to granulite facies in the south. The idea is so much ingrained in the thought and belief that it has stimulated several authors to suggest that the Dharwar is a regionally tilted block with a general tilt varying between 2 and 5 degrees, progressively exposing the palaeodepth of 10 km in the north and 25 km in the south. In that way, a depth between 60 and 65 km in the southern most part has been suggested (Ramakrishnan and Vaidhyanadhan, 2008). The concept itself is fallacious because of the fact that east–west trending isograd line that demarcates the amphibolite-granulite facies transition ('Charnockite Line' of Fermor, Fig. 3.3) cross-cut the greenstone belts having dominantly north–south trend of the tectonic grains represented by fold trends and orientation of foliation. The fact, however, is other way round. Truly speaking, the concept does not seem to have taken into consideration that the apparent increase in metamorphic grade could be a misconception ensuing from superimposition of multiple thermal events at different times. Referring to the recent isotopes studies, Jayananda et al. (2013) highlighted on the occurrence of more than one event of regional high-grade metamorphism in the Dharwar Protocontinent. The study clearly focuses on the delusion of increasing metamorphism from north to south since it was originally suggested about five decades ago (Pichamuthu, 1965).

Another disquieting fact is the division of the granite-gneiss-greenstone terrane into two different blocks: Eastern Dharwar Craton and Western Dharwar Craton (Swami Nath and Ramakrishnan, 1981). The division finds support from Rajamani et al. (1987) who presented a Plate Tectonic model in interpreting the evolution of Kolar Schist Belt based on some ill-constrained isotope data. The occurrence of the linear belt of Closepet Granite has been conceived as a 'convenient landmark' representing 'stitching granite' in the 'Suture Zone' to demarcate an important 'geosuture' in this part of the Archaean crustal block (Ramakrishnan and Vaidhyanadhan, 2008). Except such a 'grand' postulation, there is hardly any elucidation of tectonic criterion to divide this Archaean gneiss-granite-greenstone belt apart from the suggestions like 'the nature and abundance of supracrustal belts, crustal thickness, lithological assemblage, grade of regional metamorphism and degree of melting' (Swami Nath et al., 1976; Jayananda et al., 2006). According to Naqvi and Rogers (1986), the boundary between the two suggested blocks is uncertain, if

Indian Shield. https://doi.org/10.1016/B978-0-12-809839-4.00004-7

at all it exists. The most apparent difference could be the fact that the eastern part of the Dharwar Belt contains abundant 'post-tectonic' younger granites unlike the western part that show scarcity of such rocks. There is also suggestion that a major shear zone occurring west of the Closepet Granite outcrops could be the dividing 'line' between the two blocks. Drury et al. (1984), on the other hand, recognized several 'zones of high strain' out of which the one between the Chitradurga and Javanhalli belts has been considered as the possible tectonic divide between the two hypothetical blocks. In short, apart from the fact that mere occurrence of 'Ductile Shear Zones' (strong or weak) does not prove the presence of any Suture Zone marking a Plate boundary, the contradictory views postulated about the possible dividing 'line' between the two assumed crustal blocks make the concept scientifically untenable.

Remaining unconcerned about the fallacy of dividing the Archaean Dharwar Crust into western and eastern blocks, a new suggestion has been made dividing the Archaean crustal block into three or more tectono-metamorphic provinces based mainly on some geochronological data (Jayananda et al., 2013). Available geological field data do not seem to justify any such division of the Archaean crustal block of the Dharwar Province.

BOX 4.1

Greenstone belts are zones of metamorphosed mafic/ultramafic volcanic rocks with associated sedimentary rocks that occur in narrow basins within the Precambrian granite and gneiss bodies. The name comes from the dominance of the green coloured minerals, like chlorite and amphibole in the rocks. The greenstone belts in the Dharwar Province are traditionally described as the 'schist belts', for example, Kolar Schist Belt.

BOX 4.2

Supracrustal rocks (supra is from Latin meaning 'above') are those that were deposited on the existing basement rocks (usually granites and gneisses) of the Crust, hence the name. They may be further metamorphosed from both sedimentary and volcanic rocks. The supracrustal rocks are also described as the cover rocks/succession.

Keeping all these in mind, the present discussion is planned to highlight objectively the most characteristic features of the Dharwar Protocontinent, the occurrence of a number of Archaean schist belts, separated by wide zones of gneiss-granite terrane. Quantitatively, not more than 20% of the entire Archaean terrain host schist belts that evolved as greenstone belts of diverse dimensions and in all possibility at different times. Overlying this Archaean crustal block with profound unconformities occur several Proterozoic essentially intracratonic (Platform-type) cover successions of which the largest crescent-shaped outcrop of Proterozoic Cuddapah Basin occur in the east. There are also a number of other contemporaneous basins in the north like the Bhima and Kaladgi Basins.

4.2 EXTENT OF THE DHARWAR PROTOCONTINENT

Barring the northern margin, the boundaries of the Dharwar Protocontinent is pretty well defined. In the north, it extends up to the Narmada Basin beneath the covers of the Deccan Traps. Its western boundary that is thought to continue into Madagascar (assuming the pre-fragmented situation prior to the

Gondwana break-up) is presently truncated at the eastern coastline of the Arabian Sea. The northeastern boundary of the Protocontinent is bounded by the Godavari 'graben' comprising the Proterozoic Pranhita-Godavari and the Phanerozoic Gondwana basins (Fig. 3.3). Unlike these well-defined boundaries, the conventionally known eastern and the southern boundary of this large-sized Protocontinent requires redefinition in the light of some recent studies. Regarding the eastern boundary, contradicting the earlier concept that the 'Nellore-Khammam Schist Belts' that occur east of the large arcuate outcrop of the Cuddapah rocks constitute a part of the Dharwar Province (Protocontinent), recent studies indicated that these ensembles formed a part of a much younger (possibly Mesoproterozoic) geological entity totally unrelated to the Dharwar (Dobmeir and Raith, 2003; Ramakrishnan, 2003).

In the southern part of the Peninsular Indian Shield, the irregular, broadly east–west lying Charnockite Line of Fermor (1936) is traditionally considered to demarcate the southern margin of the north–south trending Dharwar granite-greenstone ensembles separated from the dominantly charnockite-bearing granulite terrane in the south. However, the southern limit of this Protocontinent as emphasized in several recent studies (Ramakrishnan and Vaidhyanadhan, 2008) seems to favour an extension up to the Palghat-Cauvery Shear Zone (Fig. 3.3) further south of Fermor's Charnockite Line.

4.3 PENINSULAR GNEISS: THE PROBLEM OF BASEMENT AND BASEMENT–COVER RELATIONSHIP

Bruce Foote (1876) was the first to mention that the 'gold-bearing' schistose rocks of the Dharwar belt overlies a basement which he described as the 'fundamental granitoid gneiss'. About four decades later, Smeeth (1915) while describing the gneisses as the 'Peninsular Gneiss Complex', interpreted these rocks as intrusive into the 'Dharwar schists'. The perception is evidently based on the misunderstanding that the gneissic rocks formed by tectono-metamorphic transformation of pre-existing 'granitic' masses at higher temperature-pressure condition could intrude into the greenstone belt rocks. The misplaced thought was clarified by Fermor (1936) who made the suggestion that the intrusive masses were nothing but the products of remelting of the pre-existing gneissic basement rocks. With time, the term Peninsular Gneiss Complex introduced by Smeeth (1915) got shortened into Peninsular Gneiss (Radhakrishna, 1983).

The Peninsular Gneiss of Dharwar Protocontinent like most of the gneiss complexes in the world shows great diversity of the suite of rocks comprising gneiss-granite ensembles formed by different processes at different times. A vast majority of rocks of this group show complex pattern of banding providing evidence of multiple folding and ductile shearing (Fig. 4.1a).

The uniformly planar geometry of thin bands occurring in the Ductile Shear Zones appears similar to those that form during the plastic deformation under the directed stress system (mylonitization). Several such bodies have been erroneously described in literature as migmatites or migmatitic gneisses.

Compositionally, the protoliths of most of the Peninsular Gneiss are meta-igneous rocks. The suggestion was made by Jayaram and Rajagopalan (1984) mainly based on the study of REE and other element distribution in the rocks. Most of the basement gneisses are of tonalite-trondhjemite-granodiorite (TTG) in composition, and have a large number of enclaves of different dimension made of wide variety of screens and inclusions of other rock types. But unlike the normal grey-coloured TTG gneisses, some of the Peninsular Gneiss appears much darker in shade with only few thinner leucocratic bands (Fig. 4.1b). In spite of the little known chemistry of these rocks, we cannot completely rule out the possibility that the protoliths of these rocks were mafic or dioritic in composition (Jayaram and Rajagopalan, 1984).

(a) (b)

FIG. 4.1

(a) Intricately folded outcrop of Peninsular Gneiss showing effects of polyphase deformation and reconstitution in the southern part of the Protocontinent; (b) Peninsular Gneiss showing *dark grey colour*, interbanded with a number of thin leucocratic bands in the southern part of the Dharwar Protocontinent.

The diverse components of the Peninsular Gneiss are so intimately intermixed that it is impossible to map each typological part separately. The only variants of the Peninsular Gneiss that has been mapped separately are the poorly foliated to massive, diapiric bodies of trondhjemite (grading to granodiorite and rarely to potassic granite). Classic examples of bodies of trondhjemite cross-cutting the schist belts occur at Holenarasipur Schist Belt (mapped as Holekote Trondhjemite), Kolar Schist Belt, Chitradurga Schist Belt, and several other schist belts. Cross-cutting relationship between the Peninsula Gneiss and younger K-granite is also observed close to the southern part of the Closepet Granite outcrop near Mysore.

There are reports of minor occurrences of fuchsite quartzite with kyanite and/or sillimanite within the bodies of Peninsular Gneiss. Correlation of these rocks with the cover schist rocks is considered unlikely (Naqvi and Rogers, 1986). On the other hand, the presence of isolated bodies of aluminous minerals like kyanite and sillimanite in Peninsular Gneiss may hint about the possible concentration of aluminous clay like sediments in the milieu gneisses and granites, providing unequivocal evidence of development of erosional surface with patches of aluminous clay deposits within the exposed parts of the Peninsular Gneiss. The time of deposition of these aluminous clays and their lithostratigraphic relationships with the other cover succession remains unknown. However, these assemblages are not widely reported. In fact, metasedimentary components in the Peninsular Gneiss suite are very few. It may not be out of point to mention about the reported occurrence of massive kyanite and/or sillimanite bodies along surfaces of unconformity, marking Archaean-Proterozoic boundary in several parts of the Indian Shield.

Before delving into the details of basement–cover relationship, it would be an erudite exercise to suggest that all the greenstone belts in the Dharwar Protocontinent had developed essentially over sialic (gneiss-granite) basement (Naha et al., 1993). Proof for this comes from the occurrence of tonalite pebbles in conglomerates of occurring in some parts of the schist belt rocks. However, there is little information available about the field relationship between the 'banded' Peninsular Gneiss and the rocks of the schist belts, primarily because of lack of outcrop along the contacts of these rocks. Even if the contact is preserved, it is difficult to delineate the surface of stratigraphic break between the two ensembles. Mainly because of these, the basement–cover relationship remains inconspicuous in most of the places, especially in the southern and eastern parts of the Protocontinent. The contact between the gneisses and the supracrustal rocks is marked in some places by quartz-pebble conglomerate notably where younger schist belts occur above the Peninsular Gneiss as in southern border of the Bababudan belt and along the western margin of the Chitradurga belt (Fig. 4.2). Quartz-pebble conglomerate also occurs at the base of schist belts in the Chikmagalur, Siddhegudda, and Kudremukh areas.

FIG. 4.2

Quartz-pebble conglomerate from the Chitradurga Schist Belt.

Image courtesy: Manoranjan Mohanty.

The oldest age recorded in Peninsular Gneiss ranges from 3.4 to 3.2 Ga; with crustal inheritance age up to 3.8 Ga has been noted in some samples (Beckinsale et al., 1980; Bhaskar Rao et al., 1992; Meen et al., 1992; Peucat et al., 1993). Apart from these oldest ages, the Peninsula Gneiss shows record of thermal reconstitution during 3.0 and 3.1 Ga. Most of these ages are recoded in the massive trondhjemite bodies that intruded into the schist belt rocks (e.g. Holenarasipur Schist Belt) as observed in Halekote (Stroh et al., 1983), Chikmagalur (Taylor et al., 1984), and several other places. According to Naqvi and Rogers (1986), this thermal event is the most recognizable event in the Peninsular Gneiss, which represents metamorphic age in the earlier formed gneisses. The end-Archaean granitic intrusions took place in two phases. The older one is between 2.7 and 2.6 Ga intruding into 3.14 Ga old tonalitic to granodioritic gneisses in the eastern part of the Dharwar Protocontinent (Nutman et al., 1996; Balakrishnan et al., 1999). Several 2.64–2.60 Ga potassic plutons have also intruded into different parts of the Peninsular Gneiss (Jayananda et al., 2006; Chadwick et al., 2007). The youngest intrusion is 2.56 Ga old detected in the migmatitic (mylonitic?) gneisses bordering the western part of the Closepet Granite (Chardon et al., 2011). The Ion Microprobe U–Pb age obtained from the greenstone-amphibolite facies transition shows nearly concordant 2.53 Ga age that provides the firm temporal link between the amphibolite facies metamorphism and the granite emplacement at higher crustal levels. Conformable whole-rock Rb–Sr isochron age of 2.50 ± 0.05 Ga was obtained from banded charnockites and incipient charnockite in amphibolite facies gneiss in southern Karnataka by Hansen et al. (1997).

BOX 4.3

Traditionally, different abbreviations for 'ages' (time before present) are used in geological literatures. Ages are abbreviated from Latin: Ga (giga-annum) or billion years; Ma (mega-annum) or million years.

The diverse age data from different parts of the Peninsular Gneiss provide proof that the vast stretch of Peninsular Gneiss 'hosting' the schist belts has a polyphased evolutionary history covering a time span of over about 1 billion years.

The crustal inheritance age of 3.8 Ga as noted above may be considered to mark the age of the oldest protolith of the Peninsular Gneiss. The 3.4 Ga indicates the age of development of the gneissic character and the associated metamorphism. Such old ages have not been reported from any of the supracrustal schist belt rocks. The oldest age for the development of the schist belt is 3.285 ± 0.17 Ga, reported from the stratiform meta-anorthosite complex from the Holenarasipur Schist Belt (Bhaskar Rao et al., 2000). The age indicates a hiatus of about 500 million years between the formation of the oldest pristine Crust and the development of the oldest greenstone belt in the Dharwar Protocontinent.

4.4 CHARNOCKITE-GRANULITE IN SOUTHERN PART OF DHARWAR PROTOCONTINENT

Several bodies of charnockite and other granulite facies rocks occur in close association with garnet-bearing gneisses (Fig. 4.3), banded iron formation, calc-silicate rocks, quartzites, and amphibolites in the southern part of the Dharwar Protocontinent. The northern limit of these occurrences is traditionally marked as Charnockite Line (Fermor, 1936). Contradicting the traditional usage, the recent studies

FIG. 4.3

Garnet-bearing paragneiss from the Biligirirangan massif in the southern Karnatak.

help to reinterpret the 'Fermor Line' as a metamorphic isograd boundary marking a transition between the upper amphibolite facies and the hypersthene-bearing granulite facies metamorphism in the south (Janardhan, 1983; Srikantappa et al., 2003). In other words, the concept implies a continuity of the Archaean Dharwar rocks further south of the defined boundary.

The proof of the presence of a single continuous belt of rocks also comes from the occurrences of the trails of the coevally metamorphosed Archaean granite-greenstone belt rocks showing a metamorphic state compatible with the enclosing granulite bodies along with the Peninsular Gneiss, meta-basaltic dykes of different generations, and their granulitic equivalents well within the massifs of Biligirirangan, Coorg, and Nilgiri Hills in the south (Rama Rao, 1940; Janardhan, 1983). An interesting information favouring continuity of Dharwar belt further south comes from the study of trace element chemistry of charnockites and the associated granites (Mahabeleshwar and Peucat, 1988; Mahabeleshwar et al., 1995). Further, the reports of late Archaean ages both in the charnockitic granulites and in the Peninsular Gneiss from the southernmost part of the Peninsular India further confirm the existence of a single continuous Archaean granulite terrane south of the Charnockite Line (Friend, 1981; Friend and Nutman, 1992; Jayananda and Peucat, 1996; Chaudhary et al., 1992).

In several outcrops especially in the northern part, south of the east–west trending 'isograd line', the charnockite occurs either in the form of small lenses or as diffused stains of charnockitic mass described as 'patchy charnockite' within the intensely deformed tonalite-trondhjemite-granodiorite gneisses (Fig. 4.4a). The continuity of Peninsular Gneiss into the massive granulite-charnockite ensemble is detected by the presence of relict banding within granulite facies rocks (Fig. 4.4b). In addition, the continuous occurrence of the linear bodies of metasedimentary and metavolcanic rocks within the granulite charnockites from the northern most outcrops of the patchy charnockite to the southernmost limit of the Protocontinent is a clear proof of transformation of the granite-gneiss-greenstone belts into granulites.

(a) (b)

FIG. 4.4

(a) Development of patches of charnokite lenses and screens (dark brown bodies) in Peninsular Gneiss.
(b) Relict traces of banding in granulite from southeastern part of the Dharwar Protocontinent.

The ages of granulite facies metamorphism in the southern part of the Dharwar Protocontinent range between 2500 ± 50 and 2540 ± 17 Ma (Hansen et al., 1997). Grew and Manton (1986) reported

2507 and 2520 Ma age from the study of allanite in pegmatoid veins cutting across amphibolite in the Kabbaldurga Charnockite. Significantly, these ages are conformable with the age of emplacement of the Closepet Granite (discussed later).

4.5 GEOLOGY OF SCHIST BELT ROCKS

Though occupying less than one-fourth of the total exposed surface area of the Archaean Crust, the schist belts are considered important as they host several important mineral deposits, notably gold, and because of that the schist belts had been the focus of attention for over a century. However, the modern geological information on these rocks was made available about one and half century ago. The present discussion is planned to highlight the geology of a few schist belts that typify the Dharwar Protocontinent mainly based on their lithotectonic character and metallogenic traits.

4.5.1 NUGGIHALLI SCHISTS BELTS

This is a narrow belt located near Nuggihalli (13°00′: 76°28′) in the southwestern part of the Dharwar Protocontinent, extending in the northwest direction for a distance of over 60 km with a maximum width of 2 km near Nuggihalli. The ultramafic suite of the Nuggihalli belt represents a differentiated intrusion of serpentinized peridotite with lenses of altered dunite, tremolite-actinolite-cummingtonite-anthophyllite-talc-chlorite magnetite schist (metapyroxenites), and gabbroic amphibolite ± garnet with layers of anorthosite (Fig. 4.5a). Importance of this schist belt lies in the occurrence of chromite associated with peridotite, ultramafic schists, and amphibolite. There are some small lenses of anorthosite with concentration of vanadium-bearing titaniferous magnetite. Copper mineralization hosted in the gabbroic anorthosite reportedly occurs in the schist belt. There are also occurrences of a number of ancient workings for gold near the southern tip of the belt. Gold mineralization is confined to lamprophyres.

(a) (b)

FIG. 4.5

(a) Layered ultramafic body and (b) association of schistose ultramafic rocks (left) occurring close to the massive trondhjemite rocks (right) in the Nuggihalli Schist Belt.

Images courtesy, Sisir Mondal.

The gneissic rocks that occur close to the schist belt show development of uniformly planar foliation at the contact suggesting strong ductile deformation as normally observed in the mylonitic gneisses. There is no clear-cut evidence of unconformity between Peninsular Gneiss and the ultramafic schist of the Nuggihalli belt. On the other hand, the occurrence of schistose ultramafic rocks close to the massive bodies of trondhjemite suggests intrusive character of these plutonic bodies (Fig. 4.5b).

4.5.2 HOLENARSIPUR SCHIST BELT

The 'trident-shaped' Holenarasipur Schist Belt (Fig. 4.6) (Ramakrishnan and Vaidhyanadhan, 2008) is considered the most 'critical, complicated and oldest (3.2-3.5 by) supracrustal' belts in the Dharwar Protocontinent (Naqvi et al., 1978). The supposedly older westernmost succession of this schist belt includes mafic-ultramafic rocks at the base overlain by metasedimentary rocks. Compositionally, the mafic-ultramafic rocks are similar to the peridotite, pyroxenite, and basaltic komatiite. At places, these are interbedded with fuchsite quartzite.

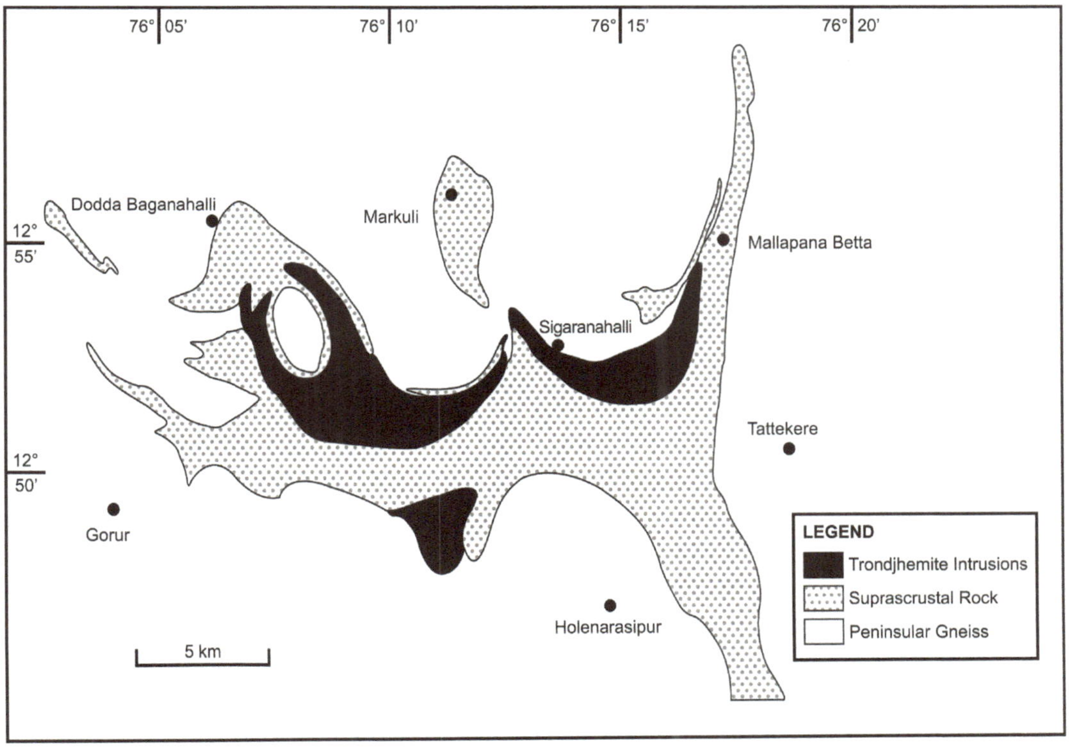

FIG. 4.6

Schematic map showing trident-shaped outcrop of in the northern part Holenarasipur Schist Belt.

The lowly metamorphosed ultramafic bodies have been widely used as convenient building materials for carving exquisite figurines and other ornamentations at the Belur and Halebid temples near Bangalore. Spinifex texture developed in the ultramafic flow is preserved in some of the images and in the scattered weathered pieces of altered ultramafic lava used for the carving of statues (Fig. 4.7a and b).

FIG. 4.7

(a) The exquisitely carved figurine in Belur Temple near Bangalore is made of altered ultramafic rocks of the Holenarasipur schist belt; (b) spinifex texture developed in the ultramafic flow (image from a weathered ultramafic lava used for carving of statues).

The overlying succession that initiate with patchy occurrence of conglomerate (named Tattekere Conglomerate, Fig. 4.8) includes a sequence of amphibolite, quartzite, and banded magnetite quartzite. Garnet-/chloritoid-bearing mica schist occurs in association with meta-basalts and quartzite. Metamorphosed felsic volcanics also occur sporadically interlayered with amphibolite/meta-basalt. The banded iron formation forms a prominent band on top of hills in the eastern arm of the schist belt.

The trident-shaped outcrop pattern of the Holenarasipur Schist Belt is thought to have developed due to the deflection of schistosity during the intrusion of plutonic 'boss-like' bodies that range in composition between trondhjemite and granodiorite (Bouhaller et al., 1993). A pre-3.00 Ga age has been suggested for the Holenarasipur Schist Belt rocks based on the age of the intrusive bodies like the Halekote Trondhjemite (Stroh et al., 1983) into the supracrustal rocks.

FIG. 4.8

Tattekere Conglomerate that separates the older, western schist belt rocks from the younger, eastern rocks.

Image courtesy: Manoranjan Mohanty.

4.5.3 WARANGAL SCHIST BELT

Warangal Schist Belt constitutes an important supracrustal succession in the northeastern part of the Dharwar Protocontinent. The lithological constituents of this easternmost greenstone belt comprise fuchsite-bearing quartzite, quartz-pebble conglomerate, cordierite-sillimanite gneiss, cordierite-anthophyllite-bearing rocks, sapphirine-kornerupine-spinel rocks, calc-silicate rocks, marbles, and banded iron formation. Mafic and ultramafic ensembles have localized but ample occurrences. Very little information is available on the lithotectonic and petrochemical characteristics of the rocks of the Warangal Schist Belt. According to Ramakrishnan and Vaidhyanadhan (2008), the Warangal rocks can be correlated with the 'oldest rock association', described as the 'Sargur Group' by the authors.

4.5.4 SALEM SCHIST BELT

The Salem has been conceived as one of the oldest 'group' of schist belt rocks in.the Dharwar Protocontinent comparable in age with the Nuggihalli and the lower part of the Holenarasipur belts. The approximate age of the evolution of this Archaean greenstone belt is thought to be between 3100 and 3300 Ma (Ramakrishnan and Vaidhyanadhan, 2008). However, the charnockite and the associated granulite from near Salem yielded Rb/Sr isochron age of 2476 ± 115 Ma by Crawford (1969). Considering the high error factor, we may presume this as the end-Archaean age of metamorphism of the older schist rocks contemporaneous with the emplacement of youngest Archaean intrusive granitoids.

The supracrustal lithology of the schist belt is poorly known because of the extensive metamorphic alteration of different rocks to granulites during end-Archaean. The age of magnesite formation is not known. Main lithology hosting magnesite deposits includes dunite that has been locally altered to serpentinite along with criss-crossed veins of magnesite (Fig. 4.9). Tectonically, the Salem Schist Belt is also known as Dharmapuri Rift Zone occurring between two NNE–SSW-trending Lineaments extending between Gudiyattam in the north and Bhavani in the south (Fig. 4.10) (Grady, 1971).

FIG. 4.9

Magnesite occurring as cross-cutting veins in altered dunite near Salem, Tamil Nadu.

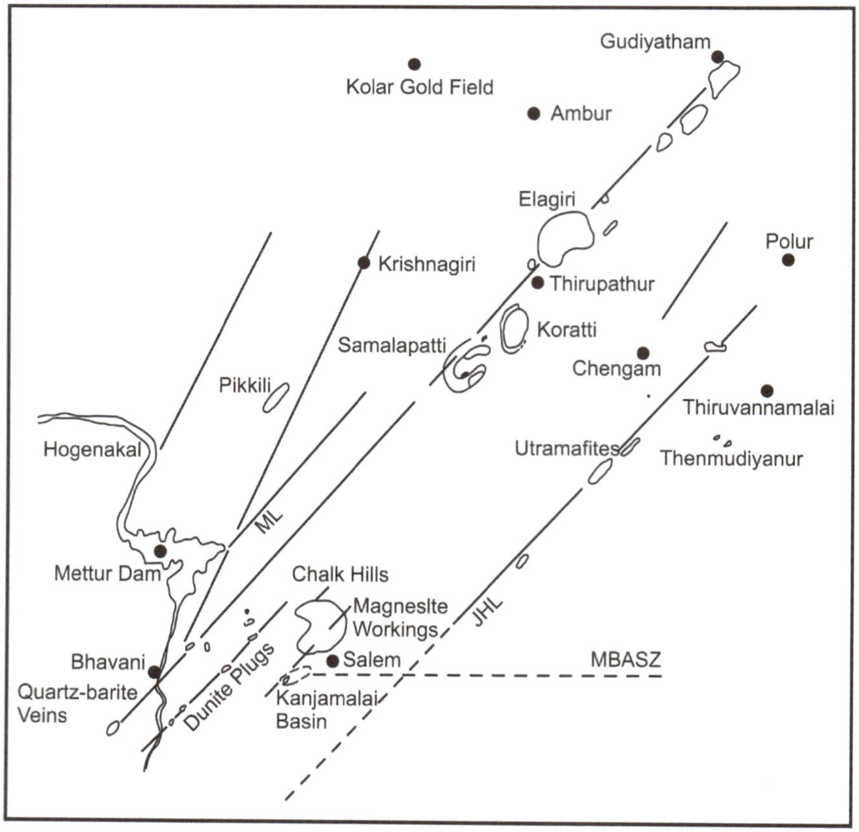

FIG. 4.10

Lineament bound Salem Schist Belt. MBASZ, Moyar-Bhavani-Attur shear zone; ML, Mettry; JHL, Javadi Hills.

Reproduced from Grady (1971) with permission.

The Salem Schist Belt, though considered as an Archaean schist belt, appears to have undergone much younger tectonothermal events related to the Quaternary deformation noted in different parts of the Indian Shield. Evidence for this comes from the occurrence of neotectonically uplifted weathered zones at Yercaud Hill much above the ground level of Salem (Fig. 4.11).

FIG. 4.11

Occurrence of bauxite deposit at the top of Yercaud Hill, providing evidence of neotectonic (Quaternary) uplift of the weathered surface much above the base level at Salem.

4.5.5 KOLAR SCHIST BELT

This is the eastern most major gold-bearing schist belt in the eastern Dharwar Protocontinent located 80 km east of Bangalore. It is an 80 km long, north–south trending schist belt consisting predominantly of amphibolites, banded iron formation, graphitic schists, and felsic schists (Fig. 4.12).

The schist belt is divided into an eastern block and a western block by a massive, fine-grained tholeiitic meta-basalt occurring in the central part of the belt. Gold mineralization occurs in the quartz-carbonate veins in the shear zones within the central massive tholeiitic meta-basalt. The felsic schists commonly known as Champion Gneiss occur along the eastern margin of the Kolar Schist Belt. These include several discontinuous bodies comprising granite-gneiss, amphibolite, banded iron formation, and vein quartz. Thin bands of graphitic schist occur along the western margin. Komatiitic meta-basalt (amphibolite) occurs on both the eastern and the western sides of the belt but more commonly along the western margin of the belt. The central zone where gold mineralization is confined within a zone strong ductile shearing. Gold mineralization in the Kolar Schist Belt shows two types of occurrence: (1) as the amphibolite-hosted quartz-carbonate vein type which is economically most important, and (2) as stratiform sulphide type hosted in the banded iron formation within amphibolite. The quartz-vein-type mineralization is of epigenetic origin. The first-order control of mineralization is tectonic, which provided structures for channel ways for the migration of fluids. These ores appear to be syngenetic within the protoliths of amphibolite. Regarding the age of the Kolar Schist Belt, Bhalla et al. (1978) recorded 2889 ± 96 Ma Rb/Sr isochron age from a sample collected from the Champion Gneiss.

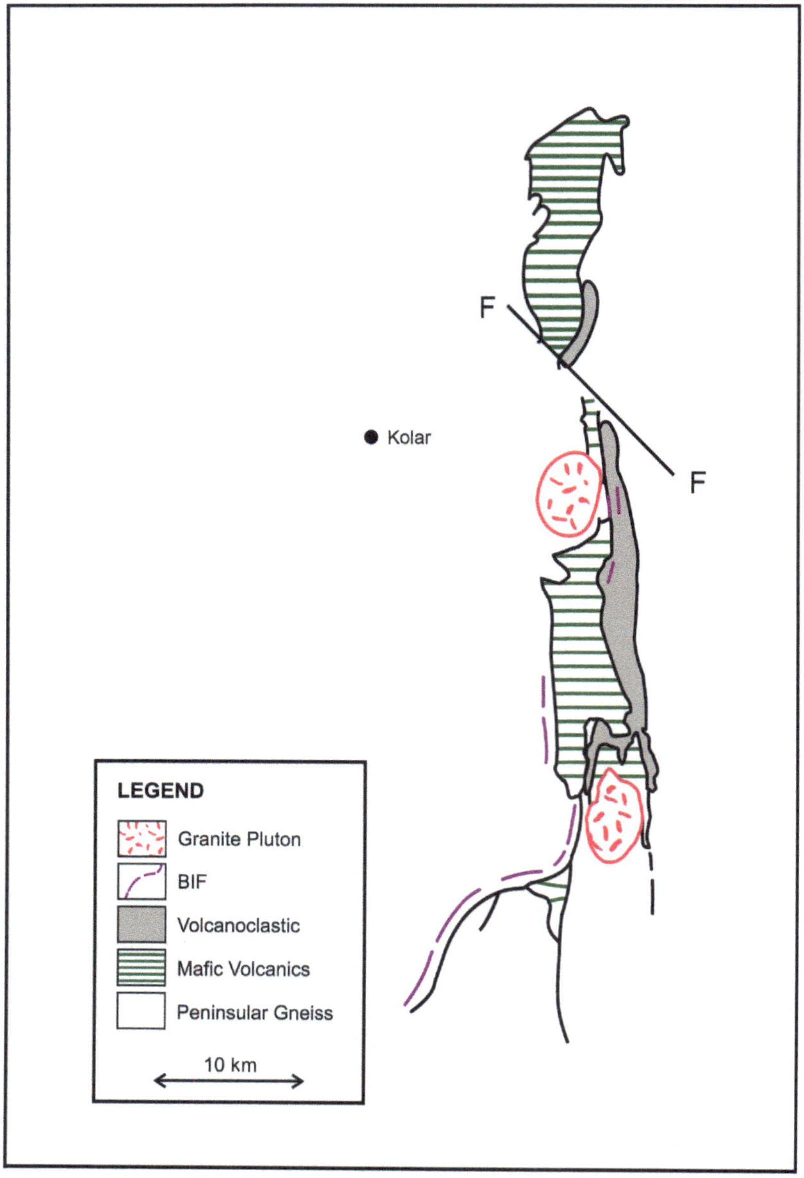

FIG. 4.12

Schematic geological map of the Kolar Schist Belt.

4.5.6 HUTTI SCHIST BELT

The Hutti Schist Belt, also known as the Hutti-Maski Schist Belt, is considered the northernmost extension of the Kolar Schist Belt occurring west of the outcrops of the Closepet Granite. It is a hook-shaped schist belt about 65 km in length and about 8 km in width, and consists predominantly of a mafic volcanic rocks with subordinate metasedimentary rocks including banded iron formation, greywacke, and polymictic conglomerate intercalated with quartzite, phyllite, limestone, and calc-silicate rocks. The volcanic rocks range in composition from basalt (the dominant suite) to acid and intermediate lavas like rhyolite porphyry. The metasediments of the Hutti Schist Belt are exposed in the eastern part of the belt and are represented by andalusite-bearing mica schist and garnet-cordierite gneiss. The schist belt is surrounded on all sides by granitic bodies of different composition, locally showing intrusive relationship.

4.5.7 CHITRADURGA SCHIST BELT

The Chitradurga Schist Belt also described as Chitradurga-Gadag Superbelt (Ramakrishnan and Vaidhyanadhan, 2008) is a 450 km long linear of Archaean greenstone belt, which vary in width between 5 and 25 km from Gadag in the north to Dodguni in the south (Fig. 4.13).

The geology of the Chitradurga belt is shrouded with controversies that could be due to the use of variety of terminology by different authors. Further, there is a general lack of understanding in the style and geometry of structure of the different units that have rendered elucidation of stratigraphic sequence complicated. Lithology wise, the belt is composed of several types of volcanic and sedimentary rocks, but on the whole, it is devoid of assemblages like ultramafic flows and layered igneous complexes similar to what have been described from some southern schist belt like the Nuggihalli Schist Belt (Ramakrishnan and Vaidhyanadhan, 2008). The volcanic rocks show wide range of composition varying from komatiitic peridotite to rhyolite. The mafic volcanics include pillowed metabasalt (Fig. 4.14a) associated with cherts hosting deposits of copper and gold (as at Ingaldhal) along with minor lead and antimony. The banded iron formation is an important component of the chemogenic sediments that include some minor carbonates (Fig. 4.14b). Amongst the terrigenous clastics, greywacke is recognized at two different stratigraphic levels. The greywacke occurring in the central part of the belt differs significantly from those of the Gadag region in the north in regards to the abundance of rock fragments, plagioclase/K-feldspar and Na_2O/K_2O ratios, and Fe_2O_3, MgO, Cr, Ni, Zr, Rb, and Sr abundances. The Gadag greywackes, on the other hand, are predominantly of volcanic source.

A stratigraphically important formation of the Chitradurga Schist Belt is the Talya Conglomerate that separates two different litho-units, mapped as older Bababudan Group and younger Chitradurga Group. The Talya Conglomerate comprises diamictites (i.e. matrix-supported conglomerates) interbedded with mudstone and sandstone. The presence of good number of faceted and bullet-shaped clasts suggests a possible reworked glacial origin of the conglomerates (Ojakangas et al., 2014) (Fig. 4.15). The Talya Conglomerate consists of clasts of basement gneiss and granitoids and orthoquartzite of the older schist belt rocks. The conglomerate is interpreted as products of debris flow type glacial deposits.

Nutman et al. (1996), based on SHRIMP U/Pb dating of zircon grains from the schistose acid volcanic rock from the Chitradurga Group, suggested 2614 ± 8 Ma age. The authors interpreted the age as

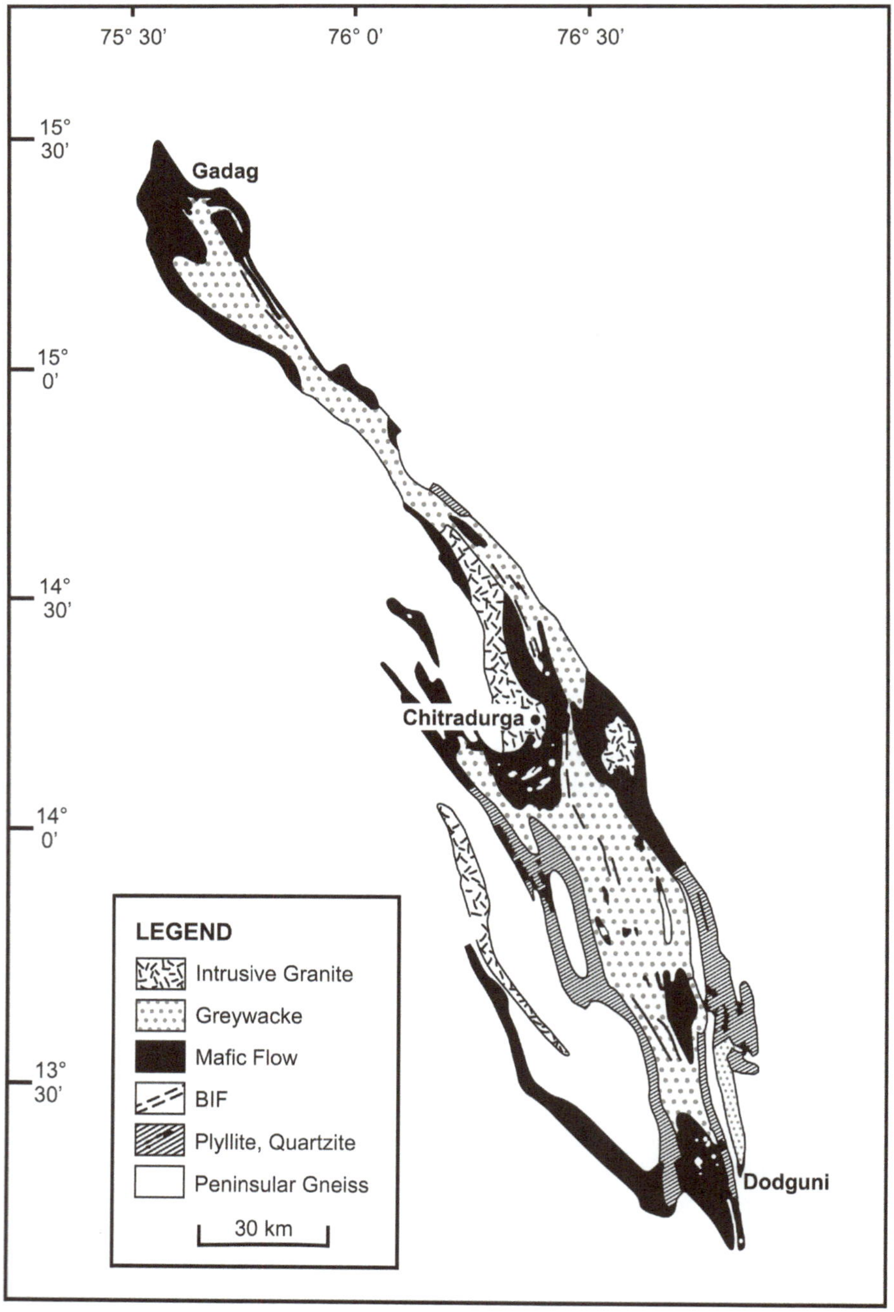

FIG. 4.13

Geological map of Chitradurga Schist Belt.

Reproduced from Radhakrishna and Vaidyanadhan (1997) with permission.

(a) (b)

FIG. 4.14

(a) Pillowed meta-basalt and (b) coaxially refolded banded iron formation in the Chitradurga Schist Belt.

Image courtesy, H.N. Bhattacharya.

FIG. 4.15

The Talya Conglomerate comprising diamictites that show the presence of good number of faceted and bullet-shaped clasts of possible reworked glacial origin.

Image courtesy, H.N. Bhattacharya.

the time of melt crystallization of the acid volcanic rocks. Very similar ages of 2648 ± 40, 2598 ± 19, and ca. 2600 Ma have been reported from the younger granites occurring near Chitradurga town by Chadwick et al. (2007). The ages suggest a little time difference between the deposition of acid volcanics and the emplacement of granitic bodies during the basin closing phase of the Chitradurga Schist Belt.

4.5.8 SHIMOGA-BABABUDAN SCHIST BELT

The westernmost schist belt that extends from Kudremukh in the south to the south of Belgaum in the north has been described as Western Ghats-Shimoga-Bababudan Schist Belt (Ramakrishnan and Vaidhyanadhan, 2008). We would, however, like to call this as the Shimoga-Bababudan Schist Belt for the sake of simplicity. This is the largest schist belt in the Dharwar covering about 30,000 km^2 between Kudremukh in the south and Belgaum in the north (Fig. 4.16). The schist belt is linked with the Chitradurga Schist Belt in the east through small occurrences of 'schists' enclosed within the Peninsular Gneiss.

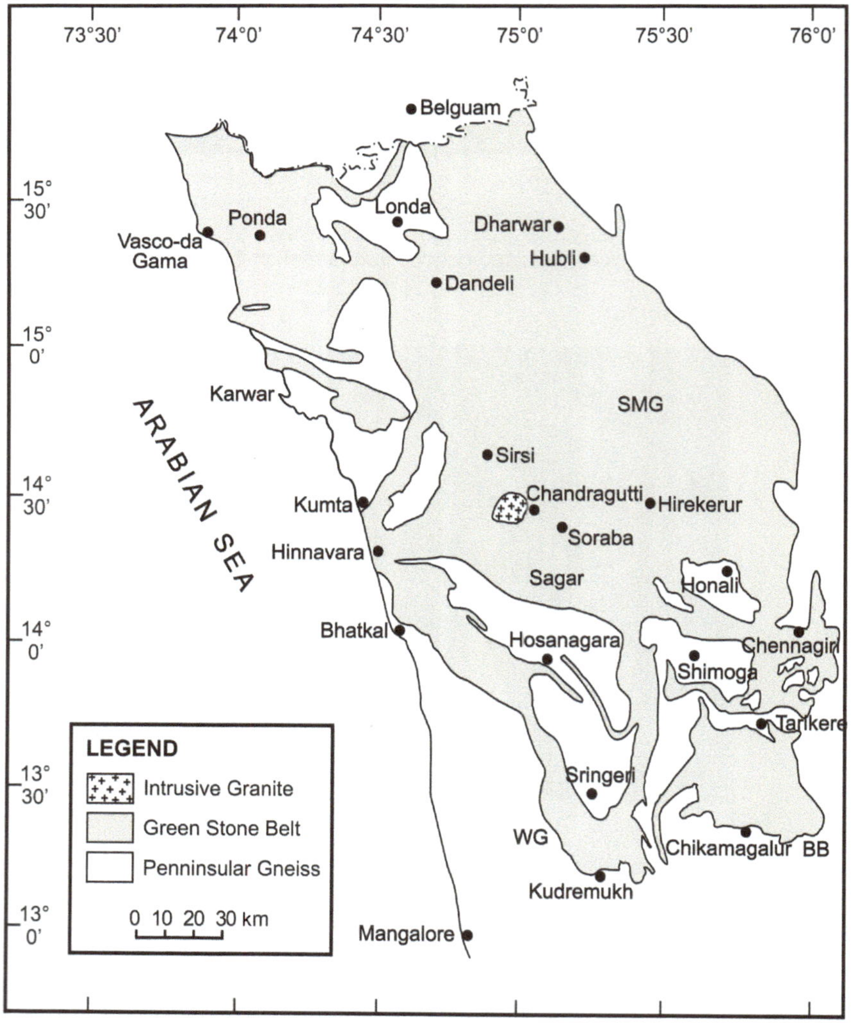

FIG. 4.16

Generalized geological map of Shimoga-Bababudan Schist Belt.

The main Bababudan outcrop that occurs in the southeast constitutes the type area of Bababudan rocks. Beginning with a thin but prominent horizon of quartz-pebble conglomerate marking a profound 'erosion' unconformity between the Peninsular Gneiss and the schist belt, the Bababudan rocks consist essentially of dark green meta-basalts, at places pillowed, and containing differentiated 'sills' (or flows?) of ultramafic rocks with lenses of chromite and titanomagnetite. Locally, sulphide mineralization is recorded within these layered bodies. The meta-basalts are interbedded with quartzite and phyllite, which locally overlie pebbly quartzite.

The Bababudan rocks continue in the north forming a part of the much larger outcrop of the Shimoga Belt, separated by a narrow belt of Peninsular Gneiss in the eastern part of the Tarikere Valley. Along the southern margin of the main body occurs basal conglomerate (quartz-pebble conglomerate) that extends discontinuously for about 40km. The quartz-pebble conglomerate is explored locally for the occurrence of uranium, copper, and gold mineralization. The quartz-pebble conglomerate grades upward into quartzite and amygdular meta-basalt, represented by chlorite-actinolite schist. The association of chlorite phyllite, banded iron formation, metagabbro, and metapyroxenite along with mafic flows forms the main bulk of the younger succession. Banded iron formation constitutes the prominent lithology capping some important hill ranges, such as at Kudremukh and nearby hills (Fig. 4.17).

FIG. 4.17

Banded iron formation constitutes the prominent lithology capping the Kudremukh Hill. The name 'Kuduremukha' literally means 'horse face' (in local language) and refers to a particular picturesque view of a side of the mountain that resembles a horse's face.

Overlying the Bababudan ensemble occurs the rocks that are correlated with the Chitradurga Group of the type area. The Chitradurga rocks in this belt show wide variability in lithology within short distances. In addition, there are occurrences of a number of basement domes making stratigraphic correlation quite problematic. Studies made in the southern part of the Shimoga Belt indicated successive occurrences of polymictic conglomerate, phyllite, greywacke, orthoquartzite, limestone, and a few bands of banded iron formation. Both greywacke and banded iron formation are locally rich in manganese oxide ores that are being mined in large scales. A major mafic-to-felsic volcanic complex occurs locally at the top of the lower succession of the Chitradurga rocks of the Shimoga Belt. The upper part of the Chitradurga Group begins with a persistent but thin unit of banded iron formation in association with carbonaceous phyllite. The upper sequence indicates a dramatic change in the depositional environment from unstable shelf to shallow lagoon to deep offshore muds.

The felsic volcanics dated by Rb–Sr, Pb–Pb, and SHRIMP U–Pb methods yielded 2600 Ma age which constrains the upper limit of the Dharwar 'Orogeny' (Ramakrishnan and Vaidhyanadhan, 2008).

4.5.9 SANDUR SCHIST BELT

The Sandur Schist Belt forms a distinctive outcrop pattern covering a part of the linear band of the Closepet Granite in the north. The irregular shape of the outcrop is said have resulted due to complex deformation during the repeated the thrusting of the rocks. The gross lithologic ensemble of this schist belt appears similar to that of the type Chitradurga Belt. The lower succession of the Sandur Schist Belt in the southwestern part is dominated by mafic and ultramafic schists with some quartzite. The sequence of rocks match with the Bababudan rocks of the type area. The overlying rocks include a siliciclastic association of greywacke and quartzite, limestone (showing development of cyanobacterial structure), mafic volcanics, and banded iron formation. The lithologic association is comparable with the lower succession of the Chitradurga Group. A unique feature of the Sandur Schist Belt is the presence of 'gneissic' veins as 'intrusive bodies' (Ramakrishnan and Vaidhyanadhan, 2008). The unusual feature may be interpreted as the slices of partially melted gneissic basement enclosed in the Sandur Schist Belt rocks.

Correlating the mafic volcanics of the Sandur Belt with those of the Kudremukh as stratigraphically coeval, the age of the basic volcanics of the Sandur Belt (Drury et al. (1984)) have been estimated to be ~2.9 Ga. The gneissic basement on which this belt is resting has yielded ~3.1 Ga age (Taylor et al., 1984). Considering the available radiometric age data, the age of the Sandur Schist Belt is presumed to be between 3.1 and 2.6 Ga. The U–Pb age of felsic volcanic from the Sandur Schist Belt is 2658 ± 14 Ma (Nutman et al., 1996).

4.6 LITHOSTRATIGRAPHY OF THE SCHIST BELT ROCKS

Robert Bruce Foot was the first to introduce the term Dharwar system for all the Archaean schistose rocks presumably buried in a number of 'synclinoria' within granite-gneiss rocks (Bruce Foote, 1876). Later, Smeeth (1915) considered the Dharwar schists as the oldest rocks dividing these into Upper Dharwar (Chloritic Division) and Lower Dharwar (Hornblendic Division). About three and half decades later, Rama Rao (1940) proposed a threefold succession of the Dharwar supracrustals dividing the Dharwar system into three units: Upper Dharwar, Middle Dharwar, and Lower Dharwar.

Several other stratigraphic successions have been proposed for the supracrustal rocks of the Dharwar Schist Belts (Naqvi et al., 1980; Pichamuthu and Srinivasan, 1983; Radhakrishna, 1983; Swami Nath and Ramakrishnan, 1981). The lithostratigraphic succession of the Dharwar Schist Belts proposed by Pichamtuhu and Srinivasan (1983) is shown in Table 4.1. Before describing the different classifications, it may be worth mentioning the sane note of caution sounded by Naqvi and Rogers (1986) about the lithostratigraphic successions erected by different authors. The common feature in all the proposed 'succession' is the assumption that the various schist belts are fragments of formerly continuous stratigraphic ensembles. In some of the classifications, a time-stratigraphic significance is attributed to the rocks based on the broad lithological similarities but occurring in the isolated belts. In such cases, the correlation based on the assumption of continuity and synchronicity would not be stratigraphically valid. The classic example is the recognition of a 'Bababudan Group' in the upper section of the Holenarasipur Schist Belt, based on lithological similarity with that in the 'type' Bababudan Basin (Bouhaller et al., 1993; Ramakrishnan and Vaidhyanadhan, 2008). A pre-3.00 Ga age has been suggested for the 'Bababudan Group' that developed in the upper section of the Holenarsipur Schist Belt. The age relationship is based on the age of the intrusive Halekote Trondhjemite (Stroh et al., 1983) into the supracrustal rocks. On the other hand, the age of Bababudan Group in the 'type' basin is said to range between 2.7 and 2.8 Ga (Chadwick et al., 1985).

Table 4.1 Stratigraphic Succession of the Dharwar Schist Belts

Group	Subgroup	Characteristic Associations
Dharwar Group	Chitradurga Subgroup	Mobile belt association; sedimentary volcanic source of deposits with conglomerate-greywacke-BIF association; mafic volcanics indicate island arc-type tholeiite
	Dodguni Subgroup	Typical platformal (stable shelf) depositional environment; occurrence of stromatolites indicate oxygenated atmosphere; carbonate and banded Mn formation; mafics are LREE-enriched
	Bababudan Subgroup	Metavolcanic (Na-rich tholeiites) and metasediments; graphite schists and iron silicates suggest nonoxygenic atmosphere; U-bearing quartz pebble conglomerate.
	Nuggihalli Subgroup	Dominantly igneous character; Komatiite lava implying thin Crust; Schists are derived from mafic-ultramafic source; no evidence of older basement

(After Pichamuthu and Srinivasan, 1983)

A major change in the stratigraphic framework has been proposed by Swami Nath and Ramakrishnan (1981) who introduced a new concept about the existence of pre-Dharwar schist rock. These authors introduced a new 'lithostratigraphic' ensemble which they named the Sargur Group. The main basis of recognizing a supposedly older schist ensemble is the identification of a 'deformed angular unconformity' between the redefined Dharwar Supergroup and the slivers of schist belt rocks embedded in the Peninsular Gneiss (Ramakrishnan and Vaidhyanadhan, 2008). The proponents of the revision suggested two possible options that could be regarded as the Sargur Group: (1) linear mafic-ultramafic belts that progressively narrow down and break-up into enclaves towards south (of the Holenarasipur Belt) or (2) as groups of enclaves scattered throughout 'Craton' but with a greater concentration in the south. Apart from such weakness of not defining the prototype of the older sequence, the suggested revision also faces criticism from structural point of view. The main objection on this count is the fact that apparent angularity noted at Sigegudda (Ramakrishnan and Vaidhyanadhan, 2008) is not truly between the two depositional surfaces but between the bedding foliation in one group and the gneissic foliation in the other (Naha et al., 1993). In short, the proposition of Sargur Group as pre-Dharwar succession appears unwarranted as it defies the standard 'code of stratigraphic correlation and nomenclature'.

4.7 CLOSEPET GRANITES: A UNIQUE TECTONIC FEATURE IN THE GEOLOGY OF ARCHAEAN CRUST

The Closepet Granite is a unique linear, arcuate body consisting of granite types having diverse composition. The southern part extending from the Cauvery River to Kalyandurga is the better-studied main body, of which the northern part extends from Rayadurga to the Deccan Trap boundary (Moyen et al., 2003). The Closepet Granite has been considered as a 'Stitching Granite' marking a Suture Zone between two tectonic belts. The outcrops of Closepet Granite form a well-marked, about 400 km long, band between 20 and 30 km wide band of linear hill range in the midst of the peneplained terrane of gneisses running north–south through Closepet and Channapatna. The granites are locally foliated in marginal portions, while the adjacent rocks are in many cases show chemical alteration. A series of porphyries, felsites, and other differentiates traverse through the outcrops of the Closepet Granite (Krishnan, 1956).

Petrologically, Closepet Granite is a composite body of predominantly potash-rich adamalite, quartz monzonite, and granite. There is a considerable variation of composition from north to south, thought to be due to different structural levels of the body from deeper in the northern region to shallower in the southern part (Moyen et al., 2003). Commonly appearing as porphyritic granite, the development uniformly planar foliation on the either side of the body granite indicates ductile shearing during emplacement of the granite. It is thought to be a sheet-like 'pluton' forming a part of the Peninsular Gneiss (Moyen et al., 2003). The SHRIMP U/Pb age of zircon grains in the Closepet Granite at Kabbaldurga is 2513 ± 5 Ma (Friend and Nutman, 1991). The quartz monzonite yielded 2518 ± 5 Ma (Jayananda et al., 1995), and the Rb/Sr whole-rock isochron age at Toranagallu is 2452 ± 50 Ma (Bhaskar Rao et al., 1992).

4.8 INTRACRATONIC PROTEROZOIC BASINS: CUDDAPAH BASIN

The Cuddapah Basin with its signature crescent-shaped map pattern is the oldest and the largest Proterozoic intracratonic basin in the Indian Subcontinent with the aggregate total stratigraphic thickness of over 12 km (Nagaraja Rao et al., 1987; Ramam and Murty, 1997; Singh and Mishra, 2002). The concave outline of the eastern margin of the Cuddapah outcrop matches with the corresponding part of the eastern coast line of India (Fig. 4.18). Area wise, the basin occupies about 44,000 km^2 and exposing the thick sedimentary cover succession along with some volcanics and rare intrusions. The entire western part the basin overlies the Archaean granite-gneiss-greenstone rocks. The pronounced cross-cutting relationship between the Cuddapah and Archaean Dharwar rocks is described in the early literatures as 'Eparchaean Unconformity' (Krishnan, 1956) marking the Archaean-Proterozoic Boundary. On the eastern side occurs some 'high-grade schists' belonging to Nellore-Khammam Schist Belt that is now correlated with the Eastern Ghats Granulite belt.

A characteristic feature of the Cuddapah sequence (including the Kurnool Group sediments) is the lack of younger cover, unlike similar other cratonic basins like the Kaladgi and Bhima in the north and northwest of the Dharwar Protocontinent.

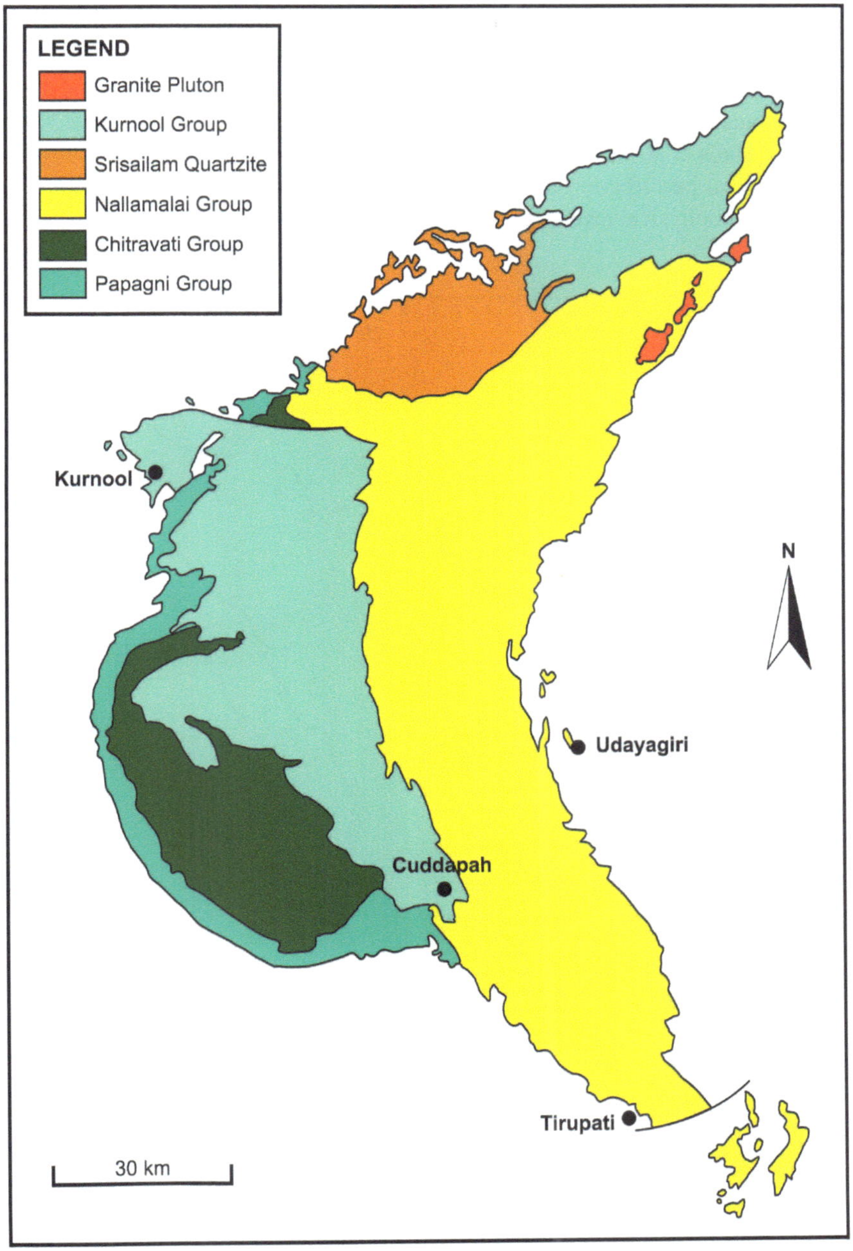

FIG. 4.18

Geological map of Cuddapah Basin showing different lithostratigraphic formations.

Reproduced from Nagaraja Rao et al. (1987) with permission.

4.8.1 BASIN EVOLUTION AND TECTONICS

Following a major tectonic hiatus after the cratonization of the Archaean Crust, the Protocontinent was invaded by mafic dyke swarms (Fig. 4.19) most of which was in the western part. The intrusion of mafic dykes took place in three different stages during the period between 2369 and 2177 Ma (French and Heaman, 2010; Halls et al., 2007; Pandey et al., 1997). The dyke swarms are thought as the initial reflection of the thermal perturbation in the Mantle prior to the opening of the basin (Chatterjee and Bhattacharji, 2001). Geophysical investigations suggest the presence of a large asymmetrical, lopolithic

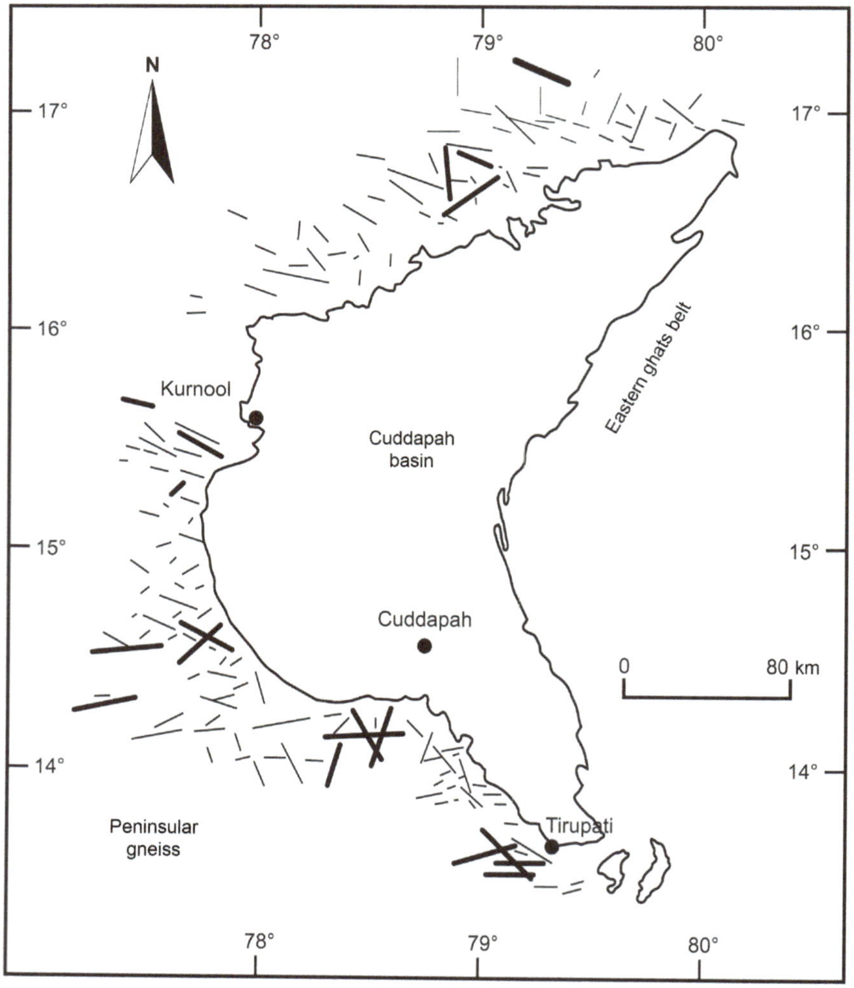

FIG. 4.19

Map of Cuddapah Basin showing location and orientation of mafic dykes around the basin. Well-studied dykes are shown in bold lines.

Reproduced from Chatterjee and Bhattacharji (2001) with permission.

mafic pluton along the base of the basin. The evolution of Cuddapah Basin is presumed to be linked with the cooling of the heat source in the deeper Crust (Bhattacharji and Singh, 1984).

The most significant feature of the Cuddapah Basin is the basin asymmetry in terms of the depth of basin fills from east to west. The thin western beds show eastward increase in thickness to about 6 km. The eastern margin of the Cuddapah Basin is marked by a prominent Ductile Shear Zone that has been interpreted as the zone thrusting. According to Singh and Mishra (2002), the eastern boundary of the Cuddapah Basin represents a 'cryptic' Suture Zone evolved during the amalgamation two different crustal provinces.

Structurally, there is a wide a variation in the nature of deformation from gentle, monoclinally tilted beds in the west to the arcuate pack of westerly overturned folds associated with thrusting and ductile shearing in the eastern beds. The eastern folded zone is known as Nallamalai Fold Belt (Narayanaswamy, 1966; Saha and Patranabis-Deb, 2014) (Fig. 4.20). This eastern fold belt is interpreted as an 'indentation' tectonic feature resulting from the collision of Dharwar Protocontinent with the Eastern Ghats tectonic front.

FIG. 4.20

Northwest vergent folds showing evidence of thrusting in the northern part of the Nallamalai Fold Belt.

Reproduced from Saha and Patranabis-Deb (2014) with permission.

4.8.2 LITHOSTRATIGRAPHY OF CUDDAPAH BASIN

The Cuddapah Basin includes two different lithostratigraphic ensembles, the older Cuddapah Supergroup overlain by the younger Kurnool Group. The Cuddapah Supergroup consists predominantly of clastic and chemogenic sedimentary rocks with minor intercalations of alkali to subalkali basaltic flows, mafic to ultramafic sills, and ash-fall tuffs in the lower part of the succession (Anand et al., 2003; Chatterjee and Bhattacharji, 2001; Nagaraja Rao et al., 1987; Ramam and Murty, 1997). The earliest systematic study of the Cuddapah rocks was made by W. King in 1872 (quoted by Krishnan, 1956), who proposed a four-fold division of the Cuddapah depositories, overlain by a younger but similar sedimentary succession. King's stratigraphic succession remained the only useful document for all subsequent studies and discussions for about a century (Krishnan, 1956). Since the late 1960s, several workers initiated studies in different parts of this intracratonic, Platform-type sedimentary basin covering different aspect of basin analysis including lithostratigraphy. The major domino effect of these studies is the suggestion of several new stratigraphic successions contradicting to some extent the time-honoured stratigraphic framework suggested by W. King in 1872. The sheer multiplicity in the proposals of lithostratigraphic succession of the basin prompted Ramakrishnan and Vaidhyanadhan (2008) to comment on the 'stratigraphic riddle' in correlating rocks having similar lithological attributes and sedimentary structures in rocks deposited at different times. In the various proposition of stratigraphic framework by different authors. Additional problems cropped up because of the observed movements noted in the outcrops along the criss-cross faults, much of which took place during the strong neotectonic (predominantly Quaternary) activities in the region (Fig. 4.21).

FIG. 4.21

A sharp scarp faces north of Tirupati, at the foot of the Tirumala Hills. The feature is a clear indication of neotectonic (Quaternary) deformation near the southeastern tip of the Cuddapah Basin.

In the light of the apparent gaps in stratigraphic correlation of the rocks of this huge cratonic basin, the study that deserves attentive consideration for understanding the lithostatigraphy of the Cuddapah Basin is that of Nagaraja Rao et al. (1987) who produced an excellent lithostructural map of the Cuddapah Basin, providing enough scope for reinterpretation. The lithostratigraphic succession suggested by the authors is appended in Table 4.2.

4.8.3 AGE OF CUDDAPAH BASIN EVOLUTION AND POST-TECTONIC INTRUSIONS

There is no definite information on the precise age of development of the Cuddapah Basin. Based on the study of uranium mineralization in the Vempalle and Tadpatri Formations of the Cuddapah Supergroup, it is suggested that the deposition of these rocks was during 1756 ± 29 Ma (Zachariah et al., 1999).

Table 4.2 Lithostratigraphic Succession of the Cuddapah Basin

	Group	Formation	Lithology
Cuddapah Supergroup	Krishna Group	Srisailam Formation (300 m)	Quartzites and slates
	Nallamai Group	Cumbum (Pullampet) Formation (2000 m)	Phyllite, Slate, quartzite, dolomite
		Bairenkonda (Nagari) Formation (1500–4000 m)	Quartzite, Shales, conglomerate with intrusives
	Chitravati Group	Gandikota Formation (300 m)	Quartzite and Shale
		Tadpatri Formation (460 m)	Shales, quartzite, dolomite, tuff, intrusives
		Pulivendla Formation(1–75 m)	Conglomerate and quartzite
	Papaghni Group	Vempalle Formation (1900 m)	Stromatolitic dolomite, chert breccia, quartzite, lavas, and pyroclastics
		Gulcheru Formation(28–210 m)	Conglometare, arkose, quartzite, and shale
Dharwar gneisses, granite, and schist belts			

(After Nagaraja Rao et al., 1987)

According to these authors, this is the minimum age for carbonate sedimentation and dolomitization in the Cuddapah Basin. A similar conclusion was drawn earlier by Bhaskara Rao et al. (1995) based on an age of 1817 ± 24 Ma for the Pulivendla mafic sill. More precise date for the mafic sill intruding the Vempalle Formation comes from the U–Pb (baddeleyite) dating of 1885.4 ± 3.1 Ma (French et al., 2008). Earlier, Anand et al. (2003) based on ^{40}Ar–^{39}Ar laser-fusion determinations on phlogopite mica, from the mafic-ultramafic sill-type intrusion in Tadpatri Formation, had suggested that the initial phase of extension and volcanism in the Cuddapah Basin was at 1.9 Ga. An interesting study was made by Sai et al. (2016) who looking into the dominance of red beds in different Cuddapah formations inferred that the Gulcheru red beds (the oldest Cuddapah formation overlying the basal conglomerates) were deposited at around 2.1 Ga, a period that corresponds to the global oxygenation event and a period that witnessed deposition of red beds in the platform type of Proterozoic basins in the world. In the absence of any definitive high-precision age of the basin evolution, we may assume that the opening of the Cuddapah Basin was around this date or a little later. Lithostratigraphic succession of the Cuddapah Basin is given in Table 4.2.

There is no information about the time of basin closing. However, information on folding and deformation event in the eastern most Nallamalai Fold Belt comes indirectly from the correlation of the deformation in the eastern part of the Cuddapah Basin with the event of collision between the Dharwar Protocontinent in the west and the Eastern Ghats Granulite Belt in the east. The timing of this event, as suggested by Valdamani et al. (2012) is 1.78 Ga. We may, therefore, logically conclude that 'inversion' of the Cuddapah Basin was earlier to this date.

The Cuddapah Basin subsequent to the folding and shearing in the eastern Nallamalai Fold Belt was affected by some post-tectonic granite intrusions. One of these is the intrusion of the Vellaturu Granite in the northeastern margin at around 1575 Ma (Crawford and Compston, 1973). In this connection, it may be worth mentioning about a unique feature known as 'Iswarakuppam Dome' in the central part of the Nallamalai Fold Belt. The broadly north–south dome-shaped structure does not show parallelism with the fold trends in the Nallamalai Fold Belt. Nagaraja Rao et al. (1987) inferred that the feature could be the result of diapiric uprise of a (unexposed) granitic body. We presume that this diapirism could be linked with the formation of the Vellaturu Domes in the northeast.

The deposition of the Kurnool Group ushered in a new phase of sedimentation in the Cuddapah Basin during the Neoproterozoic after a prolonged hiatus (Collins et al., 2014). The basal conglomerate of the Kurnool Group is known for occurrence of diamond. The source rock for the diamond is the diamond-bearing kimberlite pipes located at Wajrakarur and some other areas surrounding the Cuddapah Basin. The emplacement of these kimberlites, according to Chalapathi Rao et al. (2009), was during 1.1 Ga. We may, therefore, conclude that the deposition of the basal conglomerate of the Kurnool Group could only have taken place during early Neoproterozoic.

4.9 INTRACRATONIC PROTEROZOIC BASINS: BHIMA BASIN

The northeast–southwest oriented Bhima Basin consisting dominantly of carbonate cover is situated in the northwest of the Cuddapah Basin. Like other Proterozoic basins that occur in and around the Dharwar Protocontinent, the Bhima Basin evolved over the Archaean granite-gneiss-greenstone sequence with a profound unconformity. The basin shows faulted contact with the basement at several places. The Bhima Group was deposited in near-shore narrow basins over the Archaean basement. The succession begins with a thin conglomerate horizon passing upward to the gritty arkosic sandstone (Kale et al. 1990). The succeeding formation includes repeated occurrence of limestone with some shale and cherty layers.

King (1872) was the first to coin the term 'Bhima series' for the depositories of the basin and divided it into the Lower Muddebihal Sandstones and the Upper Talikote Limestones. This was subsequently revised by Bruce Foote (1876) who divided the 'Bhima series' into Lower and Upper formations. The recent formal lithostratigraphic nomenclature was adopted by Janardhana Rao et al. (1975), who assigned this sequence the status of a 'Group', dividing it into the Rabanpalli Formation and Shahabad Formation. The lithostratigraphic details of the Bhima Group have been provided by Kale et al. (1990) who recognized two formations, namely, the Rabanhalli Clastics and Shahabad Limestone. Truly speaking, the two formations seem to represent two major sedimentological facies types showing gradational contact between them. According to Kale and Peshwa (1995), the Rabanhalli Clastics and Shahabad Limestone are products of coeval sedimentation under two different depositional environments. Based on this correlation, the aggregate thickness of the basin depositories has been reduced from the earlier estimate of 270 m to about 150 m.

4.10 INTRACRATONIC PROTEROZOIC BASINS: KALADGI BASIN

The east–west oriented Kaladgi Basin occurs in the northern part of the Archaean Dharwar Protocontinent unconformably overlying the Archaean gneiss-schist rocks in the south. The northern part of the basin is covered under the Deccan Trap. The exposed basin covers about 830 km^2 accommodating over 4500 m thick sediments of the stand-shale-carbonate association with minor cherts (associated with fault breccia) and conglomerates deposited in continental, transitional, and shallow-marine environments. Bruce Foote (1876) was the first to map the area, who divided the sequence into Lower Kaladgi and Upper Kaladgi 'series'. About 100 years later, Viswanathia (1977) revised the stratigraphy describing lower formations as the Bagalkot Group underlying the Badami Group. Later, Jayaprakash et al. (1987) provided further details on structure and stratigraphy of the basin. The geological map prepared by the authors shows occurrence of the two groups in separate basins (Fig. 4.22). The older Bagalkot Group that occurs in the central part is flanked by the younger Badami Group both in the north and in the south.

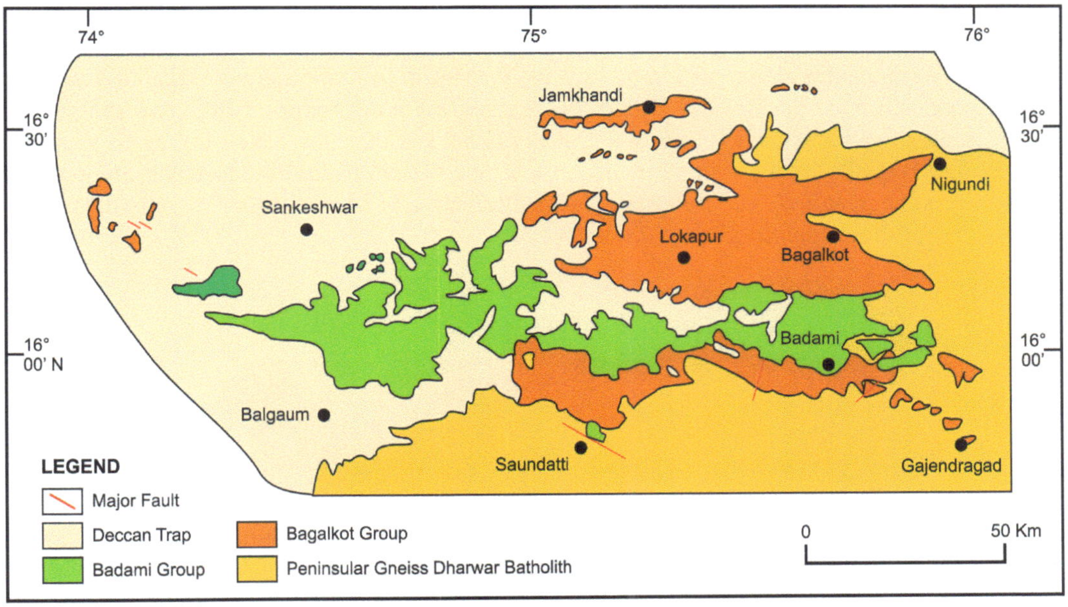

FIG. 4.22

Simplified geological map of the Kaladgi Basins.

Reproduced from Jayaprakash et al. (1987) with permission.

The Bagalkot Group starts with a conglomerate bed overlying the Archaean basement rocks with a pronounced erosion unconformity in between. A repeated fining upward sequence capped by carbonates overlies the basal conglomerate. The lower carbonate sequence shows development of stromatolitic dolomite in the middle part. That is the thickest in the entire succession. Overlying the lower unit occurs the upper succession that is lithologically similar to the lower sequence but with thick beds of argillite at the top. Another characteristic feature of the upper sequence is the development of banded haematite quartzite in the succession. The Bagalkot Group is affected by intrusion of pegmatite and quartz veins and a few dolerite dykes.

The Bagalkot Group occurring in the middle of the basin shows folding having west northwest–east southeast axial trace. Rarely the folds are superposed by the north–south trending cross folds (Jayaprakash et al., 1987). Though some east–west (or more precisely east southeast–west and northwest) trending folds have been described from the Bagalkot rocks, very little is known about the geometry of the folds described from the rocks of the Bagalkot Group. Both northerly and southerly 'vergent' folds have been described. Available information, however, preclude the possibility that the folds originated by any crustal shortening process. Mukherjee et al. (2016) have explained that the folds have resulted due to the process of gravity gliding along inclined surfaces. The occurrence of seismites in the succession along with evidence of penecontemporaneous deformation in different sedimentary layers suggests the possibility of movement along active faults (Kale and Phansalkar, 1991), which may be indirectly support the suggestion made by Mukherjee et al. (2016) about gravity gliding on tilted surfaces. The map pattern shown in Fig. 4.22 indicates that the 'deformed' Bagalkot rocks are flanked by 'subhorizontal beds' of the Badami Group (Fig. 4.23). We may, therefore, rationally presume that the deformation in the Bagalkot rocks predated the deposition of the Badami Group. From the present disposition of the two sedimentary sequences showing occurrence of the two groups of rocks at the same level, it may be suggested that the older Bagalkot rocks were vertically lifted by a system of faults. Moreover, the presence of quartz veins, pegmatites, and dolerite dykes is indicative of some form of thermal disturbance possibly in the Upper Mantle (Figs. 4.24 and 4.25).

FIG. 4.23

Horizontally layered Badami sandstone beds hosting several cave temples.

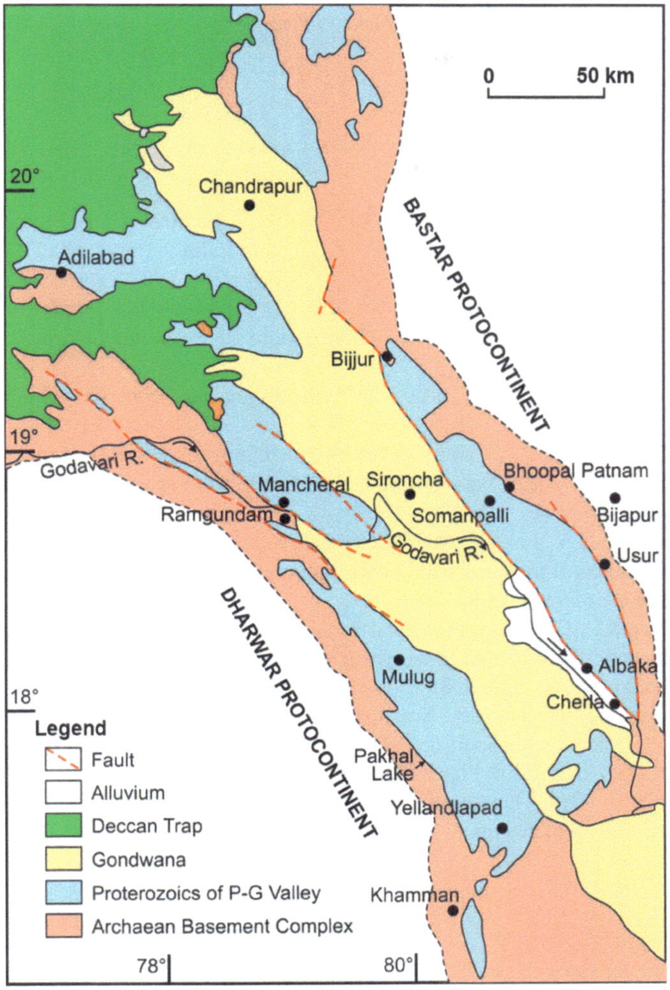

FIG. 4.24

Simplified geological map of the Pranhita-Godavari Valley showing distribution of different lithotectonic component. Map compiled from different sources.

FIG. 4.25

Deformation in the eastern lowermost succession of the Pranhita-Godavari basin, showing southwest vergent folds indicating top-to-southwest transport.

Reproduced from Saha and Patranabis-Deb (2014) with permission.

REFERENCES

Anand, M., Gibson, S.A., Subbarao, K.V., Kelley, S.P., Dickin, A.P., 2003. Early Proterozoic melt generation processes beneath the Intra-cratonic Cuddapah Basin, Southern India. J. Petrol. 44 (2), 2139–2171. https://doi.org/10.1093/petrology/egg073.

Balakrishnan, S., Rajamani, V., Hanson, G.N., 1999. U-Pb ages for zircon and titanite from the Ramagiri areas, southern India: evidence for accretionary origin of the eastern Dharwar craton during Late Archaean. J. Geol. 107, 69–86.

Beckinsale, R.D., Drury, S.A., Holt, R.W., 1980. 3,360 MY-old gneisses from South Indian craton. Nature 283, 469–470.

Bhalla, N.S., Gupta, J.M., Chabris, T., Vasudeva, S.G., 1978. Rb/Sr geochronology of rocks from the Kolar Schist Belt, South India. In: Windley, B.F., Naqvi, S.M. (Eds.), Origin and Evolution of Archaean Continental Crust. Elsevier, Amsterdam, pp. 79–84.

Bhaskar Rao, Y.J., Sivaraman, T.V., Pantulu, C.V.C., Gopalan, K., Naqvi, S.M., 1992. Rb–Sr ages of late Archaean metavolcanics and granites, Dharwar craton, South India and evidence for early Proterozoic thermotectonic event(s). Precambrian Res. 59, 145–170.

Bhaskar Rao, Y.J., Pantulu, G.V.C., Reddy, D., Gopalan, K., 1995. Time of Early Sedimentation and Volcanism in the Proterozoic Cuddapah Basin, South India: Evidence From Rb-Sr Age of Pulivendla Mafic Sill. Mem. 33. Geol. Soc., India, Bangalore, pp. 329–338.

Bhaskar Rao, Y.J., Kumar, A., Vrevsky, A.B., Srinivasan, R., Anantha Iyer, G.V., 2000. Sm-Nd isochron ages of two meta-anorthosite from Holenarsipur: constraints on antiquity of Archaean supracrustal rocks of Dharwar craton. Proc. Indian Acad. Sci. (Earth Planet. Sci.) 109, 57–65.

Bhattacharji, S., Singh, R.N., 1984. Thermo-mechanical structure of the southern part of the Indian Shield and its relevance to Precambrian basin evolution. Tectonophysics 105, 103–120.

Bouhaller, H., Choukroune, P., Ballevre, M., 1993. Diapirism, bulk homogenous shortening and transcurrent shearing in the Archaean Dharwar craton-Holenarasipur area, southern India. Precambrian Res. 63, 43–58.

Bruce Foote, R., 1876. The geological features of the south Mahratta country and adjacent districts. Mem. Geol. Surv. India 12, 139–164.

Chadwick, B., Ramakrishnan, M., Viswanatha, M.N., 1985. Bababudan—a late Archaean intracratonic volcanosedimentary basin, Karnataka, south India. J. Geol. Soc. India 26, 769–821.

Chadwick, B., Vasudev, V.N., Hegde, G.V., Nutman, A.P., 2007. Structure and SHRIMP U/Pb zircon ages of granites adjacent to the Chitradurga schist belt: implications for Neoarchaean convergence in the Dharwar craton, southern India. J. Geol. Soc. India 69, 5–24.

Chalapathi Rao, N.V., Anand, M., Dongre, A., Osborne, I., 2009. Carbonate xenoliths hosted by the Mesoproterozoic Siddanpalli Kimberlite Cluster (Eastern Dharwar craton): implications for geodynamic evolution of southern India and its diamond and uranium metallogenesis. Int. J. Earth Sci. (Geol. Rundsch.) https://doi.org/10.1007/s00531-009-0484-7.

Chardon, D., Jayananda, M., Peucat, J.-J., 2011. Lateral constrictional flow of hot orogenic crust: insights from the Neoarchean of south India, geological and geophysical implications for orogenic plateaux. Geochem. Geophys. Geosyst. 12, Q02005. https://doi.org/10.1029/2010GC003398.

Chatterjee, N., Bhattacharji, S., 2001. Petrology, geochemistry and tectonic settings of the Mafic dikes and sills associated with the evolution of the Proterozoic Cuddapah Basin of south India. Proc. Indian Acad. Sci. (Earth Planet. Sci.) 110, 433–453.

Chaudhary, A.K., Harris, N.B.W., Van Clasteren, P.C., Hawkesworth, C.J., 1992. Pan-African charnockite formation in Kerala, South India. Geol. Mag. 129, 257–264.

Collins, A.S., Patranabis-Deb, S., Alexander, E., Bertram, C.N., Falster, G.M., Gore, R.J., Mackintosh, J., Dhang, P.C., Saha, D., Payne, J.L., Jourdan, F., Backé, G., Halverson, G.P., Wade, B.P., 2014. Detrital mineral age, radiogenic isotopic stratigraphy and tectonic significance of the Cuddapah Basin, India. Gondwana Res. 28, 1294–1309.

Crawford, A.R., 1969. Reconnaissance Rb-Sr dating of the Precambrian of southern peninsular India. J. Geol. Soc. India 10, 117–166.

Crawford, A.R., Compston, W., 1973. The age of the Cuddapah and Kurnool systems, southern India. J. Geol. Soc. Australia 19, 453–464.

Dobmeir, C., Raith, M.M., 2003. Crustal architecture and evolution of the Eastern Ghats Belt and adjacent regions of India. In: Yoshida, M., Windley, B.F., Dasgupta, S. (Eds.), Proterozoic East Gondwana Supercontinent Assembly and breakup. vol. 206. Geol. Soc., London, (Sp. Pub.), pp. 145–168.

Drury, S.S., Harris, N.B.W., Holt, R.W., Reeves, S.G.J., Wightman, R.T., 1984. Precambrian tectonics and crustal evolution of South India. J. Geol. 92, 3–20.

Fermor, L.L., 1936. An attempt at the correlation ancient schistose formation of Peninsular India. Mem. Geol. Surv. India 70, 51.

French, J.E., Heaman, L.M., 2010. Precise U-Pb dating of Proterozoic mafic dyke swarms of the Dharwar Craton, India: implications for existence of Neoproterozoic supercraton Sclaria. Precambrian Res. 183, 416–441.

French, J.E., Heaman, L.M., Chacko, T., Srivastava, R.K., 2008. 1891–1883 Ma Southern Bastar–Cuddapah mafic igneous events, India: a newly recognized large igneous province. Precambrian Res. 160, 308–322.

Friend, C.R.L., 1981. The timing of charnockite and granite formation in relation to influx of CO_2 at Kabbaldurga, Karnataka, South India. Nature 294, 550–552.

Friend, C.R.L., Nutman, A.P., 1991. SHRIMP U/Pb geochronology of the Closepet Graniteand Peninsular Gneiss. Karnataka South India. J. Geol. Soc. India 38, 357–368.

Friend, C.R.L., Nutman, A.P., 1992. Response of U-Pb isotopes and whole rock geochemistry of CO_2 induced granulite facies metamorphism, Kabbaldurga, Karnataka, south India. Contrib. Mineral. Petrol. 111, 299–310.

Grady, J.C., 1971. Deep main faults in south India. J. Geol. Soc. India 12, 56–62.

Grew, E.S., Manton, W.L., 1986. A new correlation of sapphire-granulitein the Indo-Antarctica metamorphic terrain, Late Proterozoic dates from the Eastern Ghats. Precambrian Res. 33, 123–139.

Halls, H.C., Kumar, A., Srinivasan, R., Hamilton, M.A., 2007. Paleomagnetism and U–Pb geochronology of easterly trending dykes in the Dharwar craton, India: feldspar clouding, radiating dyke swarms and the position of India at 2.37 Ga. Precambrian Res. 155, 47–68.

Hansen, E.C., Stern, R.J., Devaraju, T.C., Mahabaleswar, B., Kenny, P.J., 1997. Rubidium-strontium whole-rock ages of banded and incipient charnockites from southern Karnataka. J. Geol. Soc. India 50, 267–275.

Janardhan, A.S., 1983. Transformation of Peninsular Gneiss to charnockite in Southern Karnataka. In: Naqvi, S.M., Rogers, J.J.W. (Eds.), Precambrian of South India. Mem. 4. Geol. Soc., India, Bangalore, pp. 417–435.

Janardhana Rao, L.H., Srinivasa Rao, C., Ramakrishna, T.L., 1975. Reclassifiaction of the rocks of the Bhima basin, Gulbarga district Karnataka State. Geol. Surv. India Misc. Publ. 23, 177–184.

Jayananda, M., Peucat, J.J., 1996. Geochronological framework of Southern India. In: Santosh, M., Yoshida, M. (Eds.), The Archaean and Proterozoic Terrains in Southern India Within East Gondwana. vol. 3. Mem. Gondwana Research Group, pp. 53–75.

Jayananda, M., Janardhan, A.S., Suvasubramanian, P., Peucat, J.-J., 1995. Geochronologic and isotopic constraints on granulite formation in the Kodaikanal area, Southern India. Mem. Geol. Soc. India 34, 373–390.

Jayananda, M., Chardon, D., Peucat, J.-J., Capdevila, R., 2006. 2.61 Ga potassic granites and crustal reworking in the western Dharwar craton, southern India: tectonic, geochronologic and geochemical constraints. Precambrian Res. 150, 1–26.

Jayananda, M., Tsutsumi, Y., Miyazaki, T., Gireesh, R.V., Kapfo, K.-u., Tushipokla, H., Hidaka, Kano, T., 2013. Geochronological constraints on Meso- and Neoarchean regional metamorphism and magmatism in the Dharwar craton, southern India. J. Asian Earth Sci. 78, 18–38.

Jayaprakash, A.V., Sundaram, V., Hans, S.K., Mishra, R.N., 1987. Geology of Kaladgi-Badami basin, Karnataka. Mem. Geol. Soc. India 6, 201–255.

Jayaram, S., Rajagopalan, P.T., 1984. Geochemistry and petrogenesis of the Peninsular Gneisses, Dharwar craton, India. J. Geol. Soc. India 25, 570–584.

Kale, V., Peshwa, V.V., 1995. Bhima Basin. Geol. Soc. India, Bangalore. 142 p.

Kale, V., Phansalkar, V.G., 1991. Purana basins of Peninsular India. A review. Basin Res. 3, 1–36.

Kale, V., Mudholkar, A.V., Phansalkar, V.G., Peswa, V.V., 1990. Stratigraphy of the Bhima Group. J. Palaeontol. Soc. India 35, 91–103.

King, W., 1872. The Kadapah and Karnul Formationin the Madras Presidency. Mem. Geol. Surv. India 8, 346 p.

Krishnan, M.S., 1956. Geology of India and Burmah. Higginbothams (Private) Ltd, Madras. 555 p.

Mahabeleshwar, B., Peucat, J.-J., 1988. 2.9 b.y. Rb-Sr age of granulite facies rocks Satanut-Halagur-Sivasamundram area, southern Karnataka. J. Geol. Soc. India 27, 282–297.

Mahabeleshwar, B., Jayananda, M., Peucat, J.-J., Swamy, N.S., 1995. Archaean high grade gneiss complex from Satanut-Halagur-Sivasamundram area, southern Karnataka, Southern India, Petrogenesis and crustal evolution. J. Geol. Soc. India 45, 33–49.

Meen, J.K., Rogers, J.J.W., Fullagar, P.D., 1992. Lead isotopic compositions in the western Dharwar craton, southern India: evidence for distinct middle Archaean terrains in a late Archaean craton. Geochim. Cosmochim. Acta 56, 2455–2470.

Moyen, J.-F., Martin, H., Jayananda, M., Auvray, B., 2003. Late Archaean granites: a typology based on Dharwar craton (India). Precambrian Res. 127, 103–123.

Mukherjee, M.K., Das, S., Modak, K., 2016. Basement–cover structural relationships in the Kaladgi Basin, southwestern India: indications towards a Mesoproterozoic gravity gliding of the cover along a detached unconformity. Precambrian Res. 281, 495–520.

Nagaraja Rao, B.K., Rajurkar, S.T., Ramaligaswamy, G., Ravindra Babu, B., 1987. Stratigraphy and Structure and Evolution of the Cuddapah Basin. Mem. 6. Geol. Soc., India, Bangalore, pp. 33–86.

Naha, K., Srinivasan, R., Gopalan, K., Pantulu, G.V.C., Subba Rao, M.V., Vrevsky, A.B., Bogomolov, Y.E.S., 1993. The nature of basement in the Archaean Dharwar Craton of southern India and the age of the Peninsular gneiss. Proc. Indian Acad. Sci. (Earth Planet. Sci.) 102, 547–565.

Naqvi, S.M., Rogers, J.J.W., 1986. Precambrian Geology of India. Oxford Monographs on Geology and Geophysics No. 6. 222 p.

Naqvi, S.M., Viswanathan, S., Viswanatha, M.N., 1978. Geology and geochemistry of the Holenarasipur schist belt and its place in the evolutionary history of the Indian Peninsula. In: Developments in Precambrian Geology. vol. 1. Elsevier, pp. 109–126.

Naqvi, S.M., Narayana, B.L., Rama Rao, P., Ahmad, S.M., Uday Raj, B., 1980. Geology and geochemistry of paragneisses from Javanahalli schist belt, Karnataka, India. J. Geol. Soc. India 21, 577–592.

Narayanaswamy, S., 1966. Tectonics of the Cuddapah Basin. J. Geol. Soc. India 7, 33–50.

Nutman, A.P., Chadwick, B., Krishna Rao, B., Vasudev, V.N., 1996. SHRIMP U/Pb zircon ages of acid volcanic rocks in the Chitradurga and Sandur groups, and granites adjacent to the Sandur schist belt, Karnataka. J. Geol. Soc. India 47, 153–164.

Ojakangas, W., Srinivasan, R., Hegde, V.S., Chandrakant, S.M., Srikantia, S.V., 2014. The Talya Conglomerate: an Archean (~2.7 Ga) Glaciomarine Formation, Western Dharwar Craton, Southern India. Curr. Sci. 106, 387–396.

Pandey, B.K., Gupta, J.N., Sarma, K.J., Sastry, C.A., 1997. Sm/Nd, Pb/Pb and Rb/Sr geochronology and petrogenesis of the mafic dyke swarm of Mahbubnagar, South India: implications for Paleoproterozoic crustal evolution of the Eastern Dharwar Craton. Precambrian Res. 84, 181–196.

Peucat, J.-J., Mahabaleswar, M., Jayananda, M., 1993. Age of younger tonalitic magmatism and granulite metamorphism in the amphibolite–granulite transition zone of southern India (Krishnagiri area): comparison with older Peninsular gneisses of Gorur–Hassan area. J. Metam. Geol. 11, 879–888.

Pichamuthu, C.S., 1965. Regional metamorphism and charnokitization in Mysore State. Indian Mineral. 6, 119–126.

Pichamuthu, C.S., Srinivasan, R., 1983. A billion year history of the Dharwar craton (3200–2100 m.y. ago). Mem. Geol. Soc. India 4, 121–142.

Radhakrishna, B.P., 1983. Archaean granite-greenstone terrain of the South Indian Shield. In: Naqvi, S.M., Rogers, J.J. (Eds.), Precambrian of South India. Mem. 4. Geol. Soc., India, Bangalore, pp. 1–46.

Radhakrishna, B.P., Vaidyanadhan, R., 1997. Geology of Karnataka. Geological Society of India, Bangalore. 353 p.

Rajamani, V., Krogstad, E.J., Hanson, G.N., BalaKrishnan, S., 1987. Are Patna and Bisanattam granites (adjoining Kolar schist belt) cogenetic? J. Geol. Soc. India 30, 98–105.

Rama Rao, B., 1940. The Archaean complex of Mysore. Mysore Geol. Dept. Bull. 17, 1–101.

Ramakrishnan, M., 2003. Craton-Mobile Belt Relations in Southern Granulite Terrain. Mem. 50. Geol. Soc., India, Bangalore, pp. 1–24.

Ramakrishnan, M., Vaidhyanadhan, R., 2008. Geology of India. vol. 1. Geological Society of India, Bangalore. 556 p.

Ramam, P.K., Murty, V.N., 1997. Geology of Andhra Pradesh. Geological Society of India, Bangalore. 245 p.

Saha, D., Patranabis-Deb, S., 2014. Proterozoic evolution of Eastern Dharwar and Bastar cratons, India—an overview of the intracratonic basins, craton margins and mobile belts. J. Asian Earth Sci. 91, 230–251. https://doi.org/10.1016/j.jseaes.2013.09.020.

Sai, V.V.S., Khanna, T.C., Reddy, N.R.K., 2016. Red beds in the Cuddapah Basin, eastern Dharwar craton, India: implications for the initiation of sedimentation during the Proterozoic Oxygenation event. J. Ind. Geophys. Union 20, 342–350.

Singh, A.P., Mishra, D.C., 2002. Tectonosedimentary evolution of Cuddapah basin and Eastern Ghats mobile belt (India) as Proterozoic collision: gravity, seismic and geodynamic constraints. J. Geodyn. 33, 249–267.

Smeeth, W.F., 1915. Outline of Geological History of Mysore. No. 6 Mysore Geol. Dept. Bull., Mysore. 22 p.

Srikantappa, C., Srinivas, G., Basavarajappa, H.Y., Prakash Narasimha, K.N., Basavalingu, B., 2003. Metamorphic Evolution and Fluid Regime in the Deep Continental Crust Along the N-S Geotransect From Vellar to Dharmapuram, Southern India. Mem. 50. Geol. Soc., India, Bangalore, pp. 318–373.

Stroh, P.T., Monrad, J.D., Fullagar, P.D., Naqvi, S.M., Hussain, S.M., Rogers, J.J.W., 1983. 3,000 m.y. old Halekote Trondhjemite; a record of stabilization of the Dharwar craton. In: Naqvi, S.M., Rogers, J.J.W. (Eds.), Precambrian of South India. Mem. 4. Geol. Soc., India, Bangalore, pp. 365–376.

Swami Nath, J., Ramakrishnan, M., 1981. Early Precambrian supracrustals of southern Karnataka. Mem. Geol. Surv. India 112, 1–328.

Swami Nath, J., Ramakrishnan, M., Viswanatha, M.N., 1976. Dharwar stratigraphic model and Karnataka craton evolution. Rec. Geol. Surv. India 197 (2), 149–175.

Taylor, P.N., Moorbath, S., Chadwick, B., Ramakrishnan, M., Viswanatha, M.N., 1984. Petrography, chemistry and isotopic ages of Peninsular Gneiss, Dharwar acid volcanic rocks of and Chitradurga Granite with reference to the late Archaean evolution of Karnataka craton, southern India. Precambrian Res. 23, 349–375.

Valdamani, R., Kröner, A., Vasudevan, D., Went, I., Tobschall, H., Chatterjee, C., 2012. Zircon evaporation ages and geochemistry of metamorphosed volcanic rocks from the Vinjamuru domain, Krishna Province: evidence for 1.78Ga convergent tectonics along the southeastern margin of the Eastern Dharwar craton. Geol. J. 48, 293–309.

Valdiya, K.S., 2010. The Making of India: Geodynamic Evolution. Macmillan Publishers India Ltd., India. 816 p.

Viswanathia, 1977. Lithostartigraphy of the Kaladgi and Badami Groups, Karnataka. Indian Mineral. 18, 122–132.

Zachariah, J.K., Bhaskar Rao, Y.J., Srinivasan, R., Gopalan, K., 1999. Pb, Sr and Nd isotope systematic of uranium mineralized stromatolitic dolostone from Proterozoic Cuddapah Supergroup, South India: constraints on age and provenance. Chem. Geol. 162, 49–64.

CHAPTER

BASTAR PROTOCONTINENT

5

5.1 INTRODUCTION

The Bastar Protocontinent occupies a quadrangular area bounded by the Narmada-Son Lineament in the NW, the Pranhita-Godavari Lineament (marked by Rift basin) in the SW, the Mahanadi Lineament (marked by Rift basin) in the NE, and the Eastern Ghats Granulite Belt in the SE (Fig. 5.1). Lithologically, the Protocontinent comprises patches of Archaean supracrustal rocks (greenstone belts) developed over a vast expanse of gneiss-granite ensembles of Archaean age. The post-Archaean

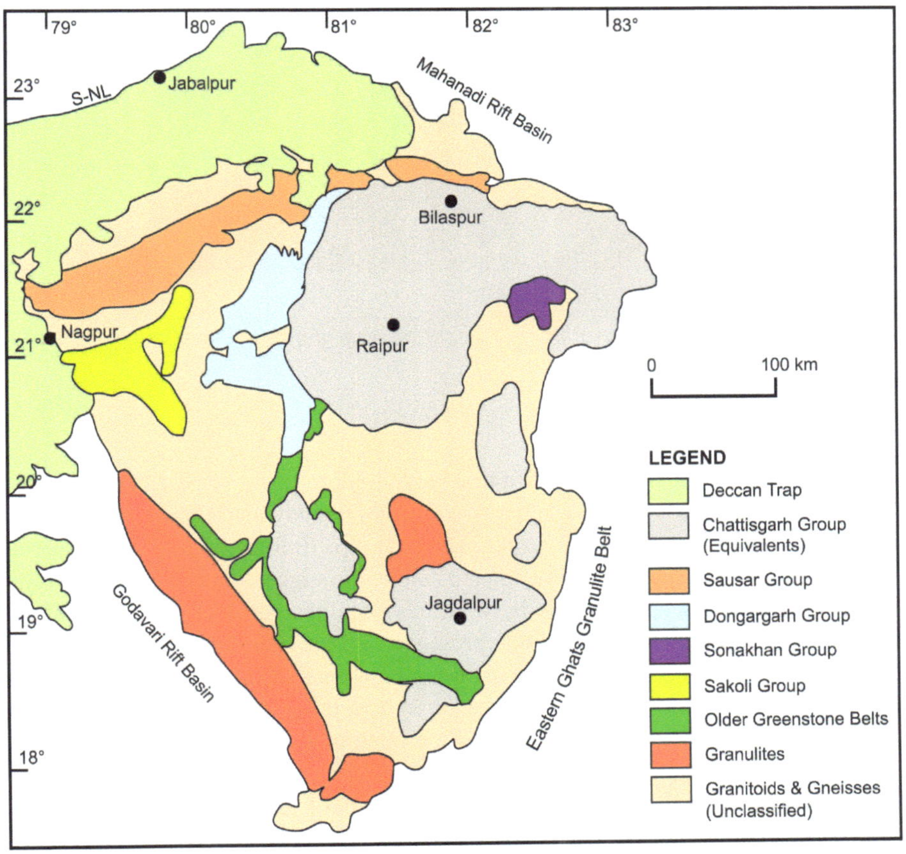

FIG. 5.1

Geological map of the Bastar Protocontinent showing distribution of major lithotectonic units.

Indian Shield. https://doi.org/10.1016/B978-0-12-809839-4.00005-9

(Proterozoic) supracrustals include the Sausar Group in the north and the intracratonic platformal successions of the Chattisgarh Group and its equivalents in the south. All these cover successions occur over the sialic Archaean basement. The Deccan Traps cover a considerable part of the Bastar Protocontinent in the north.

The oldest supracrustals that occur predominantly in the southern part show features of dismembered greenstone sequences enclosed within the sialic basement rocks. The most primitive basement rocks evolved at ~3.6 Ga. A number of episodes of reconstitution of basement rocks have been reported from the different parts of the Bastar Protocontinent. The youngest ones occur as irregular massive bodies of granites often enclosing relics of the foliated gneissic rocks (Fig. 5.2).

Some patches of granulite facies rocks occur in this Precambrian crustal block along with granites formed during different periods in the Archaean.

FIG. 5.2

An enclave of mafic body of tonalite gneiss within younger massive granite at Pithora near the Sonakhan Greenstone Belt.

Image courtesy: Erfan Mondal.

The Bastar Protocontinent is endowed with large iron-ore deposits within banded iron formation in the greenstone belts, and copper ores in the Archaean Malanjkhand Granite. Apart from that huge manganese, deposits occur in the Proterozoic Sausar Group. Before discussing the geology of the Bastar Protocontinent, it may be worthwhile to take note of the suggestion to divide the Protocontinent into two different crustal blocks separated by what has been described as the 'Central Indian Shear Zone'. However, the feature that is thought to represent a 'terrane boundary' on close examination appears like a wide zone of shear comparable in scale and complexity with that of the 'Singhbhum Shear Zone' that separates an Archaean granite-greenstone belt in the south from the younger Proterozoic fold belt in the north in the adjacent Singhbhum Protocontinent. There is absolutely no special geological, tectonic, or petrological feature that necessitates invoking the concept of a 'terrane boundary' marking a Suture Sone. Besides, the 'Magnetic Anomaly' pattern over the region covering a bit of neighbouring areas across the Protocontinent boundary confirms that the feature in the Bastar Protocontinent represents a crustal-scale shear zone and not an 'edge of the Central Indian Craton' (Rajaram and Anand, 2003).

For the sake of convenience, the discussion on the geology of the Protocontinent is divided into the following:

1. Gneisses, Granites, and Granulites.
2. The Archaean Greenstone Belts.
3. The Palaeoproterozoic Fold-Thrust Belt: Sausar Group.
4. The Late Proterozoic cratonic basins.
5. Summary of the evolutionary history of the Bastar Protocontinent.

5.2 GRANITES, GNEISSES, AND GRANULITES

The outcrops of granites and gneisses of different types and ages cover a vast area in the Bastar Protocontinent. Most of the rocks, except the oldest ones, show intrusive relationship with at least one group of supracrustals (Naqvi and Rogers, 1986). According to the early workers (Crookshank, 1963; Rajarajan, 1976), the gneissic rocks in the Bastar Protocontinent are similar to the Peninsular Gneiss of the Dharwar Protocontinent. Three different gneissic bodies are recognized, which occupy distinctly different geographic domains. These are the following:

1. Tirodi Gneiss forming the basement for the Palaeoproterozoic Sausar Group north of what has been described as 'Central Indian Tectonic Zone' (Ramakrishnan and Vaidyanadhan, 2008);
2. Amgaon Gneiss occupying the triangular area around the Sakoli Group outcrops in the northwest; and
3. Sukma Gneiss occurring in the Southern Part of the Protocontinent.

Apart from these large bodies of tonalite-trondhjemite-granodiorite gneisses with ill-defined boundaries, several patches and enclaves of gneisses also occur within the outcrops of granites of diverse composition. One such is the Baya Gneiss forming the base of the Sonakhan Greenstone Belt in the northeast.

Unlike the typical 'Archaean' banded gneiss of the Baya Gneiss 'type' (Fig. 5.3a and b), the gneissic rocks that constitute the basement for the younger Palaeoproterozoic Sausar Belt look more like a mylonite gneiss ('streaky gneiss' of Pascoe, 1973) that usually develops in sheared rocks.

In addition to the outcrops of gneissic rocks of different types and ages, there are also several bodies of massive granite that vary in composition from K-granites to granodiorite, trondhjemite, and tonalite

FIG. 5.3

(a) Dark grey Baya Gneiss and (b) Baya Gneiss showing effects of repeated reconstitution.

Image courtesy: Erfan Mondal.

at different parts of the Protocontinents. The major bodies of granite include Dongargarh, Malanjkhand, and Kanker Granites (Ramachandra and Roy, 1998). Wanjari et al. (2005) reported occurrence of at least four phases of granite that also include high-Al_2O_3 trondhjemite from the 'Amgaon Gneissic Complex' in the northern part (Sausar Belt) of the Bastar Protocontinent. These granites, according to the authors, represent lower-crustal and mid-crustal level anatexis of tonalities (tonalite-trondjhemite-granodiorite) interlayered with metasediments along the Shear Zones that supplied both heat as well as fluids.

Very little information is available about the petrochemistry of different types of gneisses in the Protocontinent. However, some information on the antiquity of these gneisses comes from the study of these gneisses by Sarkar et al. (1993) who reported single-zircon U–Pb age of 3509 + 14/−7 Ma from the trondhjemite-tonalite from Markampara in the southern part of the Protocontinent. Somewhat similar age data come from the study of Ghosh (2004) who reported 3562 ± 2 Ma $^{207}Pb/^{206}Pb$ zircon concordia age from the tonalite-trondhjemite gneiss from the central part of the Protocontinent. More recently, Rajesh et al. (2009) reported 3582.6 ± 4 Ma U–Pb zircon concordia age from the low-K tonalite-trondhjemite-granodiorite rocks from the Dalli Rajhara region in the central part of the Bastar Protocontinent. The younger well-constrained ages of granite-gneiss complex point to two different late–/end-Archaean events. These are 3018 ± 61 Ma Pb–Pb isotope age of granite intrusive into the Sukma Gneiss and 2573 ± 139 Ma Rb/Sr isochron age of the youngest (end-Archaean) intrusive leucocratic granite that include the Malanjkhand Granite hosting Andean-type Porphyry Copper-Molybdenum deposits (Stein et al., 2004).

Intermingling of granulites and gneisses is quite common in the belt of Tirodi Gneiss that occurs on either side of the outcrops of the Sausar belt in the northern part of the Bastar Protocontinent. Petrologically, these include mafic granulite, porphyritic charnockite, cordierite granulite, and amphibolites, occurring as rafts and lensoid bodies within the gneissic complex described as the Tirodi Gneiss (Sharma, 2009). The oldest age from these granulite lenses is 2672 ± 54 Ma (Roy et al., 2006), which is thought as the age of the basement (represented by Tirodi Gneiss).

Two more granulite belts are reported in the Bastar Protocontinent (Ramakrishnan and Vaidyanadhan, 2008). One of these is the Bhopalpatnam granulite belt occurring on the northern shoulder of Godavari Rift Basin (Fig. 5.1). Single-zircon U–Pb study of the Bhopalpatnam granulite shows an age range between 1.6 and 1.9 Ga (Santosh et al., 2004). Another belt is the Kondagaon granulite belt, which occurs in the middle of the Protocontinent. Charnockite and leptynite dominate especially in the western part of this granulite body. The Kondagaon granulite indicates an age of around 2.6 Ga (Vansutre and Hari, 2010).

5.3 THE ARCHAEAN GREENSTONE BELTS

Overlying the granite-gneiss ensemble whose antiquity dates back to ~3.6 Ga occur a series of greenstone belts formed at different times. Four different groups of greenstone belts of different ages are known in the Bastar Protocontinent. These are from the oldest to the youngest: (1) Sukma Greenstone Belt, (2) Bengpal Greenstone Belt, (3) Bailadila Greenstone Belt, and (4) Kotri-Dungargarh/Sakoli/Sonakhan Greenstone Belts.

5.3.1 SUKMA GREENSTONE BELT

Several discontinuous, widespread bodies of the most ancient supracrustal rocks, described as the Sukma Group (Crookshank, 1963), occur within the oldest gneissic basement locally known as the Sukma Gneiss (Fig. 5.4). The main lithological unit of this ensemble includes sillimanite-bearing

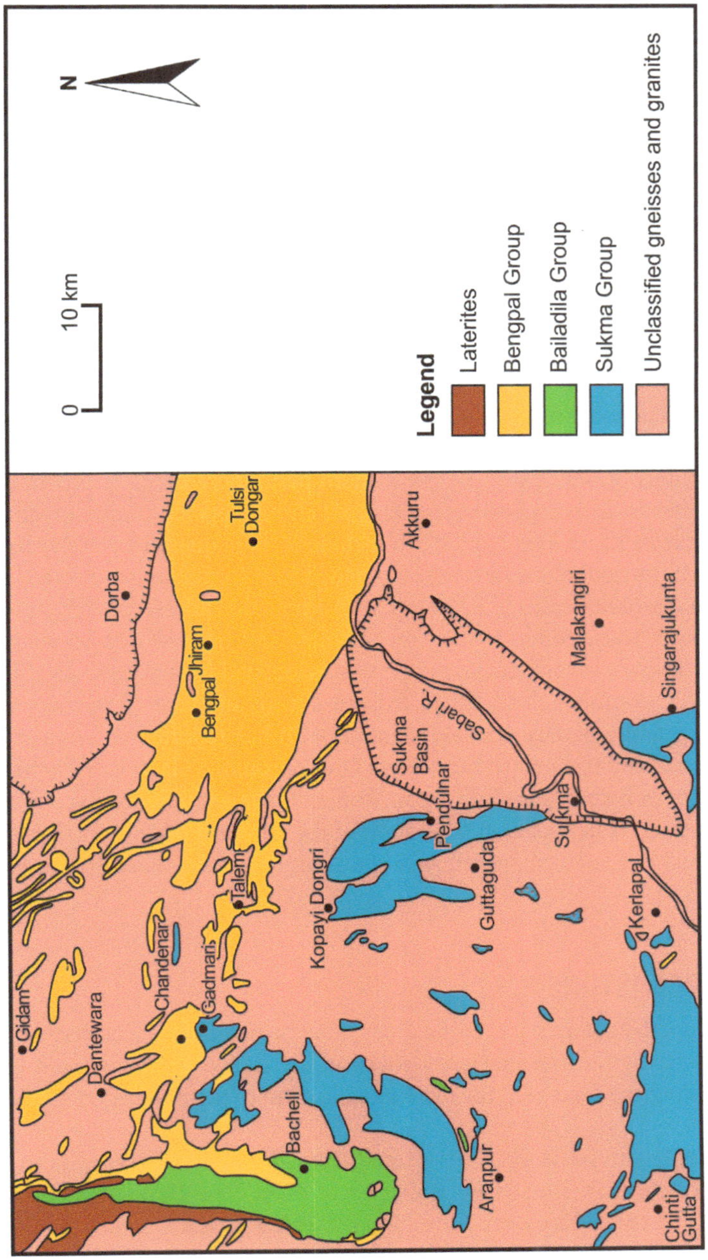

FIG. 5.4

Geological map of the Southern part of the Bastar Protocontinent, showing outcrop distribution of the older group of greenstone belts.

quartzite, calc-silicate gneiss and amphibolite, cordierite-anthophyllite schist, cordierite-biotite-sillimanite-garnet schist, and banded iron formation. The basal unit of the Sukma Group is sillimanite-bearing quartzite with or without fuchsite, corundum, and graphite. The association of high-alumina minerals in the quartzite is significant suggesting derivation from sialic source. The localized occurrences of the high-alumina deposits may be interpreted as alumina-rich palaeosols. The presence of fuchsite in the sequence further indicates that the source rocks also included chromium-bearing ultramafic protoliths, while the presence of graphite indicates a reducing 'oxygen-deficient' condition of the depositional environment. The Sukma Basin was evolved on the tonalite-trondhjemite-granodiorite gneiss. The oldest age of the basement is cited by Sarkar et al. (1993) who reported single-zircon U–Pb age of 3509 + 14/−7 Ma from the trondhjemite-tonalite gneisses of the Markampara area a little south of Sukma (Fig. 5.4).

5.3.2 BENGPAL GREENSTONE BELT

The Bengpal Greenstone Belt exposing rocks of the Bengpal Group (Crookshank, 1963) occurs in a NNW-SSE trending band that extends from Bailadila Hills in the central Bastar to Tulsi Dongar Hills close to the Eastern Ghats Granulite Belt (Fig. 5.4). In the eastern part, the outcrops of the Bengpal rocks show more continuous presence than in the western part where the outcrops of the greenstone belt rocks are relatively sparse. Lithologically, the Bengpal Group appears similar to that of the Sukma Group. The main constituents of the Bengpal Group include sericite quartzite interlayered with amygdular meta-basalt and tuff, ferruginous quartz schist, banded iron formation, and some isolated bodies of conglomerate. The meta-basalts are interlayered with immature arkose grading to orthoquartzite (Niogi et al., 1996), quartz wacke/lithic wacke, metapelites, and banded iron formation (Khan and Bhattacharyya, 1993). There are several enclaves of rocks associated with andalusite-chiastolite schist, andalusite quartzite, and chloritoid schist, which formed during an event of thermal metamorphism induced by intrusion of granitic bodies. Ramakrishnan (1990) has reported an angular unconformity between the Bengpal Group and the underlying Sukma Group from the Gadmiri-Chandenar area in the northern part of the Southern Bastar (Fig. 5.5). Apart from this, the antiquity of the Sukma Group with respect to the Bengpal Group is indicated in the nature of the outcrops of the former making a girdle-like pattern of the isolated outcrops compared with the straight-running band-like appearance of the Bengpal outcrops (Fig. 5.4).

5.3.3 BAILADILA GREENSTONE BELT

Named as Bailadila Group, the components of the greenstone belt include an ensemble of mildly metamorphosed, dominantly ferruginous sedimentary rocks in the Bastar Protocontinent (Fig. 5.6). The Bailadila Hills, representing the type area for the Bailadila Group, form the southernmost part of the Kotri-Dongargarh belt.

The iron-ore deposits in the Bailadila Hills extend from the north of Bailadila to Rowghat and Dalli Rajhara that contribute iron ores to the Bhilai Steel Plant. The Bailadila ores are considered similar to the Superior-type banded iron-ore formation occurring in association with the fine clastics, tuffs, and some minor mafic volcanics (Ramakrishnan and Vaidyanadhan, 2008). The Sukma Group forms a garland-like outcrops around the north-south trending Bailadila Hills. The predominantly east-west trend of the Sukma Group and the west north-south trending Bengpal Group have been

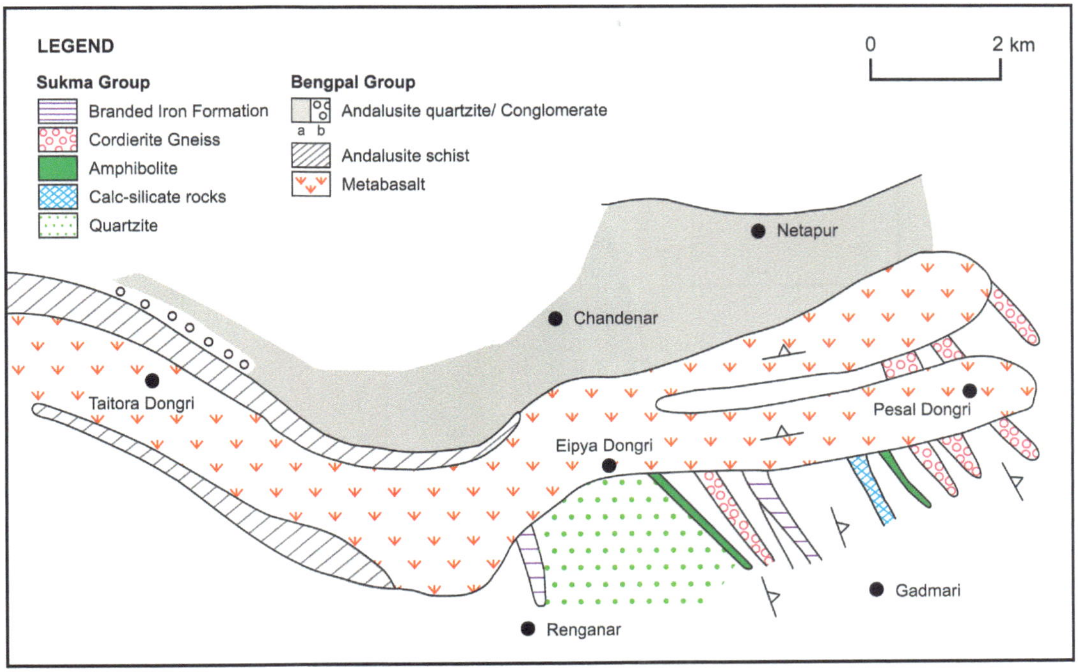

FIG. 5.5

Map showing unconformable relationship between the Bengpal Group and the Sukma Group north of Gadmiri.

After Ramakrishnan and Vaidyanadhan (2008) with permission.

swept to parallelism with the north trending Bailadila Group from the west of Geedam to the Bailadila Hills (Fig. 5.5) concurrently with the deformation of the Bailadila Group.

The following lithostratigraphic succession of the Bailadila Group (Table 5.1) is suggested by Khan and Bhattacharyya (1993).

5.3.4 THE END-ARCHAEAN GREENSTONE BELTS

The greenstone belts that occur north of the north-south running linear Bailadila Greenstone Belt may be described as the end-Archaean greenstone belts. These include the Sakoli Belt in the west, the Kotri-Dongargarh Belt in the middle, and the Sonakhan Belt in the east. Brief geological account of this youngest group of greenstone belts is appended here.

5.3.4.1 Sakoli Belt

The Sakoli Group, which covers an area of about 350 km^2 forming a triangular outcrop pattern (the Bhandara Triangle of Naqvi and Rogers, 1986) (Fig. 5.7), constitutes the westernmost ensemble of the end-Archaean greenstone belts in the Bastar Protocontinent. The ensemble consists of low-grade metamorphic rocks comprising dominantly a volcano-sedimentary sequence made of slates, phyllites, and bimodal volcanic suite including meta-basalts, along with meta-chert, conglomerate, and banded

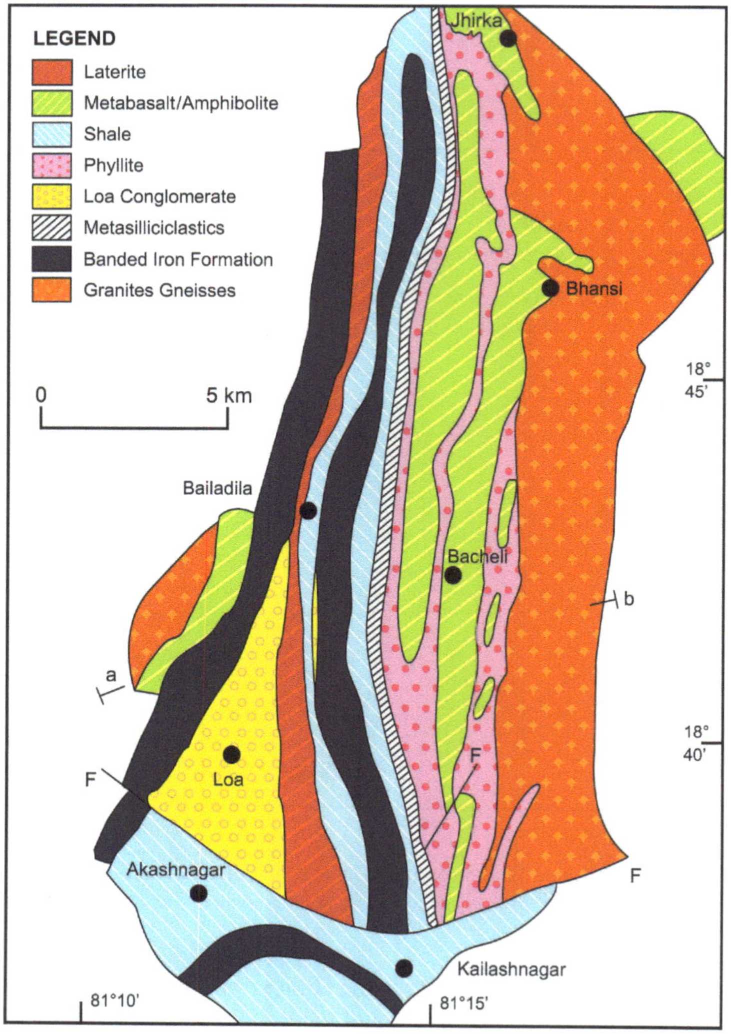

FIG. 5.6

Lithologic map of the north-south trending Bailadila Range.

Reproduced from Khan and Bhattacharyya (1993) with permission.

Table 5.1 Stratigraphic succession of the Bailadila Group

Formation	Lithology
Kailash Nagar Formation (200–500 m)	Banded hematite quartzite, banded magnetite quartzite with pockets of iron-ore deposits
Loa Formation (5–150 m)	*Upper horizon*: Intensely lateritized ferruginous shale *Lower horizon*: Polymictic conglomerate in silt and sand
East Ridge Formation (206–639 m)	Ferruginous tuffaceous shale interbedded with ferruginous chert, banded ferruginous chert/jaspilite, and carbonaceous phyllite
Bacheli Formation (240–335 m)	Graded bedded wackes and cross bedded arenite with intercalations of shale
Bansi Formation (600–1000 m)	Chlorite-mica phyllite with or without porphyroblasts of andalusite. Amydular basalt

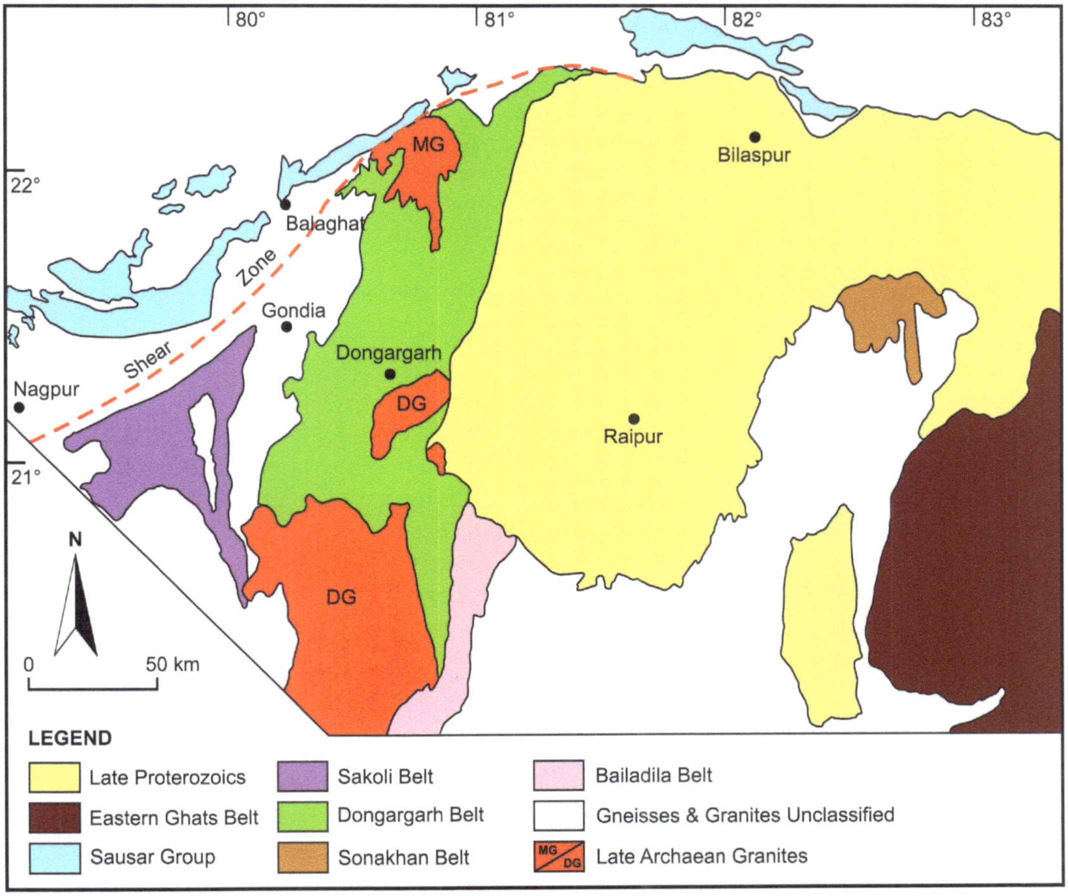

FIG. 5.7

Geological map of the northern part of the Bastar Protocontinent showing distribution of the three end-Archaean Greenstone belts south of the prominent shear zone. MG, Malanjkhand Granite; DG, Dongargarh Granite.

iron formation. Amongst the constituent lithologic units, the slate and phyllite dominate the rock assemblages covering about 80% of the area. The rest 20% of the outcrops belong to the felsic to mafic bimodal volcanics. The dominantly gneissic rocks described as the Amgaon Gneiss constitute the basement of the Sakoli Group.

The following stratigraphic succession has been suggested for the Sakoli Group by Bandopadhyay et al. (1995) (Table 5.2).

There is hardly any detailed study about the structural geometry of the Sakoli Basin, which is stated to be very complex. One suggestion is that the development of complex structure has resulted because of repeated folding of the initial complex geometry of the Sakoli Basin (Ramakrishnan and Vaidyanadhan, 2008). The Sakoli Group does not host any major mineral deposit except the reported occurrence of gold in the felsic volcanic suites (Mishra et al., 2001).

Table 5.2 Stratigraphic succession of the Sakoli Group

Formation	Lithology
Pawni Formation	Slate, phyllite, arkose, quartz arenite, and conglomerate
Bhiwapur Formation	Mainly metapelites (± chloritoid, andalusite, garnet, and staurolite) with interbands of felsic volcanic/tuffs and exhalative sediments (coticules, tourmalinites, chlorites, banded iron formation, rare basalts), and Zn-Cu-gold and scheelite mineralization
Dhabetekri Formation	Mainly meta-basalts with minor ultramafics, cherts, metapelites, and banded iron formation
Gaikhuri Formation	Conglomerate, coarse arenite, arkose, minor pelites (partly carbonaceous), and banded iron formation

5.3.4.2 Kotri-Dongargarh Belt

While revising the stratigraphy of the rocks of the traditionally described Sakoli belt rocks, Sarkar (1958) introduced a new stratigraphic entity in the Kotri-Dongargarh region in the north-central part of the Bastar Protocontinent, which he named the Dongargarh Group ('Group' replacing 'system' of the author). The newly recognized Dongargarh Group was earlier considered a part of the pristine 'Sakoli system' (Pascoe, 1973). Subsequent to the initial proposal to define a new group, the stratigraphic status of the Dongargarh Group was elevated to the level of 'Supergroup'. This revision, however, does not appear justified as it is in contradiction of the 'Code of Stratigraphic Nomenclature' recommended by the Geological Survey of India (1971). In the present discussion, we would prefer to refer this ensemble as the Dongargarh Group following the suggestion of Sensarma and Mukhopadhyay (2014).

Sandwiched between the Sakoli selt and the granite-gneiss (Amgaon Gneiss) on the west and the late Proterozoic Chattisgarh Basin on the east, the broadly linear outcrops of the Dongargarh Group extend for about 300 km in the north-south direction (Fig. 5.7). The Dongargarh Group consists of pyroclastics and volcanics, which also include volcano-clastic arenite and arkosic wackes. Sensarma and Mukhopadhyay (2014) proposed a revision of Dongargarh stratigraphy defining a continuous stratigraphic succession with only minor breaks between different lithostratigraphic formations (Table 5.3). The Dongargarh succession shows a weakly developed regional synformal structure (Sitagota Syncline) that has a low 20° plunge in the northerly direction (Sensarma and Mukhopadhyay, 2014). This description, however, does not match with the penetrative nature of the tectonic structures marked by the occurrence of steeply dipping foliation surfaces in the different rocks.

Table 5.3 Stratigraphic succession of the Dongargarh Group

Group	Formation
Dongargarh Group	Mangikhuta Volcanics
	Karutola Formation
	Sitagota volcanics
	Chandsuraj Formation
	Pitepani Volcanics
	Bijli Rhyolites

Some isotope data are available on the basal unit represented by the Bijli Rhyolite of the Dongargarh Group. A wide range of ages have been suggested by different workers based on the study of Rb–Sr systematics of the Bijli Rhyolite, which vary between 2180 ± 25 Ma (Sarkar et al., 1993) and 2503 ± 35 Ma (Krishnamurthy et al., 1990). Such a wide variation in ages, according to Sensarma and Mukhopadhyay (2014), could be the result of mere thermal disturbance on the isotope system at different times. On the other hand, the dating based on U–Pb single-crystal zircon study (Ghosh, 2004) indicates ~2530 Ma age of the oldest rhyolite in the Kotri belt. Further, the granites intrusive into the components of the Dongargarh Group show a close but slightly younger age range. Several isotope age data are known from the different parts of the belt, such as (i) 2465 ± 22 Ma (Dongargarh Granite) (Rb–Sr whole rock age, Krishnamurthy et al., 1988), (ii) 2405 ± 63 Ma (Malanjkhand Grey Gneiss) (Rb–Sr whole rock age, Panigrahi et al., 1993), (iii) 2467 ± 38 Ma Malanjkhand Granite (Rb–Sr whole rock age, Panigrahi et al., 1993), (iv) 2478 ± 9 Ma (*Re*–Os age, Stein et al., 2004), and (v) 2490 ± 8 Ma (U–Pb single-zircon age, Panigrahi et al., 2002). Taken together, all the above cited ages confirm a virtual conformable end-Archaean evolution of the granites that intruded into the Dongargarh Group, marking the culmination of the Archaean crustal evolutionary history of the belt. Further, considering the age of the Bijli Rhyolite, the oldest formation of the Dongargarh Group, we may presume a very short span for the evolution of the end-Archaean greenstone belt of Dongargarh.

5.3.4.3 Sonakhan Group

The Sonakhan Greenstone Belt covers an area of about 1200 km^2 extending in NNW-SSE direction for about 40 km from Sonakhan (21°23′35″: 82°48′55″) in the north to Remra (21°17′N: 82°46′E) in the south having maximum width of 40 km in the central part. Stratigraphically described as the Sonakhan Group, the belt exposes a bimodal volcanics-sedimentary sequence that unconformably overlies the basement gneissic complex locally known as the Baya Gneiss. The late Proterozoic Chhattisgarh Basin covers the entire greenstone succession in the north and east.

The Sonakhan Group comprises three formations: (i) Baghmara Formation, (ii) Arjuni Formation, and (iii) Bilari Formation (Das et al., 1990). The oldest Baghmara Formation consists dominantly of meta-ultramafites, meta-basalt, metagabbro, pyroclastics of intermediate to basic composition, ignimbrite, rhyolite, acid tuff, pebbly tremolite-actinolite schist, and banded iron formation. The next younger is the Arjuni Formation, which constitutes a thick sedimentary pile interspersed with minor volcanics and overlies the basal Baghmara Formation. The polymictic conglomerate member the Arjuni Formation is known as the Jonk Conglomerate, which is dominantly a clast-supported conglomerate body. The Jonk Conglomerate includes mixtures of boulders and large-sized pebbles and blocks of protolithic rocks. The very nature of the Jonk Conglomerate suggests its fluvioglacial origin (Figs. 5.8 and 5.9). The youngest Bilari Group comprises both felsic and mafic intrusive and extrusive bodies like the Arangi mafic volcanics, meta-basalt, and pyroclastics along with rhyolite. The larger bodies of sills with NW-SE trend occur within the Baghmara Formation.

5.3.5 THE PALAEOPROTEROZOIC SAUSAR FOLD-THRUST BELT

The Sausar Fold-Thrust Belt, which extends from the north of Nagpur to the northeast of Bilaspur (Fig. 5.1), forms a narrow, about 32 km wide, arcuate zone spanning over 215 km between Ramakona (21°44′00″: 78°50′37″) in the east and Baihar (22°05′00″: 80°32′42″) in the west. Lithostratigraphically,

FIG. 5.8

Outcrop of Jonk Conglomerate showing dominantly clast-supported boulders and large pebbles. The conglomerate includes a large chunk of strongly foliated gneissic body.

Photo courtesy: Erfan Mondal.

FIG. 5.9

Assorted well-rounded boulders and pebbles, some of which appear as bullet-shaped body, provide impression of glacial environment of deposition.

Image courtesy: Erfan Mondal.

the supracrustal rocks of this fold-thrust belt are known as the Sausar Group, the different components of which show intimate intermixing with gneisses and granites of different ages (Tirodi Gneiss and younger granites) (Fig. 5.10). Sausar Group is made up primarily of metamorphosed rocks of the sand-shale-carbonate association hosting rich manganese ore deposits (Fermor, 1909, 1936). The calcareous sedimentary formations are better developed in the north and west, whereas the argillaceous formations dominate in the south and east of the belt.

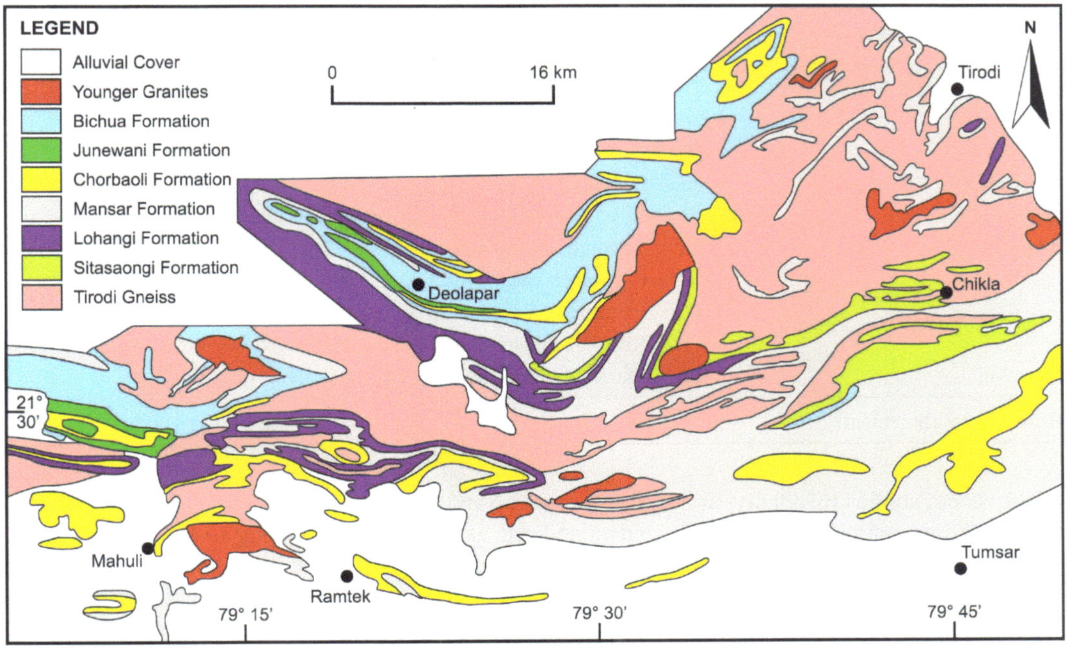

FIG. 5.10

Geological map of the western part of the Sausar Group between Tirodi in the east and northwest of Mahili in the west showing a very complex geometry of the outcrop pattern.

Compiled from different sources.

The stratigraphic scheme of the Sausar Group has been a matter of debate for a long time. Ramakrishnan and Vaidyanadhan (2008) highlighted several reasons for the polemic:

1. A complex deformation pattern resulting from the development of large recumbent folding and thrusting that greatly inhibited proper understanding of the normal stratigraphic order of the constituent rocks.
2. The absence of indubitable field evidence for determining the basement-cover relationship over the major part of the area.
3. In spite of the reported occurrences of polymictic conglomerate in different parts, there is little evidence to prove that these are 'basal' conglomerates formed during the basin opening of the Sausar Group.
4. The widespread sedimentary facies variation noted in highly dismembered rock sequences.

However, in spite of all these difficulties in erecting the stratigraphic succession of Sausar Group, the scheme proposed by Narayanaswamy et al. (1963) is being thought as a 'workable' stratigraphic succession (Table 5.4) for the study of the Sausar supracrustal rocks that occur in a highly tectonized belt in the northern part of the Bastar Protocontinent. It may be worth remembering that the stratigraphic succession shown in Table 5.4 does not differ significantly from that, which was originally proposed by L.L. Fermor and W.D. West about seven decades ago (Krishnan, 1956).

Table 5.4 Stratigraphic succession of Sausar Group

Formation	Lithology
Bichua	Dolomite, serpentine-bearing marble, calc-silicate granulite
Junewani	Muscovite-biotite-quartz schists, granulite, and gneisses
Chorbaoli	Quartzite, micaceous and feldspathic quartz schist, (local) conglomerate
Mansar	Mica schist, graphite phyllite, and quartzite hosting manganese ores
Lohangi	Three interdigitating members:
	Lohangi member: calcareous dolomite and marble
	Utiketa member: calc-silicate granulite and gneisses
	Kadbi Khera member: quartz-biotite granulite and gneisses
Sitasaongi	Quartz-muscovite-feldspar schist and intercalated quartzite
Tirodi Gneiss (basement)	

The recent studies (Mohanty, 2006; Mohanty and Nanda, 2016; Mohanty et al., 2015) highlighted some important points, which help to understand the Sausar lithostratigraphy to a great extent. The first is the suggestion that the Sausar Group does not only comprise the metasedimentary rocks but also include volcanic rocks especially in the basal part (Mohanty, 2006). Further, based on some well-preserved sedimentary features in the basal succession of the Sausar Group, a suggestion has been made about the glaciogenic origin for the diamictites that occur in the lower part of the succession (Fig. 5.11a and b) (Mohanty et al., 2015). Added to these is the discovery of palaeosols from the contact of the Sausar Group with the basement, the Tirodi Gneiss (Mohanty and Nanda, 2016).

The Sausar Fold-Thrust Belt is bounded between two very prominent northerly dipping shear zones. The southern Shear Zone separates the high-grade Sausar metasediments in the north from the dominantly low-grade end-Archaean greenstone belts in the south (Jain et al., 1991).

FIG. 5.11

(a) Diamictite showing assorted large pebbles in arenitic matrix in the Mahili area and (b) angular clast of 'vein quartz' within the diamictite.

Reproduced from Mohanty et al. (2015) with permission.

In early literatures, the Sausar rocks have been described as stacks of complexly folded sequence (Pascoe, 1973). The complexity of structure finds expression in the outcrop pattern of this fold-thrust belt (Fig. 5.12a and b). According to Fermor (1909), the reversal of rock sequence from the northern to the southern part is because of the development of large recumbent folding. The major recumbent fold that occurs as a large fold-thrust nappe is called the Deolapar Nappe (West, 1936). The development of thrust slices in the form of nappes in the Sausar Belt has also been compared with the Alpine tectonics by the early 19th-century geologists (Pascoe, 1973).

FIG. 5.12

(a) Packs of isoclinal folds developed in manganese-bearing metasediments of Chikla Mine area showing rounded antiforms and sharp 'pinched in' synforms. (b) Stacked outcrop scale isoclinal recumbent fold in calc-silicate rock, near the base of 'Deolapar Nappe'.

(a) Reproduced with permission from Mohanty (2010); (b) Reproduced with permission from Chattopadhyay et al. (2003).

Some confirmation of the intricate deformation structures in smaller scale comes from the studies of Mohanty (1988, 2010), Chattopadhyay et al. (2001), and Chattopadhyay and Bandopadhyay (2004). The commonly occurring deformation structures reported by these workers include stacks of sheared and dissected isoclinal folds both in the outcrop and in the map scale (Fig. 5.12a and b).

Some controversy surrounds the description of the large-scale 'fold-thrust' structure described as the 'Deolapar Nappe' by West (1936). However, Chattopadhyay et al. (2003) seem to confirm indirectly the earlier suggestions about the 'nappe' structure interpreting the large-scale structure as an allochthonous block representing a thrust nappe formed over a low-angle thrust plane. Development of similar detachment structure has also been indicated by Mohanty (2010) who prepared a map of a large overfolded fold with detached lower limb in the Ramtek area (Fig. 5.13).

A mention has already been made about close intermingling of the bodies of granulites and the granite-gneisses described as the Tirodi Gneiss (Sharma, 2009). The granulite bodies having different petrological compositions generally occur as rafts and 'lens-shaped' outcrops of different sizes on either side of the Sausar Belt rocks. The oldest age of these granulite lenses is 2672 ± 54 Ma (Roy et al., 2006), which is considered the age of the Tirodi Gneiss representing the basement on which the overlying Sausar Group was deposited.

The information on the geochronologic age of the Sausar Group is scanty, and the available interpretations appear quite messy. Objectively speaking, based on the youngest ages (~2.5 Ga) of the basement gneiss-granite association, it may be said that the Sausar Group, which unconformably overlies these basement rocks, is Proterozoic in age. Several isotope age data are known from

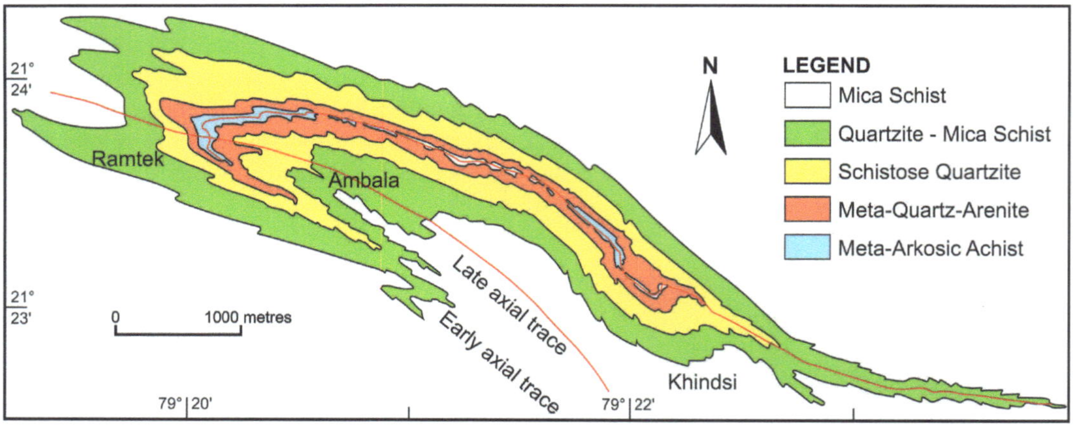

FIG. 5.13

Simplified geological map of the Ramtek area, showing large-scale axial folding of an early fold around a later isoclinal fold with detached lower limb.

Reproduced from Mohanty (2010) with permission.

different parts that suggest that the Archaean Crust-forming processes in the Bastar Protocontinent did not culminate precisely at 2500 Ma, but the tectonothermal activity continued beyond this landmark date. Examples are (i) 2465 ± 22 Ma age of the Dongargarh Granite (Rb–Sr whole rock age, Krishnamurthy et al., 1988), (ii) 2405 ± 63 Ma age of the Malanjkhand Granite of grey gneiss type (Rb–Sr whole rock age, Panigrahi et al., 1993), (iii) 2467 ± 38 Ma age of the Malanjkhand Granite (Rb–Sr whole rock age, Panigrahi et al., 1993), (iv) 2478 ± 9 Ma age of Malanjkhand Granite (Re–Os age, Stein et al., 2004), and (v) 2490 ± 8 Ma U–Pb single-zircon age of granitic rocks (Panigrahi et al., 2002). Taken together, all the above cited ages provide proof that the granite-gneiss complex formed the basement of the Sausar Group did not get fully cratonized at least before ~2450 Ma. Though the depositional age of the overlying Sausar Group is not known precisely, the occurrence of palaeosols over the peneplained basement (Mohanty and Nanda, 2016) provides an indication of a prolonged period of erosion, weathering, and nondeposition subsequent to the date of cratonization of the youngest Archaean basement, which in the case of Bastar Protocontinent could not have been earlier than ~2450 Ma. At the same time, considering that the peneplanation is a very slow process, we may rationally presume that the Sausar Basin did not open at least before 2350 Ma. Some information on the possible age of the basin formation comes from the reported evidence of 'snowball' Earth from the basal succession of the Sausar Group (Mohanty et al., 2015). Globally, this Palaeoproterozoic snowball event known as the Huronian glaciation, spans between 2400 and 2100 Ma. In this context, the occurrence of sedimentary manganese deposits in Sausar Group assumes importance. Successive phases both of the deposition of manganese and of glaciation have been recorded in the Palaeoproterozoic in different parts of the globe (Roy, 1981, 2006). Incidentally, this was also the period of massive growth of cyanobacteria in shallow seas, which helped in enhancing oxygenation of the atmosphere. In the Indian Shield, drawdown of atmospheric CO_2 coupled with wide cyanobacterial activity has been noted in the Aravalli Supergroup, which based on geochronological data was during 2.2–2.1 Ga (Roy and Purohit, 2015). In the absence of any direct geochronologic age of the Sausar Group, we have reasons to assume that the basin opening of the Sausar was also at the same time as that of the Aravalli Supergroup.

5.3.6 THE LATE PROTEROZOIC CRATONIC BASINS

Several isolated bodies of late-Proterozoic cratonic basins, described in early literatures as the Purana Basins, occur in the Bastar Protocontinent. Of these, the Chattisgarh Basin, covering an area of 36,000 km^2 area in the Bilaspur-Raipur region, is the largest situated on the northeastern edge of the Archaean cratonic block of Bastar. Apart from this, there are also a number of smaller 'outlier' type outcrops of cratonic Platform type depositories in the Protocontinent (Fig. 5.14). The deformation structures noted in some localized blocks are the results of 'postinversion' cratonic deformation processes.

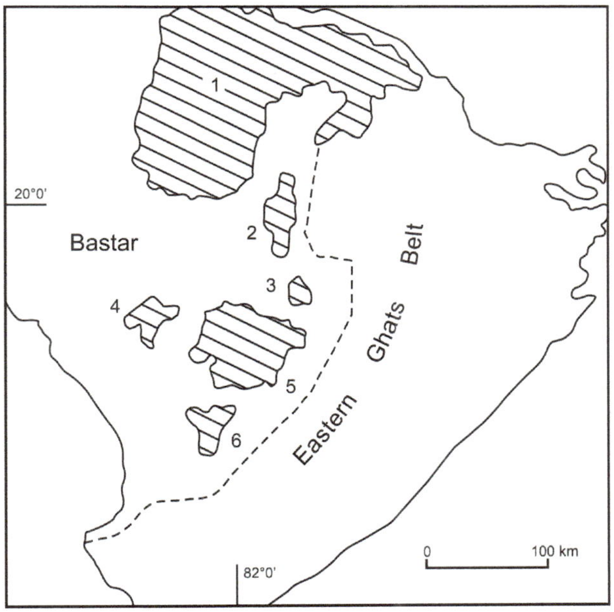

FIG. 5.14

Geological sketch map showing distribution of late-Proterozoic cratonic basins (Purana Basins) in the Bastar Protocontinent. 1, Chattisgarh; 2, Khariar; 3, Ampani; 4, Abujhmar; 5, Indravati; 6, Sabari.

5.3.6.1 Chattisgarh Basin

The Chattisgarh Basin is the largest late-Proterozoic cratonic basin in the Bastar Protocontinent. The basin contains about 2500 m thick sediments that belong to the shale-sand-carbonate association deposited in multiple sedimentary cycles in three different subbasins (Fig. 5.15). The presence of fan-delta deposits at the base, followed by thick sandstone-mudstone succession, welded tuff (ignimbrite) in the shale units and carbonate platform grading into pelagic environment suggests deposition essentially in cratonic basin environment without major hiatuses (Deb, 2004). The vast limestone and dolomite resources of the basin feed the Bhilai Steel Plant and some nearby cement factories. The lithostratigraphic scheme of the Chattisgarh Basin is given in Table 5.5.

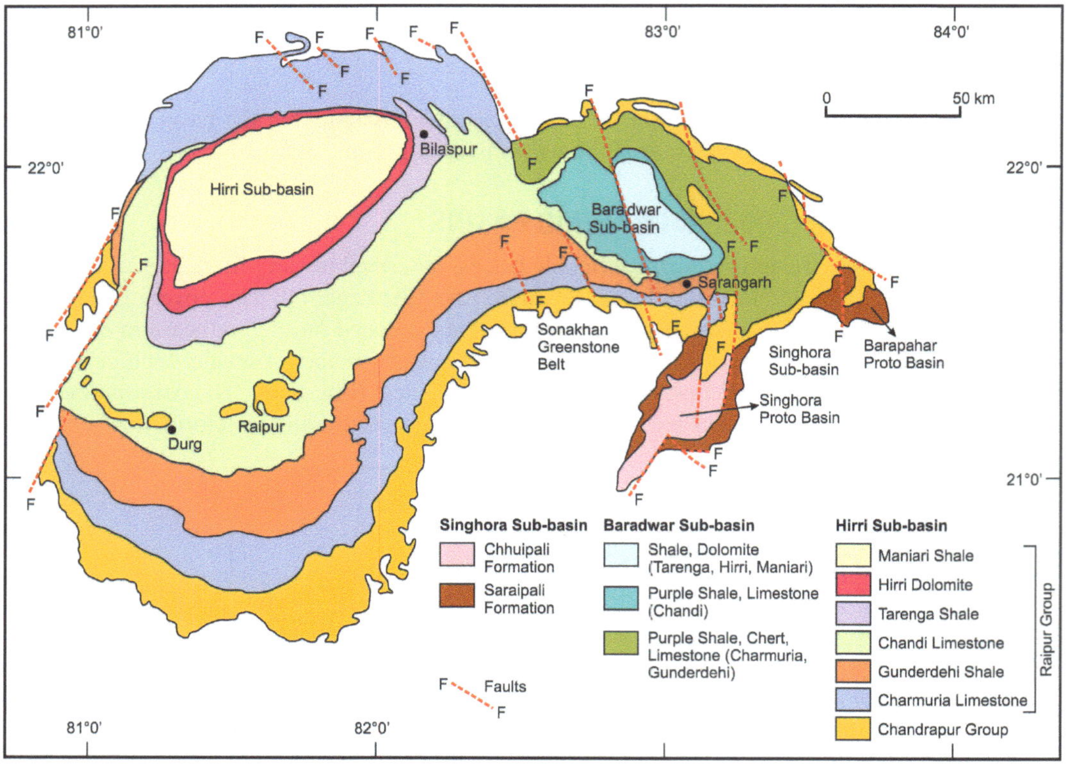

FIG. 5.15

Generalized geological map of the Chattisgarh Basin.

Table 5.5 Lithostratigraphy of the Chattisgarh Basin

Group	Formation (thickness in m)	Lithology
Raipur Group	Maniari Shale (70)	Purple shale, dolomite, gypsum
	Hirri Dolomite (70)	Grey dolomite
	Tarenga Shale (180)	Dolomitic shale, shale-chert beds, purple shale, limestone
	Chandi Limestone (670)	Stromatolitic dolomite, limestone, glauconitic sandstone, shale
	Gunderdehi Shale (430)	Shale with limestone interbeds, arenite—shale, ignimbrite
	Charmuria Limestone (490)	Phosphatic limestone with shale interbeds, cherty limestone, phosphatic dolomite, chert-like interbeds
	Bijepur Shale (100)	Green and brown calcareous shale with sandy interbeds
Chandrapur Group	Kansapathar Sandstone (200)	Glauconitic sandstone
	Chaporadih Shale (200)	Shale with arenite interbeds
	Lohardih Conglomerate (20)	Purple arkose, gritty arenite, and basal conglomerate
Singore Group	Chhuipali Shale (300)	Shale with chert, limestone, dolomite, siltstone
	Bhalukona Stone (20)	Sandstone, siltstone, minor shale
	Saraipali Shale (60)	Variegated shale with siltstone, limestone, porcellenite, felsic tuff
	Rehatikhol Conglomerate (20)	Feldspathic arenite, arkose, and basal conglomerate
Crystalline basement of the Bastar Protocontinent		

Over the major part of the basin, the beds show horizontal to subhorizontal orientation with rolling dips in areas showing mild folding. A few open antiformal folds have developed close to the faults that generally developed in the marginal parts of the Basin. The eastern projections of the basin are deformed at their margins during the 'collision'-type impact of the Eastern Ghats Granulite Belt. Geochronological data based on K/Ar systematics on the glauconites from Chandrapur Group suggest sediment deposition age in the time span between 1250 and 1300 Ma (Kruezer et al., 1977). The basin with the irregular rectangular-shaped configuration of outcrop is surrounded by a number of kimberlite pipes.

5.3.6.2 Khariar Basin

The Khariar Basin covers about 1500 km^2 area containing over 1000 m thick sediments in the eastern part of the Bastar Protocontinent. This irregular, oval-shaped, broadly N–S trending basin is juxtaposed against the faulted margin close to the Eastern Ghats Granulite Belt in the east. On the western side, the basin is underlain by the Archaean gneiss-granite complex containing enclaves of supracrustals like the banded iron formation, mica schist, amphibolite, and pillowed meta-basalt. As in the case of the Chattisgarh Basin, several intrusions of kimberlites are also noted near the basin margins.

Lithostratigraphy of the Khariar Basin following Das et al. (2001) is indicated in Table 5.6.

Table 5.6 Lithostratigraphy of the Khariar Basin

Group	Formations
Piari Group (600–1000 m)	Ling Dongri Sandstone (60–120 m)
	Tarjhar Shale (205 m)
	Galighat Sandstone (150–300 m)
	Disconformity
	Neor Sandstone (40–160 m)
	Kulharighat Limestone (980–120 m)
	Devdhara Sandstone (10–80 m)
Basement gneisses and granites	

The Piari Group rocks are involved in open-to-tight asymmetrical folding trending NW-SW. The intensity of deformation increases in the easterly direction near the contact of the Eastern Ghats Granulite Belt showing evidence of strong shearing and faulting (Ramakrishnan and Vaidyanadhan, 2008).

No precise age data are available for the rocks of the Khariar Basin. However, suggestion has been made about the Lower to Upper Riphean age (i.e. between 1400 and 800 million years ago) based on the study of the stromatolite structures developed in the carbonate formations of Piari Group (Ramakrishnan and Vaidyanadhan, 2008).

5.3.6.3 Ampani Basin

The Ampani Basin (Fig. 5.14) occurs on a small plateau south of Khariar. The basin includes about 280 m thick sandstone-shale sequence occupying an outcrop area of about 220 km^2. The lithostratigraphy of the basin as worked out by Balakrishnan and Babu (1987) includes a thin body of basal conglomerate with interbands of subarkose (180 m), siltstone (20 m), and purple shale interbanded with impure limestone (70 m). Like the Khariar Basin, the Ampani Basin shows gentle folding over the major part of the outcrop. The intensity of deformation increases towards the eastern margin close to the western boundary of the Eastern Ghats Granulite Belt, where the overturned isoclinal folds have developed along with shear zones.

5.3.6.4 Indravati Basin

This is an irregular, rhomb-shaped basin covering an area of about 900 km^2. The basin shows about 500 m thick sequence of shale-sand-carbonate association and covers an area of about 900 km^2 area (Fig. 5.16). The carbonate succession includes cement-grade limestone and dolomite, which are of economic importance. The following stratigraphic succession (Table 5.7) has been erected by Ramakrishnan (1987). The basin infillings of these late-Proterozoic deposits are classified as the Indravati Group.

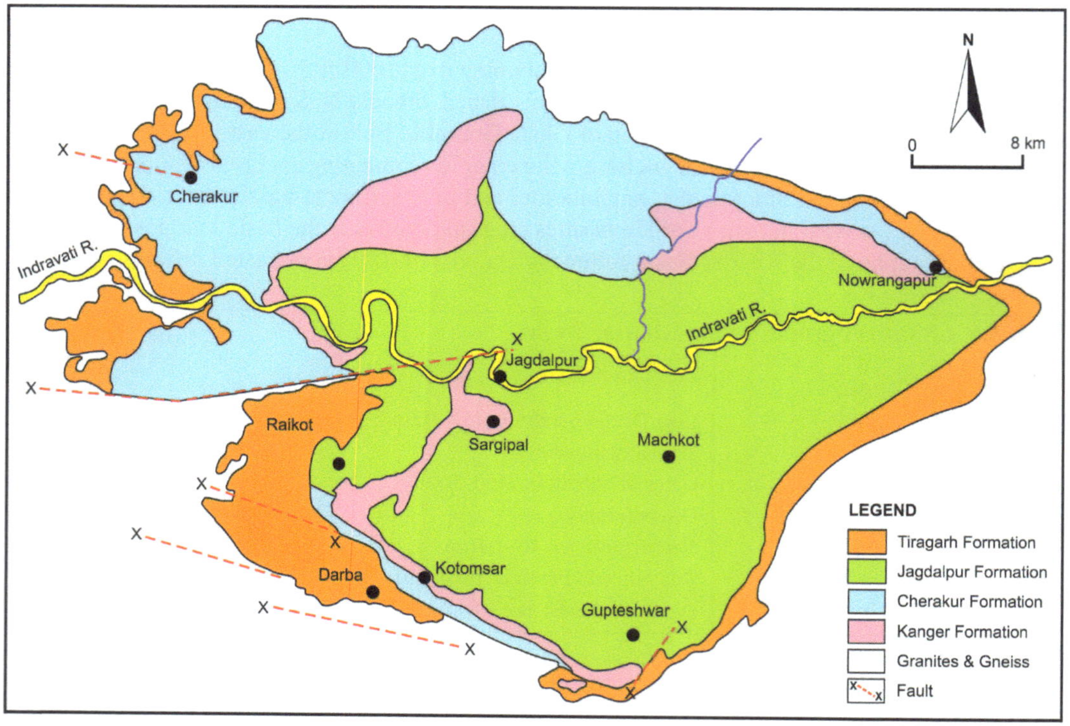

FIG. 5.16

Geoloical map of the Indravati Basin.

Reproduced from Ramakrishnan (1987) with permission.

Table 5.7 Lithostratigraphy of the Indravati Group

Jagdalpur Formation (200–250 m)	Purple shale with stromatolitic dolomite
Kanger Limestone (50–200 m)	Purple limestone and laminated shale
Cherakur Formation (50–60 m)	Purple shale and siltstone with interbeds of arkosic sandstone, chert pebble with conglomerate and 'grit'
Tiratgarh Formation (50–60 m)	Quartz arenite with minor laminated shale, sandstone, and pebble conglomerate
Granites gneisses and other Archaean supracrustals	

Structurally, the Indravati Group comprises a flat-lying sequence with gentle undulating dips, except in the southeastern margin with the Eastern Ghats Granulite Belt where the rocks show deformation with the development of faults and shear zones.

5.3.6.5 Sabari Basin

The triangular-shaped small basin near Sukma is named, named Sabari Basin because the River Sabari cuts across the basin almost through the middle. The lithologies of the basin includes a basal conglomerate with a thick-bedded orthoquartzite, overlain by creamy and grey-coloured limestone. The youngest succession includes purple- to grey-coloured shales.

5.3.7 SUMMARY OF THE EVOLUTIONARY HISTORY OF THE BASTAR PROTOCONTINENT

The Bastar Protocontinent includes well-preserved records of Precambrian evolutionary history that spans over two and a half billion years before present. A critical analysis of the Precambrian geology of this Protocontinent indicates four different stages of crustal growth beginning with the development of the earliest sialic Crust over which the successive supracrustal ensembles were deposited. The growth history of the Precambrian Crust was initiated with the development of granite-gneisses, which are compositionally comparable with the tonalite-trondjhemite-granodiorite gneisses that characterize the oldest basement of the ancient Shield rocks in different parts of the world. The available geochronological data indicate that the oldest sialic Crust in the Bastar Protocontinent formed at ~3.56 Ga before present. The older supracrustal formations that evolved over this sialic basement included greenstone belts formed during the mid-Archaean to the late-Archaean time. The most primitive of these is the Sukma Group. No precise date is available about the formation of the Sukma Group. But conventionally, it is taken to have developed before the intrusion of the 3.00 Ga old granitic rocks. In spite of the antiquity, the lithologic formations of Sukma Group show dominance of aluminous silicate minerals in the succession. Apart from that, there are also some localized pockets of high-alumina deposits that may be interpreted as alumina-rich palaeosols. The next greenstone sequence is the Bengpal Group, which overlies the Sukma Group with a profound angular unconformity in between. Lithologically, the Bengpal formations show considerable similarity with the older Sukma Group rocks. The Bailadila Group, which evolved next in the succession, includes an ensemble of mildly metamorphosed, dominantly ferruginous sedimentary rocks. The Bailadila Hills, representing the type area for the Bailadila Group, host rich iron-ore deposits that are considered similar to the Superior-type banded iron-ore formation occurring in association with the fine clastics and tuffs, with minor mafic volcanics.

The greenstone belts that occur north of the north-south running linear Bailadila Greenstone Belt are described as the end-Archaean greenstone belts. These include (1) the Sakoli Belt in the west, (2) the Kotri-Dongargarh Belt in the middle, and (3) the Sonakhan Belt in the east. The lithogic ensembles occurring in these belts include Sakoli Group, Dongargarh Group and the Sonakhan Group.

The Sakoli Group consists of low-grade volcano-sedimentary sequence with meta-chert, conglomerate, and banded iron formation. The complex structural geometry of the Sakoli Group is thought to have developed due to repeated folding of the intricately framed initial basin geometry. Introduced as a new stratigraphic entity in the Kotri-Dongargarh Belt, the newly defined Dongargarh Group includes metapelites; quartz arenite, with minor ultramafic rocks; felsic volcanics; and the banded iron formation.

Available isotope data confirm the end-Archaean age of the Dongargarh Group. The intrusion of granites marks the culmination of the Dongargarh tectonothermal cycle.

The third end-Archaean greenstone belt in the east is stratigraphically described as the Sonakhan Group, which includes a bimodal volcano-sedimentary sequence overlying the basement gneissic complex locally known as Baya Gneiss. An important component of the Sonakhan Group is the Jonk Conglomerate, which includes mixtures of boulders and large-sized pebbles and blocks and chunks of protolithic rocks simulating the characters of fluvioglacial deposits (Figs. 5.8 and 5.9).

Separated from the outcrops of the Archaean granite-gneiss complexes and greenstone belts by a narrow tectonized belt where the Proterozoic Sausar Group occurs intimately intermingled with granite-gneiss complexes of different ages along with granulites. Sausar Group is made up primarily of metamorphosed rocks of the sand-shale-carbonate association hosting rich manganese ore deposits. The basal succession of the Sausar Group comprises diamictites of glaciogenic origin.

The Sausar Group occurs as stacks of complexly folded sequence showing the development of thrust slices in the form of nappes, which have been compared with the Alpine Nappes of the Central Europe. Based on scanty data on geochronology collating with the different depositional features, the age of Sausar Group is tentatively set at 2.2–2.1 Ga.

Several isolated bodies of late-Proterozoic cratonic basins (Purana Basins in early literatures) occur in Bastar Protocontinent of which the Chattisgarh is the largest. situated on the northeastern edge of the Archaean cratonic block of Bastar. There are also several other smaller cratonic basins like Khariar, Ampani, Abujmarh, Indravati, and Sabari, apart from a number of smaller 'outlier' like outcrops. The basins contain thick sediments belonging to the shale-sand-carbonate association and were deposited in multiple sedimentary cycles along with welded tuff (ignimbrite) in the shale units. Carbonate is absent in some of the basins. The vast limestone and dolomite resources that occur in some of the major basins, feed the steel plants and cement factories.

Majority of the basins show horizontal to subhorizontal orientation of beds and have rolling dips. Folding and shearing are noted at the eastern margins of some basins, which resulted from 'collision'-type impact of the emplacement of the Eastern Ghats Granulite Belt in the east of the Bastar Protocontinent.

No precise age data are available for these cratonic basins. However, based on the study of the stromatolite structures developed in the carbonate formations, suggestion has been made about the depositional age of these basins during 1400 to 800 million years ago. Geochronological study based on K/Ar systematics on the glauconites suggests that the sediment deposition age is between 1300 and 1250 Ma.

Majority of the cratonic basins with irregular to rectangular-shaped configuration of the outcrops is surrounded by a number of kimberlite pipes. The association of the two suggests possibility that the cratonic-platformal basins evolved due to underplating of mafic alkaline magmatism in different parts of the Bastar Protocontinent like those in the Dharwar Protocontinent. Based on the emplacement age of the southern Indian kimberlites at 1.1 Ga (Chalapathi Rao et al., 2009), we may conclude that the development cratonic-platformal basins of the Bastar Protocontinent predated the onset of Neoproterozoic at ~1 Ga before present.

REFERENCES

Balakrishnan, P., Babu, Mahesh, 1987. . The geology of the Ampani outlier. Kalahadi District, Orissa. Geol. Soc. India, Mem. 6, 281–286.

Bandopadhyay, B.K., Roy, A., Huin, A.K., 1995. Structure and tectonics of parts of the Central Indian Shield. Mem. Geol. Soc. India 38, 155p.

Chalapathi Rao, N., Anand, M., Dongre, A., Osborne, I., 2009. Carbonate xenoliths hosted by the Mesoproterozoic Siddanpalli Kimberlite Cluster (Eastern Dharwar craton): implications for the geodynamic evolution of southern India and its diamond and uranium metallogenesis. Int. J Earth Sci (Geol Rundsch). https://doi.org/10.1007/s00531-009-0484-7.

Chattopadhyay, A., Bandopadhyay, B.K., 2004. Fold-thrust tectonics of the Sausar Fold Belt, Central India, and its bearing on the adjacent high pressure metamorphic rocks. Geol. Surv. India, Sp. Publ. No. 84, 319–330.

Chattopadhyay, A., Bandopadhyay, B.K., Khan, A.S., 2001. Geology and structure of the Sausar Fold Belt: a retrospection and some new thoughts. Geol. Surv. India 64, 251–263.

Chattopadhyay, A., Huin, A.K., Khan, A.S., 2003. Structural framework of Deolapar Area, Central India and its implications for Proterozoic nappe tectonics. Gondwana Res. 6, 107–117.

Crookshank, H., 1963. Geology of southern Bastar and Jeypore from the Bailadila range to the Eastern Ghats. Mem. Geol. Soc. India 87, 150p.

Das, N., Dutta, D.R., Das, D.P., 2001. Proterozoic cover sediments of the south eastern Chhattisgarh State and adjoining parts of Orissa. Geol. Surv. India Spec. Publ. 55 (2), 237–262.

Das, N., Roy Burman, K.J., Vatsa, U.S., Mahurkar, Y.V., 1990. Sonakhan schist belt- a Precambrian granite-greenstone complex. Geol. Surv. India Spec. Publ. 28, 118–132.

Deb, S.P., 2004. Lithostratigraphy of the Neoproterozoic Chhattisgarh sequence: its bearing on the tectonics and palaeogeography. Gondwana Res. 7, 323–337.

Fermor, L.L., 1909. The manganese ore deposits of India. Geol. Surv. India, Spec. Publ. 3, 55–67.

Fermor, L.L., 1936. An attempt at the correlation of the ancient schistose formation of Peninsular India. Mem, Geol. Surv. India 70, 217p.

Geological Survey of India, 1971. Code of Stratigraphic Nomenclature of India. . Misc. Publ. 20. 28p.

Ghosh, J.G., 2004. 3.56 Ga tonalite in the central part of the Bastar craton, India: Oldest Indian date. J. Asian Earth Sci. 23, 359–364.

Jain, S.C., Yedekar, D.B., Nair, K.K.K., 1991. Central Indian shear zone: a major pre-Cambrian crustal boundary. J. Geol. Soc. India 37, 521–531.

Khan, M.W.Y., Bhattacharyya, T.K., 1993. A reappraisal of the stratigraphy of Bailadila group, Bacheli, Bastar District, M.P. J. Geol. Geol. Soc. India 42, 279–311.

Krishnamurthy, P., Chaki, A., Pandey, B.K., Chimote, J.S., 1988. In: Geochronology of the granite-rhyolite suites of the Dongargarh Supergroup, central India. Proceedings of 4th National Symposium on Mass Spectrometry, Bangalore, Earth and Planetary Science. 2, pp. 1–5.

Krishnamurthy, P., et al., 1990. Magmatic rocks of the Dongargarh Supergroup, central India-their petrological evolution and implications on metallogeny. Geol. Surv India, Sp. Publ. 28, 303–319.

Krishnan, M.S., 1956. Geology of India and Burma. Higginbothams Private Ltd, Chennai, India, 555p.

Kruezer, H., Karre, W., Krustein, M., Schnitzer, W.A., Murti, K.S., Srivastava, N.K., 1977. K/Ar dates of glauconites from Chandrapur Series (Chattisgarh Basin Central India). Geol. Jarbuch, B 28, 23–36.

Mishra, T., Jose, B., Deepa, K. K., Sreehari, S. M. S. and Venkatesh, A.S. 2001. Auriferous mineralization in the Sakoli Group, central India, with particular reference to sulphide-bound gold. Appl. Earth Sci. Trans. Inst. Mining Metall. B., 110. https://doi.org/10.1179/aes.2001.110.2.103.

Mohanty, S.P., 1988. Structural evolution of the southern part of the Sausarbelt near Ramtek, Nagpur district, Maharashtra. Indian J. Geol. 60, 200–210.

Mohanty, S.P., 2006. Evidence of volcanism and glaciation from the Sausar Group central India. J. Geol. Soc. India 68, 764–768.

Mohanty, S.P., 2010. Tectonic evolution of the Satpura Mountain Belt: a critical evaluation and implication on supercontinent assembly. J. Asia Earth Sci. 39, 516–526.

Mohanty, S.P., Barik, A., Sarangi, S., Anindya, S., 2015. Carbon and oxygen isotope systematics of a Paleoproterozoic cap-carbonate sequence from the Sausar Group, Central India. Palaeogeogr. Palaeoclimatol. Palaeoecol. 417, 195–209.

Mohanty, S.P., Nanda, S., 2016. Geochemistry of a paleosol horizon at the base of the Sausar Group, central India: implications on atmospheric conditions at the Archean-Paleoproterozoic boundary. Geosci. Front. 7, 759–773.

Naqvi, S.M., Rogers, J.J., 1986. Precambrian Geology of India. In: Oxford Monographs on Geology and Geophysics. vol. 6. 222p.

Narayanaswamy, S., Chakraborty, S.C., Vemban, N.A., Shukla, K.D., Subramanyam, M.R., Venkatesh, V., Rao, G.B., Anandalwar, M.A., Nagarajaiah, R.A., 1963. Geology and manganese ore deposits of the manganese belt of Madhya Pradesh and adjoining Maharashtra. Bull. Geol. Surv. India, Series. A 22 (1). 69p.

Niogi, S., Misra, H., Hariya, Y., 1996. Geochemistry of the Dongargarh volcanics rocks Central India: implications for the Precambrian mantle. Precambrian Res. 76, 77–91.

Panigrahi, M.K., Misra, K.C., Bream, B., Naik, R.K., 2002. In: Genesis of the granitoid affiliated copper–molybdenum mineralization at Malanjkhand Central India: facts and problems. Proceedings of the 11th Quadrennial IAGOD Symposium and Geocongress, Windhoek, Namibia (Electronic Extended Abstract).

Panigrahi, M.K., Mookherjee, A., Pantulu, G.V.C., Gopalan, K., 1993. Granitoids around Malanjkhand copper deposit: types and age relationship. Proc. Indian Acad. Sci. (Earth Planet. Sci.) 102, 399–413.

Pascoe, E.D., 1973. A Manual of Geology of India and Burma. vol. 1. Geological Survey of India, Calcutta, 485p.

Rajarajan, K., 1976. Epizonal granitic complex in Bhandara and Chanda Districts, Maharashtra. Geol. Surv. India Misc. Publ. 23, Part 2, pp. 347–363.

Rajaram, M., Anand, S.P., 2003. Central Indian tectonics revisited using aeromagnetic data. Earth Planets Space 55, e1–e4.

Rajesh, H.M., Mukhopadhyay, J., Beukes, N.J., Gutzmer, J., Belyanin, G.A., Armstrong, R.A., 2009. Evidence for an early Archaean granite from Bastar craton, India. J. Geol. Soc. Lond. 166, 193–196. https://doi.org/10.1144/0016-76492008-089.

Ramachandra, H.M., Roy, A., 1998. Geology of intrusive granitoids with particular reference to Dongargarh granite and their impact on the tectonic evolution of the Precambrian in Central India. Indian Minerals 52, 15–32.

Ramakrishnan, M., 1987. Stratigraphy and sedimentary environment and evolution of the Late Proterozoic Indravati Basin. Geol. Soc. India Mem 6, 139–160.

Ramakrishnan, M., 1990. Crustal development in southern Bastar, Central Indian craton. Geol. Surv. India, Calcutta, Sp. Publ. No. 28, 44–66.

Ramakrishnan, M., Vaidyanadhan, R., 2008. Geology of India. vol. 1. Geological Society of India, Bangalore, 556p.

Roy, S., 1981. Manganese deposits. Academic Press, London. 458p.

Roy, S., 2006. Sedimentary manganese metallogenesis in response to the evolution of the Earth system. Earth-Sci. Rev. 77 (4), 273–305.

Roy, A., Kagami, H., Yoshida, M., Roy, A., Bandopadhyay, B.K., Chattopadhyay, A., Khan, A.S., Huin, A.K., Pal, T., 2006. Rb-Sr and Sm-Nd dating of different metamorphic events from the Sausar Mobile belt, central India: implications for Proterozoic crustal evolution. J. Asian Earth Sci. 26, 61–76.

Roy, A.B., Purohit, R., 2015. Lithostratigraphic, geochronological and depositional framework of the Precambrian basins of the Aravalli Mountains and adjoining areas, Rajasthan, India. In: Precambrian Basins of India: Stratigraphic and Tectonic Context. Mem. 43. Geol. Soc. London, pp. 55–65.

Santosh, M., Yakoyama, K., Acharyya, S.K., 2004. Geochronology and tectonic evolution of Karimnagar and Bhopalpatnam granulite belts, central India. Gondwana Res. 7, 501–518.

Sarkar, S.N., 1958. Stratigraphy and tectonics of Dongargarh system: a new system in the Precambrian Bhandara-Drug-Balaghat area, Bombay and Madhya Pradesh. J. Sci. Engg. Res. 1 (2), 237–268.

Sarkar, G., Corfu, F., Paul, D.K., McNaughton, N.J., Gupta, S.N., Bishui, P.K., 1993. Early Archean crust in Bastar Craton, Central India—a geochemical and isotopic study. Precambrian Res. 62, 127–137.

Sensarma, S., Mukhopadhyay, D., 2014. Stratigraphy of the ~2.5 Ga Dongargarh belt, central India: key observations and suggested revisions. Gondwana Geol. Mag. Spec. 16. Online publication: www.ggsnagpur.org.

Sharma, R.S., 2009. Cratons and Mobile Belts of India. Springer, 304p.

Stein, H.J., Hannah, J.L., Zimmerman, A., Markey, R.J., Sarkar, S.C., Pal, A.B., 2004. A 2.5 Ga porphyry Cu–Mo–Au deposit at Malanjkhand, Central India: implications for late Archean continental assembly. Precambrian Res. 134, 189–226.

Vansutre, S., Hari, K.R., 2010. Granulite belts of Central India with special reference to the Bhopalpatnam Granulite Belt: significance in crustal evolution and implications for Columbia Supercontinent. J. Asian Earth Sci. 39, 794–803.

Wanjari, Nishchal, Asthana, Deepanker and Divakara Rao, V. Remnants of early continental crust in the Amgaon Gneisses, central India: geochemical evidence. V. 2005. Gondwana Res. 8, 589–595.

West, W.D., 1936. Nappe structure in the Archaean of Nagpur District. Trans. Nat. Inst. Sci. 1, 93–102.

CHAPTER

SINGHBHUM PROTOCONTINENT

6

6.1 INTRODUCTION

Singhbhum Protocontinent occupies a triangular area bounded between the Eastern Ghats Granulite Belt in the south and Mahanadi Lineament (Rift Basin) in the southwest, the Damodar Valley Rift Basin in the north, and the youngest the sediment-filled Bengal Basin in the east (Fig. 6.1). Though named after the 'pristine' district of Bihar in 'British-India', the region covered under the Singhbhum Protocontinent includes significant parts of three adjacent States of Odisha (pristine 'Orissa'), Chhattisgarh, Jharkhand, and West Bengal (Fig. 6.1).

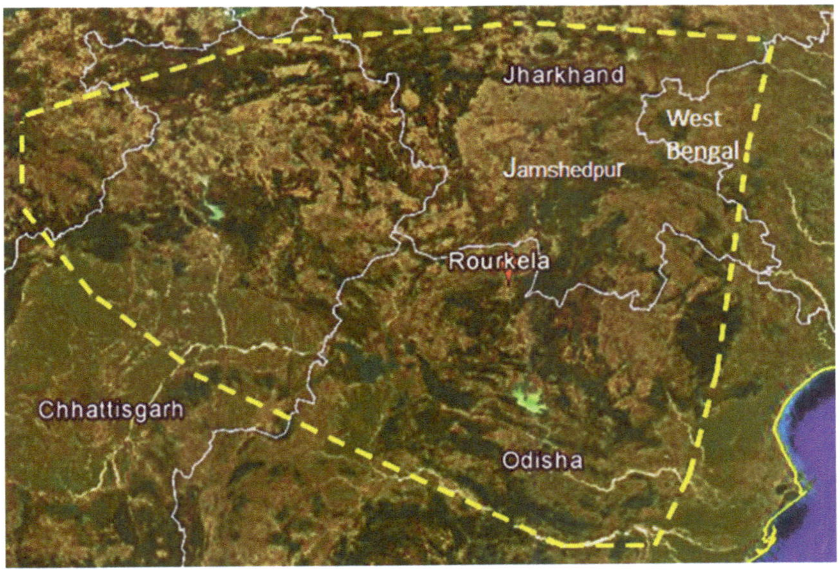

FIG. 6.1

Geographical location of the Singhbhum Protocontinent shown on Google Earth imagery.

BOX 6.1

The delimitation of the Singhbhum Protocontinent by the Damodar Valley Rift Basin in the north (Gondwana Rift Basin) is based on suggested scheme of dividing the Protocontinents of the Indian Shield by the 'joins' marked by the Gondwana Rift Basins as discussed earlier (Chapter 3).

Indian Shield. https://doi.org/10.1016/B978-0-12-809839-4.00006-0

The Protocontinent comprises three major lithotectonic blocks: the Singhbhum Granite Complex in the south, the Chhotanagpur Gneissic Complex in the north, and the North Singhbhum Fold Belt in the middle (Fig. 6.2). Like all other Protocontinents of the Indian Shield, the geological understanding of this Precambrian crustal block remained hazy and marred with disagreement and debate. The array of suggested tectonostratigraphic models proposed by different workers (Mahadevan, 2002; Ramakrishnan and Vaidyanadhan, 2008; Sarkar and Gupta, 2012; Valdiya, 2010, to cite a few) stand testimony to the misunderstanding and misconceptions that surround the geological growth history of this terrane. In view of this, it is planned to describe objectively the major components of the Singhbhum Protocontinent in the first place.

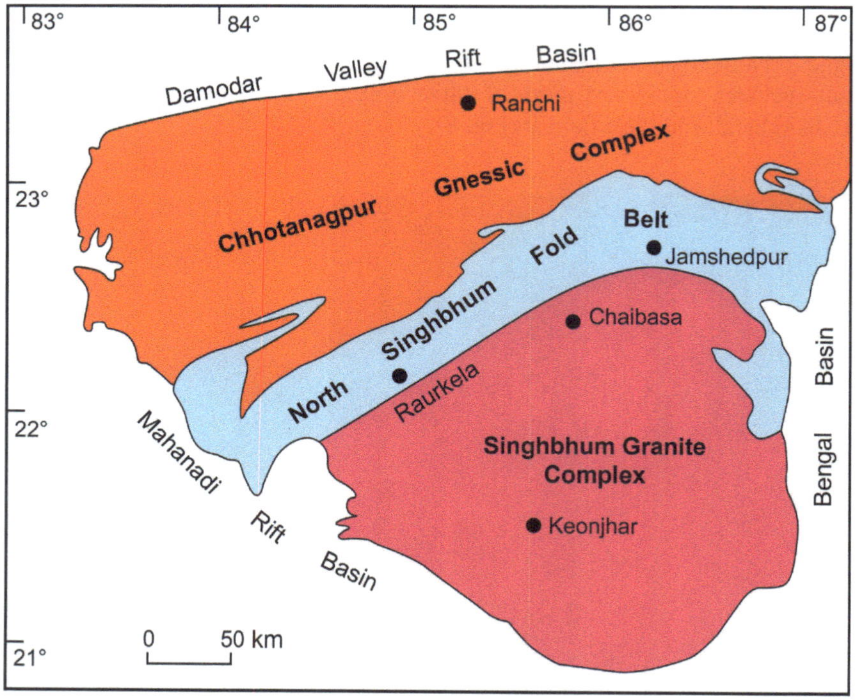

FIG. 6.2

Geological sketch map showing major lithotectonic units of the Singhbhum Protocontinent.

6.2 SINGHBHUM GRANITE COMPLEX

6.2.1 CONSTITUENTS AND GEOLOGICAL HISTORY

The Singhbhum Granite Complex (also described as the Singhbhum Archaean Craton, Roy and Bhattacharya, 2012) includes certain characteristic features that make this crustal block geologically distinctive. Firstly, it provides information on the existence of very old sialic Crust (~3800 Ma or older) that constituted the possible earliest basement over which the oldest supracrustals were deposited

(Basu et al., 1981). Another important feature of the Singhbhum Granite Complex is the history of its early cratonization by around 3100 Ma (Roy and Bhattacharya, 2012), which is very much unlike most of the Archaean basement rocks in other Protocontinents of the Indian Shield. Subsequent to that, there is no record of any tectonometamorphic and granitic activity in the rocks till the end of the Archaean barring imprints of some younger thermal events.

BOX 6.2

Supracrustal rocks (supra is from Latin meaning 'above') are those which were deposited on the existing basement rocks comprising granites and gneisses, of the Crust, hence the name. The supracrustal rocks are also described as the cover rocks/succession.

For understanding the geology of the Singhbhum Granite Complex, it would be useful to include different bodies of granitoids like those of Nilgiri, Bonai, Palla Lahara, Mayurbhanj, and Kapdipara that possibly represented several separate magmatic events, under the single head 'Singhbhum Granite' (Saha, 1994). In addition to these larger bodies, there are also a number of xenolithic enclaves of different dimensions comprising older tonalite-trondhjemite-granodiorite gneisses (Fig. 6.3). Petrographically, the granitic rocks include biotite-granodiorite grading to adamellitic granite, biotite trondjhemite, and leucogranite. Patches of chlorite- and epidotic-rich granodiorite are observed in the areas close to the shear zones bordering the Singhbhum Granite Complex (Saha, 1994).

BOX 6.3

The terms 'granitoid', 'granitic rock', and 'granites' are used for the variety of coarse-grained, felsic plutonic rock similar to granite that are composed predominantly of feldspar and quartz. Included in these categories are granite (sensu stricto), granodiorite, quartz monzonite, quartz diorite, syenite, tonalite, and trondhjemite.

Saha (1994) recognized three different phases of granitic intrusions in the Singhbhum Granite suite based essentially on the petrochemical data. The phase-I intrusions are relatively K-poor

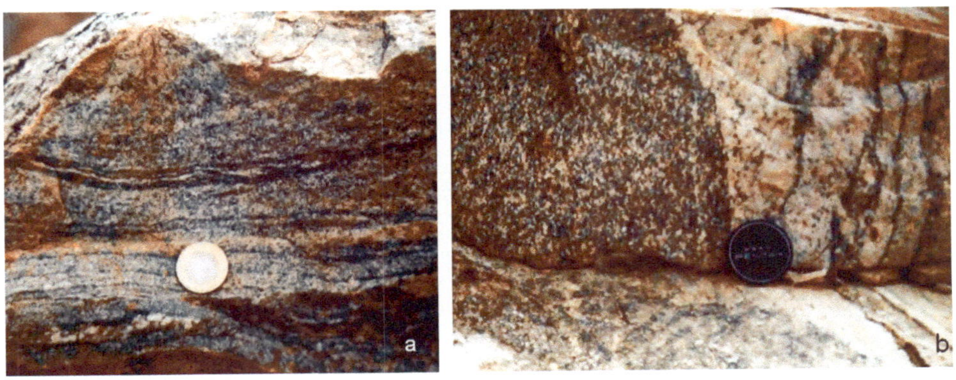

FIG. 6.3

(a) Banded gneiss representing trondhjemite-tonalite-granodiorite association locally occurs as enclaves within porphyritic granite. (b) Thick band of dioritic gneiss alternating younger bands of leucogranite.

granodiorite grading to trondhjemitic diorites. The phase-II and phase-III granitic rocks range in composition between granodiorite and adamellite-granite. The older granitic bodies that are compositionally similar to tonalite gneisses show gently sloping REE pattern with moderately enriched LREE and flat HREE without demonstrating Eu anomaly. The other varieties of granitic rocks are moderately enriched in LREE showing flat HREE pattern and negative Eu anomaly. Pb–Pb whole-rock isochron dates from phase-I and phase-II granites show ages between 3442 ± 26 and 3298 ± 63 Ma (Ghosh et al., 1996). Basu et al. (1981) reported a nine-point Sm–Nd isochron date of 3775 ± 89 Ma from the tonalitic gneisses, considered by the authors as intrusive into the oldest supracrustal rocks. This is the oldest date so far reported from the Singhbhum Granite Complex. The age of detrital zircons from the oldest supracrustal in the central part of the Singhbhum Granite Complex ranges between 3628 ± 38 and ~3550 Ma (av. of 3551 ± 13, 3555 ± 21, 3522 ± 19, and 3583 ± 25 Ma) (Goswami et al., 1995; Misra et al., 1999). Closest to the detrital zircon dates is the report of 3664 ± 79 Ma Pb–Pb isochron age from the tonalitic gneiss of the same region by Ghosh et al. (1996).

Apart from the occurrences of different types of granites and gneisses, there are several supracrustal bodies that vary between small rafts and smaller enclaves and large bodies within in the Singhbhum Granite Complex (Fig. 6.4). A multistaged evolution of the Singhbhum Granite Complex has been suggested based on the petrochemical and geochronological studies of the different lithotectonic/stratigraphic ensembles (Saha et al., 1988; Saha, 1994; Misra et al., 1999; Misra, 2006; Nelson et al., 2014). Roy and Bhattacharya (2012) explicated a coherent crustal evolutionary history of the Singhbhum Granite Complex based on the reappraisal of field relationships between the different lithological ensembles (Fig. 6.4) taking due note of the tectonometamorphic, magmatic, and sedimentation history of the different lithotectonic associations, including the granites and gneisses of different ages.

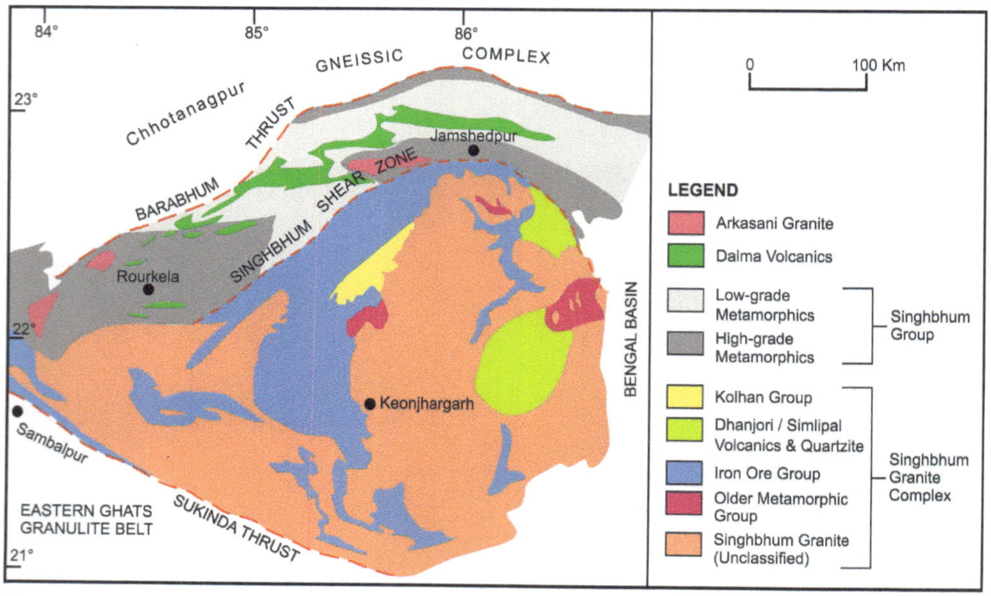

FIG. 6.4

Generalized geological map of the Singhbhum Protocontinent showing distribution of different lithotectonic units.

The oldest supracrustal ensembles are known as the Older Metamorphics. Lithostratigraphically described as the Older Metamorphic Group, it includes amphibolite, calc-silicate rock, and mica schist along with minor bodies of fuchsite-bearing quartzite. These bodies show evidence of intense ductile deformation and shearing and penetrated by intrusive granites of different ages (Fig. 6.5a and b).

FIG. 6.5

(a) Evidence of complex deformation pattern in the calc-silicate rocks of the Older Metamorphic Group showing intricately folded, detached isoclines separated by irregular gneissic rafts and bands. (b) Intrusion of massive granitoid into the calc-silicate bodies belonging to Older Metamorphic Group.

Reproduced from Roy and Bhattacharya (2012) with permission.

The banded iron-formation-bearing ensembles constitute the Iron Ore Group (the lithostratigraphic term 'group' replacing 'series' of early workers like Jones, 1934 and Dunn and Dey, 1942) that is the most commonly occurring supracrustals in the Singhbhum Granite Complex (Fig. 6.6a and b). The major outcrops of the banded iron formation occur in three different basins (Mahadevan, 2002; Valdiya, 2010). These are (1) the western Koira-Jamda Basin, (2) the eastern Gorumahisani-Badampahar Basin, and (3) the southern Daitari-Palla Lahara Basin. A common trait amongst the three basins is the occurrence of the banded iron formation hosting huge iron-ore deposits. The banded iron formation over a large part of its occurrence shows subhorizontal beds (Fig. 6.6a). Elsewhere, the tilted and folded beds

FIG. 6.6

(a) Banded iron formation showing subhorizontal layering. (b) Banded iron formation showing local warping and other karst-related deformation.

are associated with brittle deformation resulting from down-sagging or 'caved-in' gravity-induced collapse of the lithoformations (during the subsequent subsurface karst-related weathering, most probably during the Quaternary period) (Fig. 6.6b). The formation of loose 'blue dust' type iron-ore concentration (Fig. 6.6b) took place during such karst-related weathering processes.

The other important metasedimentary rocks in the Iron Ore Group include quartzite, conglomerate (locally with quartz pebble), and shale with minor bodies of calc-silicates, marble, and mafic metavolcanics. One such ensemble in the western part of the Singhbhum Granite Complex is known as the Darjing Group. This poorly known rock association consists of conglomerate, arkosic quartzite, black-shale facies rocks, and some calcareous metasediments showing amphibolite facies metamorphism. Correlation of these rocks with the Iron Ore Group is based on the evidence of intrusion of the ~2800 Ma old Tamperkola Granite (Bandyopadhyay et al., 2001) into the Darjing Group rocks. Unlike most of the banded iron formations, the Darjing rocks show evidence of multiple deformations. Closely associated with the Darjing Group both in space and time is the Kunjer Group (Rao et al., 1964), the latter having broadly similar lithologic character.

An important component of the Iron Ore Group is the ultramafic rocks hosting chromite and platinum group minerals. The chrome-bearing ultramafic rocks occur in the southern Daitari-Palla-Lahara basin, around Sukinda in Odisha and Jojuhatu in the Koira-Jamda area in Jharkhand (Chakraborty and Chakraborty, 1984; Mondal et al., 2006). The chromitite deposits of the Nuasahi and Sukinda massifs constitute a part of layered ultramafic bodies that include dunite and orthopyroxenite. Apart from chromite, the ultramafic bodies in the southern part the Singhbhum Granite Complex are the hosts of the 'lone' platinum group mineral deposits in India (Haldar, 2017).

Very little information is available on the nature of basement over which the rocks of the Iron Ore Group were deposited. The field relationship, however, provides a clear indication that the banded iron-formation-bearing 'Iron Ore Group' basins formed after the closing (cratonization) of the Older Metamorphic Group. Lithostratigraphically, overlying the Iron Ore Group and the youngest group of ca. 3100–3200 Ma old granitoids occur the Simlipal Volcanics that together with intertrappean sediments constitute the Simlipal Group (Mazumdar, 1996). The Simlipal Group forms a spectacular oval-shaped outcrop in the southeastern part the Singhbhum Granite Complex (Fig. 6.4). The Simlipal Basin has been described as a volcano-sedimentary basin evolved over a Plume head (Banerjee and Ghosh, 1994). The strong lithological similarity between the Simlipal and Dhanjori is implied in the chemical character of the two volcanic formations. Misra and Johnson (2005) reported Pb–Pb and Sm–Nd whole-rock isochron ages of 2794 ± 270 and 2787 ± 270 Ma, respectively, from the Dhanjori volcanics. In spite of the high error factor, we may presume that the Simlipal and Dhanjori volcano-sedimentary basins were coeval in the late-Archaean time, postdating the deposition of the Iron Ore Group.

The age comparable with the formational age of the Simlipal and Dhanjori volcanics is known from a number of plutonic and volcanic bodies, such as the Tamperkola Granite in the western margin of the Singhbhum Granite Complex. Pb–Pb zircon ages of this granitoid body are 2809 ± 12 and 2822 ± 67 Ma (Bandyopadhyay et al., 2001). The other one is from a small 'tonalitic' pluton south of the Simlipal basin that yielded Pb–Pb zircon ages of 2803 ± 4 and 2811 ± 3 Ma (Misra et al., 1999). The Rengali-Riamal charnockite in the southeastern part of the Singhbhum Granite Complex yielded ca. 2740 Rb–Sr whole-rock isochron age (Sarkar et al., 2000). A comparable age of c.2806 ± 6 Ma has been determined from felsic volcanic (dacitic tuff) sample collected from an isolated outcrop in the western part of the Singhbhum Granite Complex (Nelson et al., 2007).

A system of dominantly mafic dykes called Newer Dolerite occurs within the Singhbhum Granite Complex, cross-cutting both the granitic rocks and the associated supracrustal rocks (Roy and Bhattacharya, 2012). The dykes show two distinct orientations trending NN*E*-SSW and NW-SE. The strict regularity in the orientation of the two sets of dykes provides proof of their intrusion along a conjugate set of vertically oriented fracture systems (Fig. 6.7a and b).

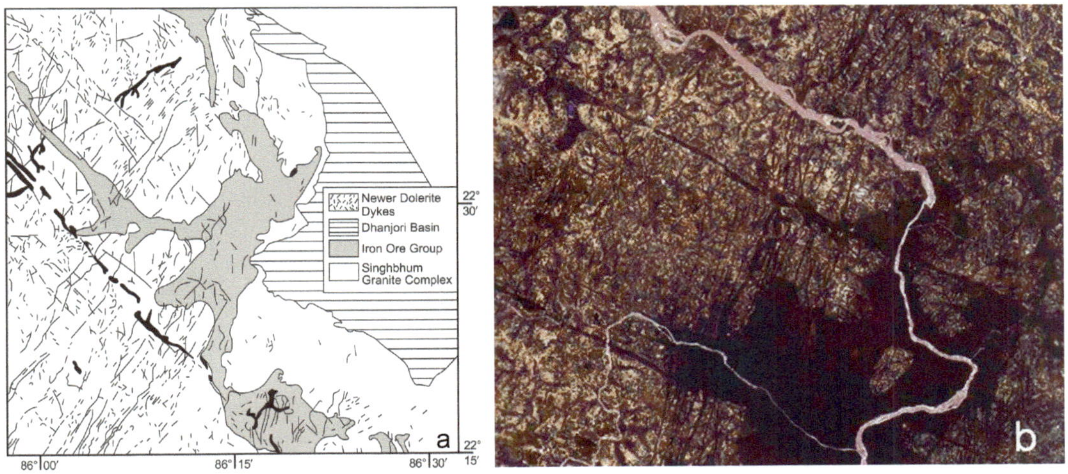

FIG. 6.7

(a) Geological map of a part of the Singhbhum Granite Complex showing dominantly conjugate pattern of intrusion of Newer Dolerite. (b) Google Earth Imagery scene between lat. 21°27′ and 21°37′ and long 85°52′ and 86°04′ showing a conjugate pattern of the Newer Dolerite dyke system in a part of Singhbhum Granite Complex.

Reproduced from Roy and Bhattacharya (2012) with permission.

The opening of the N–S trending Kolhan Basin post-dating the emplacement of the cross-cutting Newer Dolerite dykes has been suggested by Roy and Bhattacharya (2012) on the assumption of the continuity of the E-W extension-related tectonic regime after the phase of dyke intrusions. Its end-Archaean age is based on the consideration of the conformable stress pattern responsible both for the basin opening and the development of the conjugate fracture system in the Singhbhum Granite Complex at around 2600 Ma. Though the two events appear virtually contemporaneous, a pre-Kolhan age of the Newer Dolerite dykes is stipulated on the negative evidence that the dykes intruding the granitoids and older supracrustals invariably stop short of the margin of the Kolhan basin (Pascoe, 1973).

6.2.2 TECTONOSTRATIGRAPHIC AND GEOCHRONOLOGIC FRAMEWORK

The summary of the tectonostratigraphic and geochronologic framework of the Singhbhum Granite Complex is given in Table 6.1.

Table 6.1 Tectonostratigraphic and Geochronologic Framework of the Singhbhum Granite Complex

Tectonostratigraphic Units	Tectonic Framework	Age in Million Years (Ma)
Kolhan Group	Half-graben-type cratonic basin	From <2600 to ~2500
Newer Dolerite	Fracturing of rigid Crust and dyke intrusions	~ 2600
Dhanjori Group and Simlipal Volcanics, granitoid plutons	Plume-related magmatism and sedimentation	~ 2800
Iron Ore Group (including Darjing Group and Kunjer Group)	Greenstone belt evolution over stabilized sialic Crust	~ 3000
Crust building granitic activities	Late synkinematic intrusion	From ~3200 to ~ 3100
Older Metamorphic Group	Early Greenstone belt development	From ~3380 to ~3285
Formation of early sialic Crust		From >3800 to ~3400

6.3 CHHOTANAGPUR GNEISSIC COMPLEX

6.3.1 LITHOLOGIC CONSTITUENTS

The Chhotanagpur Gneissic Complex, also described as the Chhotanagpur Granite-Gneiss Complex, occupies a large area in the northern part of the Singhbhum Protocontinent between the North Singhbhum Fold Belt in the south and the Damodar Valley Rift Basin in the north (Roy, 2012) (Fig. 6.2).

The Chhotanagpur Gneissic Complex includes predominantly gneisses and granitoids of diverse petrological composition and age. The gneisses show typical banded character that looks quite similar to those constituting the ancient basement rocks in different parts of the Indian Shield (Fig. 6.8a and b). Petrochemistry of these rocks matches with the trondhjemite-tonalite-granodiorite suite. There are a

FIG. 6.8

(a) and (b) Outcrops of Chhotanagpur Gneiss showing banded character.

Image courtesy: Alokesh Chatterjee.

FIG. 6.9

A mylonite gneiss showing uniformly planar foliation. The 'eye-shaped' outcrop of a large porphyroclast of feldspar has grown pushing aside the foliation surfaces.

Image courtesy: Alokesh Chatterjee.

few gneissic bodies that are compositionally more mafic in character and are described as the diorite gneiss. The amphibolite, also described as the hornblende gneiss (Pascoe, 1973), occurs locally as enclaves of various shapes and sizes within the gneiss-granite complex. Some of the gneissic rocks show the development of uniformly layered, planar shape fabric of secondary origin that may be described as mylonite gneiss (Roy et al., 2016). These are the reconstituted gneissic rocks that occur close to the outcrops of the Bengal Anorthosite (described later). In certain instances, these reconstituted gneisses bear evidence of synkinematic reconstitution of the pristine gneisses through the development of secondary foliation that is locally disturbed by the growth of 'eye-shaped' feldspar porphyroblast (Fig. 6.9).

Several bodies of granites (with poorly known petrochemistry) occur in the form of isolated plutons in the Chhotanagpur Gneissic Complex. Most of the granite plutons occur in the Ranchi-Purulia Belt. In general, the granites are coarse-grained, porphyritic type showing tendency to weather into rounded, dome-like bosses and tors (Fig. 6.10).

FIG. 6.10

The youngest group of granite occurs in the form of isolated tors amidst flat country.

Image courtesy: Indrani Roy.

Singh and Krishna (2009) divided the granites of the Chhotanagpur Gneissic Complex into two different categories: the grey and the pink granites. The grey granites, according to the authors, have evolved from the melting of tonalite-trondjhemite-granodiorite gneisses. The granites especially of the 'pink' types are the youngest intrusive bodies within the gneissic complex (Saxena et al., 1992; Mahadevan, 1992, 2002) (Fig. 6.11a). Migmatitic nature of some of these granites is indicated by the growth of pink-coloured feldspar between the dark protolithic layers (Fig. 6.11b).

FIG. 6.11

(a) Pink granite showing dominance of pink-coloured K-feldspar. (b) Migmatitic nature of the pink granite is manifested through the growth of pink feldspar in between dark protolith layers.

Image courtesy: Alokesh Chatterjee.

Several bodies of supracrustal rocks (mainly metasedimentary and metavolcanic rocks) occur as enclaves within the gneissic rocks. These include mica schist, banded iron formation, marble, and quartzite along with patches of mafic and ultramafic rocks. Most of these metasedimentary inclusions show low-grade metamorphic character varying between greenschist and amphibolite facies. The high-grade granulite facies rocks have localized occurrence, mainly along two belts: (i) in the eastern part in the Purulia District of West Bengal and (ii) in the Ranchi-Palamau area in Jharkhand. The granulite assemblages are represented by khondalite (garnet-sillimanite ± graphite), calc-silicate granulite (scapolite-wollastonite-calcite-garnet ± quartz), charnockite (hypersthene-granite), two-pyroxene granulite with or without garnet, hornblende granulite, and leptinite; all occur in the form of dismembered bands within granitic gneisses (Sharma, 2009).

Geologically, the most significant feature of the Chhotanagpur Gneissic Complex is the occurrence of lenticular or elliptical bodies of anorthosite, locally associated with thin lenses of syenite (Mahadevan, 2002). The largest body of anorthosite occurs at the eastern margin of the Singhbhum Protocontinent. Known as the Bengal Anorthosite, this, about 40 km long 'tadpole-shaped' intrusive body (Ghosh et al., 2008) (Fig. 6.12), is enveloped predominantly by pelitic granulites and felsic gneisses (Bhattacharya and Mukherjee, 1987). The core of the anorthosite body consists of 'grey anorthosite' that is bordered by medium-grained, equigranular 'white anorthosite'. The grey anorthosite comprises coarse-grained plagioclase megacrysts that show magmatic flow-related alignment (Ghosh et al., 2008). Supracrustal enclaves within the anorthosite body include patches of sillimanite-bearing

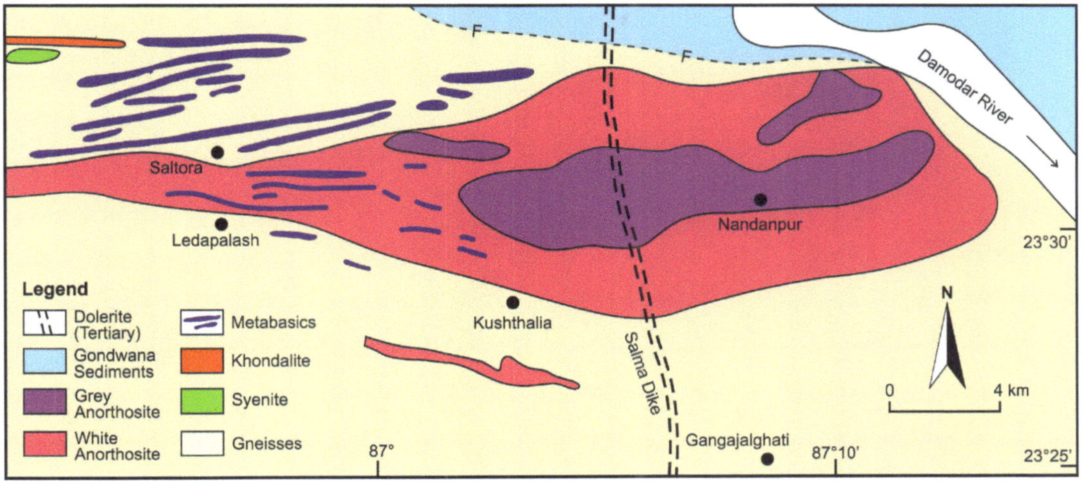

FIG. 6.12

Geological map of the Bengal Anorthosite and surrounding areas.

Reproduced from Ghosh et al. (2008) with permission.

mica schist, khondalite, quartzite, quartz-magnetite rock, and calc-silicate gneiss. A few inclusions of relic basement complex occur within the anorthosite body that is composed of quartzofeldspathic gneiss, biotite/hornblende gneiss, charnockite, and leptynite.

6.3.2 TECTONO-THERMAL EVOLUTION OF THE CHHOTANAGPUR GNEISSIC COMPLEX

Very little or virtually no information is available on the antiquity and precise age of the gneiss-granite rocks of the Chhotanagpur Gneissic Complex except some casual mention in the literature that 'the Chhotanagpur gneisses (Fig. 6.8a and b) resemble the Peninsular Gneiss of the Dharwar Protocontinent' (Pascoe, 1973). A similar suggestion was also made by Ghosh et al. (2008) based mainly on the field studies of the Chhotanagpur gneisses. Sharma (2009) who discussed this issue at length seems to validate the concept of Archaean antiquity of the Chhotanagpur gneisses. This is in conformity with the reports of Ramakrishnan and Vaidyanadhan (2008), who based on TD_M model ages and SM-ND ages favoured Archaean age of the Chhotanagpur Gneissic Complex. The type of the supracrustals like the banded iron formation, marble, and quartzite, in all probability, suggests their development in the greenstone-belt-type depositional environment. Presently, these lithologic formations are in highly dismembered state.

Apart from physical resemblance with the ancient (mid- and late-Archaean) gneisses, some of the outcrops of the banded gneisses provide indubitable evidence of reconstitution through folding and transposition of earlier formed gneissic foliation (Fig. 6.13a). There are also signs of metamorphic overprinting of high-temperature minerals like almandine garnet and (ortho-) pyroxene on the reconstituted gneissic rocks (Fig. 6.13b).

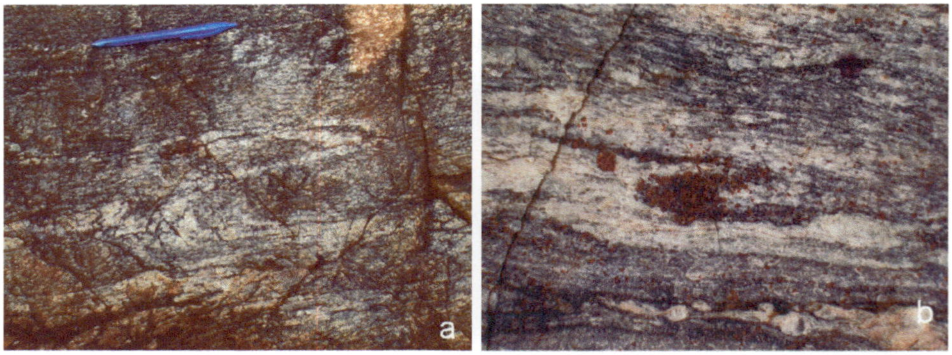

FIG. 6.13

(a) Folding and reconstitution of grey, banded gneiss with the development of 'transposed' foliation parallel to the axial surface of folds. (b) Growth of metamorphic minerals (garnet and pyroxene) overprinting on the reconstituted grey, banded gneiss.

Image courtesy: Alokesh Chatterjee.

Amongst the supracrustals, the pelitic rocks constitute the most dominant component of the Chhotanagpur Gneissic Complex especially in the southern part. Out of these, the mica schists along with some other metasedimentary bodies showing low greenschist to amphibolite facies metamorphism occur mainly in the central part of the east-west running belt. These pelitic-dominant lithologies are taken over on the either side of the belt by the youngest bodies of granite and pelitic granulites. The granites have evolved through low-temperature melting of the sedimentary protoliths (Singh and Krishna, 2009). Based on this interpretation, we may suggest that the dominantly pelitic metasediments covered the entire east-west running belt stretching from the Purulia District of West Bengal in the east to the Ranchi-Palamau area in Jharkhand in the west.

The presence of a wide shear zone within the mica schists (including the granulitic equivalents) in the southern part of the Chhotanagpur Gneissic Complex is difficult to explain. Because of the lack of any information on its relationship with either the gneisses that occur on the northern fringe or with the metasedimentary sequence belonging to the northern part (Chandil Domain) of the North Singhbhum Fold Belt. The point worth mentioning in this context is the suggestion that the pelitic schists, 'paragneisses', and other metasedimentary components in the southern part of the Chhotanagpur Gneissic Complex represent 'Older metasediments' correlatable with the North Singhbhum Fold Belt rocks (Banerji, 1991). According to the author, both the basement gneisses and the sedimentary supracrustals were overprinted by later high-grade granulite facies metamorphism and the intrusion of the plutonic bodies including the anorthosite at later dates.

Based on the available geochronological data, five different phases of tectonomagmatic activities in the Chhotanagpur Gneissic Complex have been suggested. These were during (i) 2.5–2.4, (ii) 2.2–2.0, (iii) 1.6–1.4, (iv) 1.2–1.0, and (v) 0.9–0.8 Ga (Singh and Krishna, 2009). But before discussing on these 'dates', it would be useful to look into the geological evolutionary history based on the observed field relationships between different components of the Chhotanagpur Gneissic Complex. The occurrence of typical tonalite-trondjhemite-granodiorite and diorite gneisses (Fig. 6.8a and b) in different parts of the Chhotanagpur Gneissic Complex suggests the presence of sialic basement as in all other Precambrian

Protocontinents in the Indian Shield. However, in the absence of any well-constrained isotope dates, it is not possible to precisely determine the time of formation of the earliest basement in this gneiss-granite terrane. However, their 'Mesoarchaean' antiquity is suggested based on the development of gneissic banding in these rocks similar to those generally observed in the different Archaean basement gneisses world over. Further, the occurrence of banded iron formation and other metasedimentary bodies along with the mafic and ultramafic ensembles helps to confirm the greenstone belt association of these supracrustal ensembles.

Following the development of Archaean Crust, the dominantly shale facies rocks along with other associated sediments were deposited in the shallow seas simultaneously with similar such sediments that form a part of the Proterozoic basins north of the Dalma Volcanics. Crustal shortening following the closure of the basin introduced a foliation in these pelitic bodies. The intrusion of anorthosite magma took place from the synkinematic to late-synkinematic phase of the deformation. The tectonic domain marked by north-south-directed stress seems to control the 'tadpole'-shaped outcrop pattern of the largest anorthosite body in the eastern part of the Chhotanagpur Gneissic Complex (Fig. 6.12). This phase of deformation and associated metamorphism was also responsible for the development 'spaced' 'mylonitic' foliation in the basement rocks (Fig. 6.8). The pelitic granulites evolved later followed successively by the intrusion/emplacement of the 'grey' and 'pink' granites.

Single-zircon studies indicated a concordant U–Pb crystallization age of 1550 ± 2 Ma of the anorthosite (Chatterjee et al., 2008). The postcrystallization thermal reconstitution of the anorthosite body along with associated metapelite was during 947 ± 27 and 950 ± 20 Ma (Chatterjee et al., 2008). The little older age of 995 ± 24 Ma thermal event is recorded in the pelitic granulite located north of the anorthosite body (Chatterjee et al., 2008; Maji et al., 2008).

Summarizing, we may say that unlike the southern Singhbhum Granite Complex, the evolutionary history of the Chhotanagpur Gneissic Complex is much prolonged extending (in all probability) from Mesoarchaean to the early Neoproterozoic.

A tentative lithostratigraphic-geochronologic framework of the Chhotanagpur Gneissic Complex is shown in Table 6.2.

Table 6.2 Summary of Lithostratigraphic/Geochronologic History

Tectonothermal Events	Age (Probable)	Reference
Emplacement of pink granites	850–800 Ma	Singh and Krishna (2009)
Emplacement of grey granites	1000–950 Ma	Singh and Krishna (2009)
Granulite facies metamorphism	~1000 Ma	Chatterjee et al. (2008)
Intrusion of anorthosite bodies	~1550 Ma	Chatterjee et al. (2008)
Development of greenstone belts	Late-Archaean ([a])	–
Development of early sialic crust	Mesoarchaean ([a])	–

[a] *Age 'Late-Archaean' is unconfirmed and subject to change in future.*

6.4 NORTH SINGHBHUM FOLD BELT

The North Singhbhum Fold Belt is a relatively narrow belt of folded sequence comprising dominantly of metasedimentary rocks along with volcanics and some granitoids. The entire belt is sandwiched between the Singhbhum Granite Complex in the south and the Chhotanagpur Gneissic Complex in the north (Fig. 6.2). The southern boundary of the Chhotanagpur Gneissic Complex with the North Singhbhum Fold Belt is transitional, because much of the southern part of the belt is reported to include the fold belt rocks including mica schists, and some other metasedimentary bodies, which have undergone 'migmatization' during late Palaeoproterozoic time.

Purely for the purpose of description, the entire belt of metasediments and volcanics of the North Singhbhum Fold Belt is divided into three domains (Fig. 6.14). These are (1) the Jamshedpur Domain including the Dalma Volcanics that bound the metasediments (mica schist phyllite and quartzite) in the north and west; (2) Chandil Domain, occurring between the Dalma Volcanics in the south and the Chhotanagpur Gneissic Complex in the north; and (3) the Western Domain occurring approximately west of 85°15′ longitude including the Gangpur Group. Apart from the metasedimentary ensembles and volcanics, the North Singhbhum Fold Belt includes patches of granites and gneissic rocks of diverse ages in all the three domains. The largest body is that of the Chakradharpur Granite (Fig. 6.14) that occur in the southwestern part of the of the Domain 1 rocks.

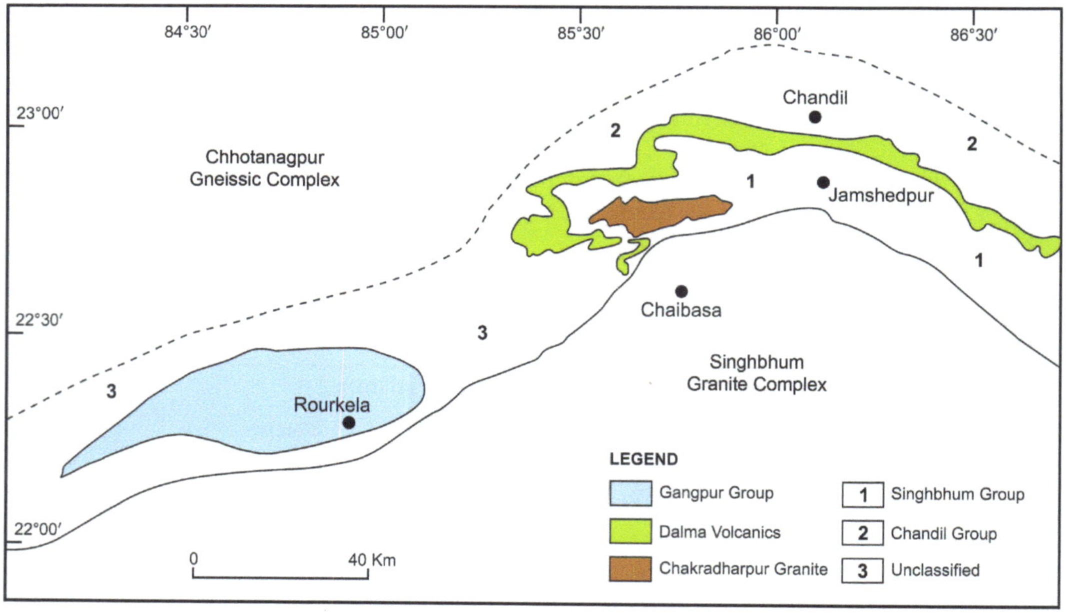

FIG. 6.14

A schematic map of the North Singhbhum Fold Belt occurring between the Singhbhum Granite Complex in the south and Chhotanagpur Gneissic Complex in the north.

For the purpose of description, the term like Singhbhum Group proposed by Sarkar and Saha (1962) is retained but only for the metasedimentary ensembles of the Jamshedpur domain situated between the Dalma Volcanics in the north and west and the Singhbhum Granite Complex in the south (Fig. 6.14).

For the similar lithological associations that occur in the north of the Dalma Volcanics, the term Chandil Group (Ramakrishnan and Vaidyanadhan, 2008) is used.

6.4.1 SINGHBHUM GROUP

The Singhbhum Group dominantly comprises metamorphosed shale-sand-bearing assemblages with some linear bodies of amphibolite that possibly represent metamorphosed mafic intrusions. This arcuate belt of metasediments is bounded between the outcrops of the Dalma Volcanics in the north and west and late Quaternary laterite cover in the east. In the south, the belt is truncated by a very prominent shear zone that passes close to the northern boundary of the Singhbhum Granite Complex. Some isolated bodies of conglomerate (the Bisrampur Conglomerate, e.g. Fig. 6.15) along with arkosic sandstone and shale occur south of the shear zone bordering the Singhbhum Granite Complex. These little deformed and virtually unmetamorphosed rock formations (conglomerate, shale, and arkosic sandstone) are considered a part of the schist-quartzite belt rocks in the north (Bhattacharya and Mahapatra, 2008).

FIG. 6.15

Bisrampur Conglomerate showing little deformed assorted pebbles and boulders including some from the banded iron.

Lithostratigraphically, the entire sedimentary succession between the Dalma Volcanics in the north and west and the Singhbhum Granite Complex in the south has been described as the Singhbhum Group by Sarkar and Saha (1962). The authors have also divided the Singhbhum Group into an upper Dhalbhum Stage and a lower Chaibasa Stage. Field studies in the central part of the area have, however, demonstrated continuity in the metasedimentary sequence without any structural and stratigraphic break (Naha, 1962; Roy, 1966). Sarkar (1982) who worked in the western part of the belt has also stated categorically that the lithostratigraphic subdivision of the Singhbhum Group by Sarkar and Saha (1962) is arbitrary.

The overall structure of the folded rocks of the Singhbhum Group rocks has been described as the Singhbhum Anticlinorium by Sarkar and Saha (1962), following the earlier suggestion of Dunn and Dey (1942). A prominent shear zone has developed along the southern limb of the anticlinorium which has traditionally described as the Singhbhum Shear Zone marked by a wide zone of mylonite and mylonitic phyllite (phyllonite of Knopf, 1931) (Fig. 6.16a and b). Sarkar and Saha (1962) described it as the Copperbelt Thrust as it truly marks a zone of thrusting along which the northern fold belt rocks are thrust over the Singhbhum Granite Complex. There are also several occurrences of copper mineralization along the belt.

A similar shear zone of lesser intensity has also developed in the north affecting the northern limb of the regional antiform and the adjacent parts of the Dalma Volcanics. Apart from thrusting and shearing along the southern limb of the anticlinorium, some additional complexity is noticed in the easternmost outcrops because of superposition of some NNE-SSW trending folds and associated faulting. These isolated folds, which affected only the northern margin of the Dalma outcrop, and the associated faults do not match with the regional structural style, and in all probability represent a distinctly younger structural feature unrelated to the Precambrian evolution of the terrain.

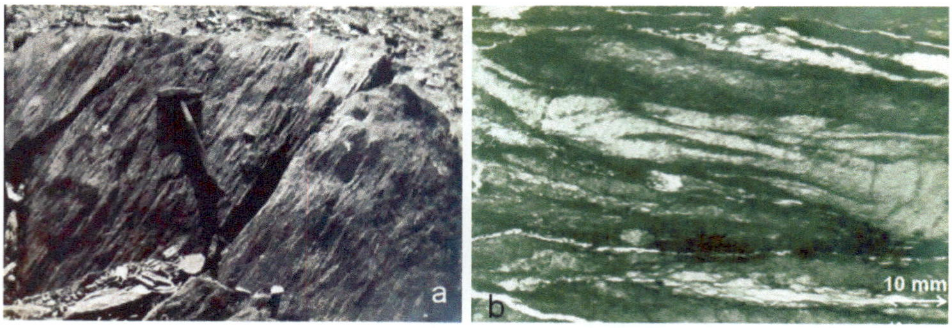

FIG. 6.16

(a) Strongly striated shear surface in quartzose mica schist from Singhbhum Shear Zone. (b) Phyllitic mylonite ('Phyllonite', Knopf, 1931) showing bands of extremely fine-grained of quartz alternating with chlorite-biotite-rich folia. Because of the extremely intense shearing and 'protoclastic' granulation, no individual minerals are identifiable.

The structural and tectonic characteristic of the major structure (also described as the Jamshedpur-Ghatshila Antiform) has been worked out in detail mainly in the central and the western part of the fold belt (Naha, 1965; Roy, 1966; Bhattacharya, 1978). The anticlinal nature of the regional fold gets confirmation through the studies of the primary sedimentary features like cross-bedding and ripple marks (Fig. 6.17a) so well developed in the arenites (quartzites) in the central part of the basin. Bhattacharya (1991) described these lowly metamorphosed metasediments as shallow marine tidal deposits. Seismic-shock-induced highly contorted beds preserving records of seismic-shock-related water escape structures (also described as 'seismites') are well preserved in these metasediments (Fig. 6.17b) (Bhattacharya and Bandyopadhyay, 1998). Prevalence of such structures in the north Singhbhum rocks indicate their deposition in a tectonically active basin.

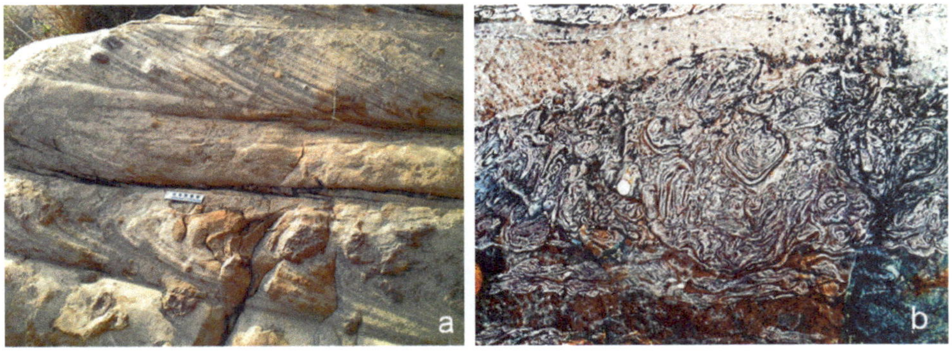

FIG. 6.17

(a) Cross bedding in the topmost bed of quartzite suggesting direction of 'stratigraphic younging' in the rocks. The lower beds show folded cross beds. (b) Highly contorted beds preserving records of seismic shock.

Image courtesy: H.N. Bhattacharya.

The small-scale folds that developed in this belt indicate open-to-tight fold geometry showing development of the schistosity of the axial planar type (Fig. 6.18a). The broadly open-to-tight small-scale folds have changed into tight and isoclinal geometry with the progressive development of crenulation cleavage in the rocks close to the Shear Zones both in the north and in the south (Fig. 6.18b).

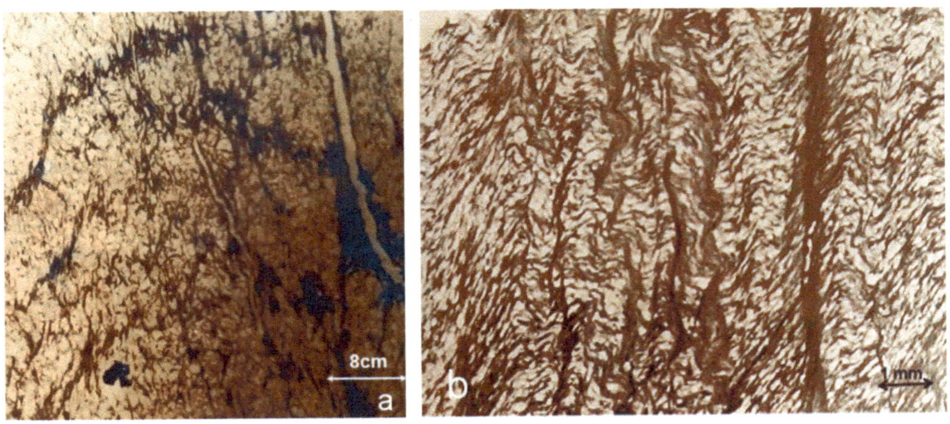

FIG. 6.18

(a) Photomicrograph of a fold showing the development of incipient schistosity through recrystallization of chlorite and biotite parallel to the axial planes of small-scale folds. (b) Crenulation cleavage formed across microfolds in the early formed schistosity in the Shear Zone rocks.

A typical Barrovian-type metamorphic character has been recorded in the belt (Naha, 1965; Roy, 1966; Lal and Singh, 1978). Studies in the central part of the Singhbhum Anticlinorium have highlighted the progressive nature of the metamorphism showing successive development of prograde regional metamorphic minerals from chlorite to sillimanite zones through biotite, almandine garnet, and kyanite/staurolite zones (Roy, 1966; Fig. 6.19). Microstructural studies have helped to provide proof of

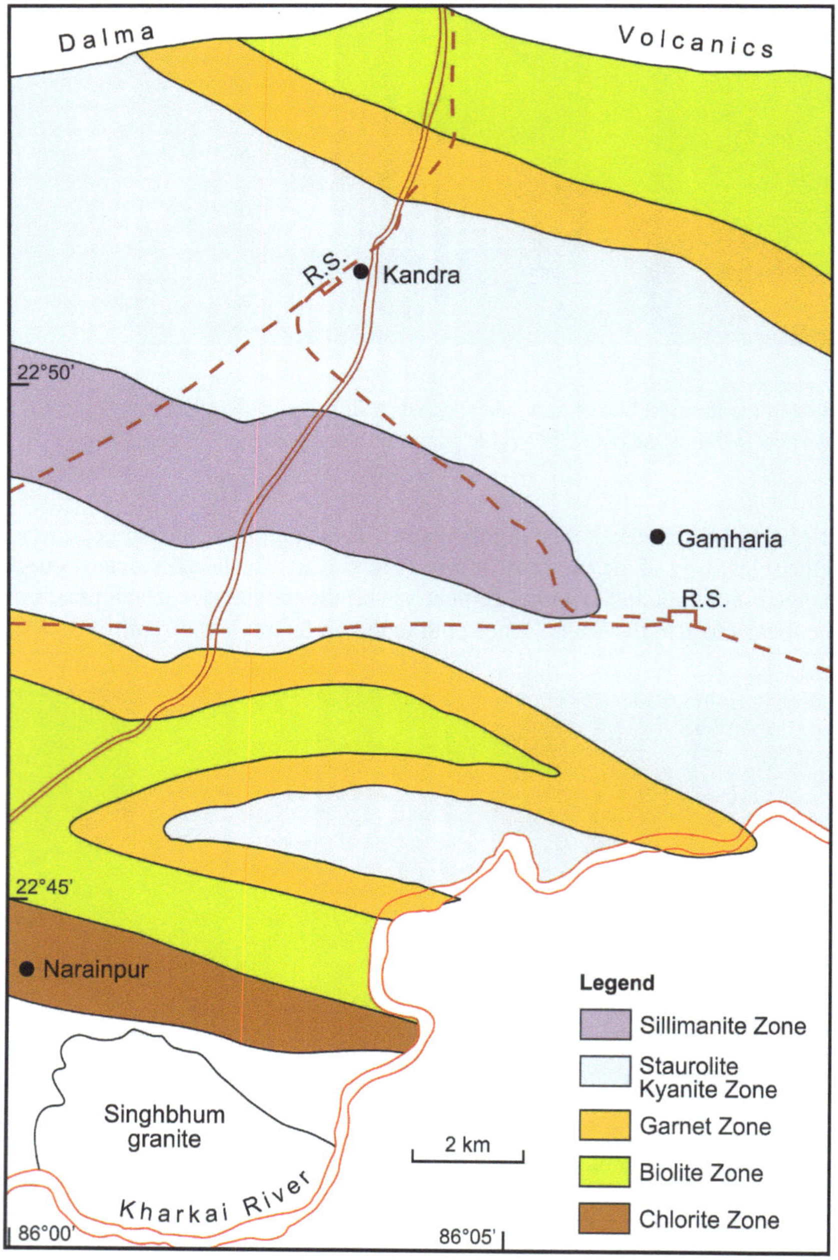

FIG. 6.19

Map showing the spatial distribution of the zones of progressive regional metamorphism in the central part of the 'Singhbhum Antiform'.

Reproduced with modification after Roy (1966) with permission.

progressive regional metamorphism in the main part of the fold belt coevally with the major folding of the belt. This was followed at a later stage by retrogressive metamorphism concomitant with the phases of shearing both along the northern and southern shear zones (Fig. 6.20a–d).

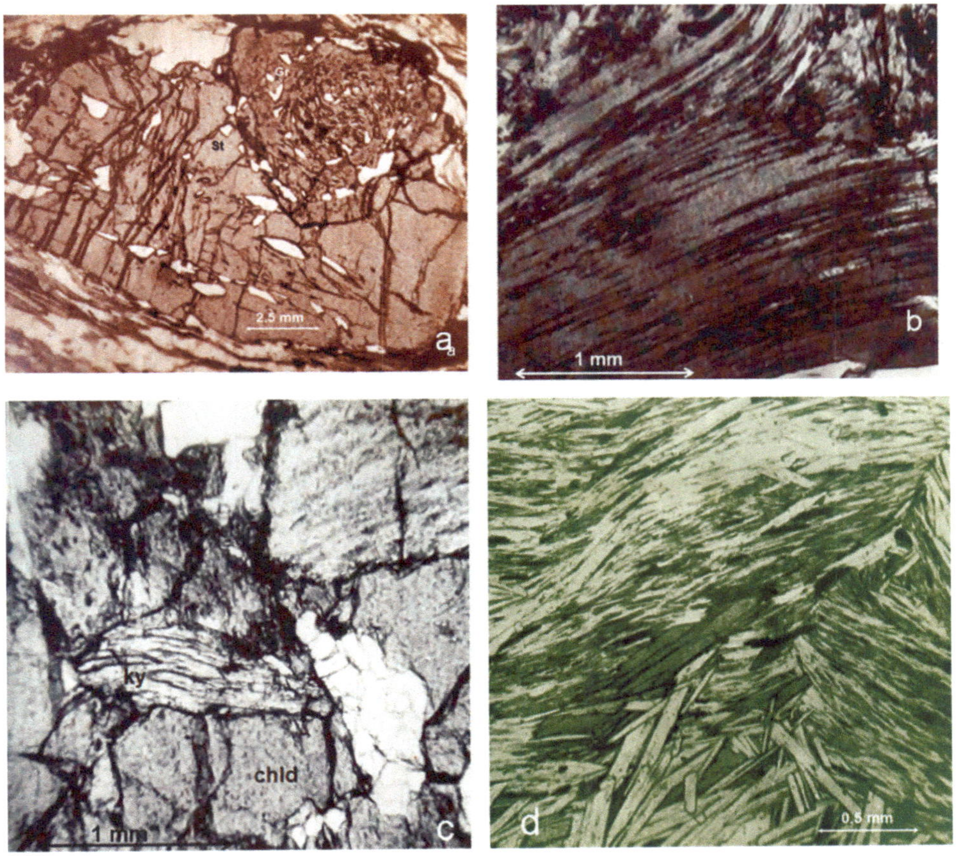

FIG. 6.20

Microstructural features hinting at the successive growth of metamorphic minerals (Roy, 1966). (a) Microtextural evidence indicating relative time of the formation of almandine garnet and staurolite; spirally oriented inclusions in garnet indicate synkinematic growth of almandine garnet, whereas the straight trails of inclusions in staurolite indicate its post-tectonic growth. (b) Development of sillimanite needles in a large flake of biotite. (c) Deformed kyanite (ky) enclosed in large, recrystallized grain of chloritoid (chld), which is a retrograde mineral. (d) Synkinematic (related to shearing deformation) recrystallization of muscovite-chlorite in the core of folded sheared rocks.

6.4.2 CHANDIL DOMAIN

The belt comprises dominantly of pelitic schists and other metasediments intermixed with tuff and related volcanic components. Several bodies of chert and carbonaceous shale in rhythmically banded form occur along with the impure limestone and/or dolomite mainly in the southern edge close to the outcrops of the Dalma Volcanics. Sporadic occurrences of base metal sulphide along with phosphorite occurrences are known locally in this belt. There are thick bands of quartzites that are thought to be chemically precipitated silica (metachert).

The northern contact of this lithoassociation with the Chhotanagpur Gneissic Complex is marked by a prominent shear zone. However, based on the suggestion that much of the southeastern part of the Chhotanagpur Gneissic Complex comprises migmatite gneisses formed from pelitic protolith, it can be justifiably said that migmatites with some relic bodies of mica schist merely represent the northern continuation of the migmatized pelites in the Chandil domain. The pelitic rocks mixed with the volcanic tuffs show low to very low grade of metamorphism. The metamorphic grade increases eastward from near Chandil (22°57′; 86°04′) reaching up to staurolite-kyanite grade of the Barrovian metamorphism.

Some occurrences of felsic volcanics like rhyolite have been reported from a number of places (Fig. 6.21) (Ray et al., 1996). In addition to rhyolite, there are isolated bodies of alkali syenite with nepheline and sodalite occurring in thin bands. Mafic-ultramafic rocks represented by chlorite schist, talc schist, amygdaloidal basaltic flows, and ultramafic intrusives are more common in the eastern sector than in the western where they form elongate 'strike ridges'.

FIG. 6.21

Uniformly layered body of rhyolite from the northern boundary of the outcrops of the Dalma Volcanics.

The pelitic-dominant Chandil domain shows some features that make it distinctive from the Jamshedpur domain in the south. Both the domains were included in the 'Singhbhum Anticlinorium' described by Dunn and Dey (1942). Ramakrishnan and Vaidyanadhan (2008) listed some features that make the Chandil domain distinctive. These are:

(i) The metapelites are generally of low metamorphic grade.
(ii) There is abundance of chert and carbonaceous shale-chert rhythmites along with bodies of limestone/dolomite.
(iii) There are large bodies of mafic-ultramafic rocks and abundance of felsic volcanics, along with syenite and nepheline syenite.
(iv) Carbonatites are seen along the Tamar-Porapahar Shear Zone in the north.

6.4.3 GANGPUR GROUP

The Gangpur Group constituting a distinctive lithostratigraphic ensemble occurs in the western part of the North Singhbhum Fold Belt. The manganese- and carbonate-bearing ensemble of the Gangpur Group has been described as an anticlinorium having an east-northeast to west-southwest trending axial direction. Krishnan (1937), who worked out a stratigraphic succession of the rocks of the region, divided the ensemble into two sequences. The younger one has been described as the Iron Ore Series. The older succession has been described as the Gangpur Series (Table 6.3).

The following stratigraphic succession of the Gangpur 'Anticlinorium' has been suggested by Krishnan (1937):

Table 6.3 Lithostratigraphic Succession of the Gangpur Series

Iron Ore Series	Phyllites, Slates, and Lavas
	Raghunathpur Conglomerate
Gangpur Series	Phyllites and mica schists
	Upper carboniferous phyllites
	Calcitic marbles
	Dolomitic marbles
	Mica schists and phyllite
	Lower carbonaceous quartzites and phyllites
	Gondites with associated phyllites
	(Basement not exposed)

Based on the occurrence of inverted 'current bedding', observed in some locality, a suggestion has been made that the major Gangpur structure represents a 'synclinorium' by Kanungo and Mahalik (1967). Banerjee (1967) who remapped the area detected a large-scale inversion of the Gangpur structure. He reported that the large-scale Gangpur fold represents an antiformal structure but with an inverted stratigraphy. He advanced a reinterpretation of the Gangpur structure proposing coaxial refolding of an earlier easterly plunging reclined fold. Later, Chaudhuri and Pal (1983) proposed that the large-scale Gangpur structure is a prototype of the geometry of the Dublabera fold (Fig. 6.22). While revising the stratigraphic succession earlier suggested by Krishnan (1937), Chaudhuri and Pal (1983) advocated a stratigraphic continuity of the different ensembles in the region forming a single continuous sequence without a many major break. These authors proposed a stratigraphic scheme completely reversing the order of the lithostratigraphic units earlier suggested by Krishnan (1937). In this connection, it may be highly educative to note the geometry of the Dublabera fold, which appears like 'mushroom'-type fold interference. If the large-scale Gangpur fold truly follows the geometry of the Dublabera fold as shown in Fig. 6.23, it may appear a far-fetched assumption to conclude that the total stratigraphy of the Gangpur is inverted. In the absence of any detailed information on the fold geometry, it may be rational to consider the stratigraphic succession erected by Krishnan (1937) as a workable stratigraphic succession for the future work.

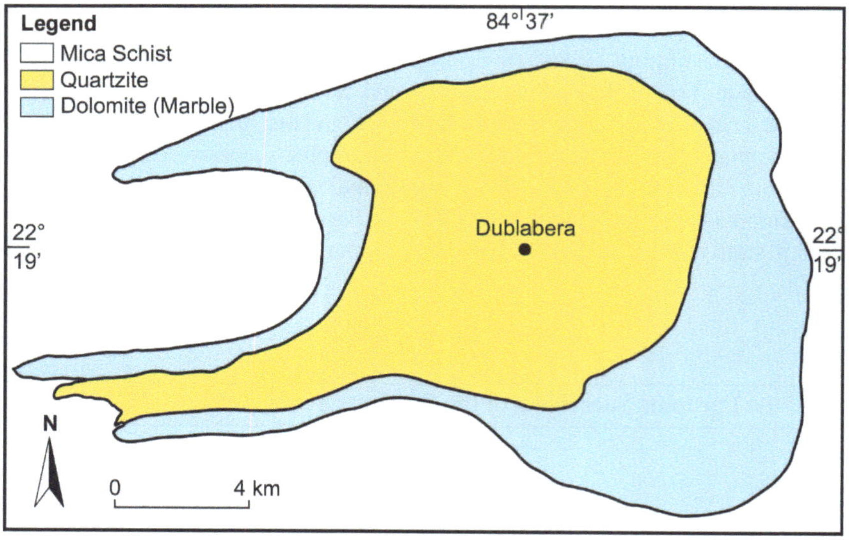

FIG. 6.22

Outcrop pattern of the mushroom-shaped 'Dublabera Fold' considered a prototype of the Gangpur 'anticlinorium'.

Reproduced from Chaudhuri and Pal (1983) with permission.

The debate continues on the understanding of the stratigraphic relationship of the Gangpur Group with the other components of the North Singhbhum Fold Belt (Fig. 6.15). However, based mainly on the metallogenic consideration, the Gangpur Group having manganese-bearing horizon associated with the calcitic and dolomitic carbonates may be correlated with the Sausar Group that occur in the adjacent Bastar Protocontinent. But unlike the evidence of 'snowball Earth' in the manganese-bearing rocks of Sausar, there is hardly any study done on the similar line on the Gangpur Group of the Singhbhum Protocontinent.

6.4.4 DALMA VOLCANICS

The Dalma Volcanics, also called as the Dalma Lavas (Dalma Group, Ramakrishnan and Vaidyanadhan, 2008), occupies the median zone in the eastern part of the North Singhbhum Fold Belt (Sarkar, 1982) (Fig. 6.23). Topographically, the Dalma Volcanics consists of rows of hill ranges that, at places, attain precipitous heights (Fig. 6.24).

The outcrops of the Dalma Volcanics do not extend beyond the halfway mark of the total outcrop length of the fold belt (Fig. 6.23). In the present discussion of the Dalma Volcanics, the outcrops of volcano-sedimentary rocks known as the Ongabira lavas have been included in the domain of the Dalma Volcanics based on the observed similarities on the structural, petrological, and geochemical characteristics (Pascoe, 1973; Sarkar, 1982; Sarkar and Gupta, 2012). The bulk of the lava pile is made of basalt flows (now in lowly metamorphosed state) preserving a strong compositional bimodality. In general, the high magnesian picritic volcanics have developed at the base; whereas low-K tholeiite basalt flows occur in the upper part of the volcanic sequence (Bose et al., 1989). The Dalma Volcanics show patches of volcanic breccia and agglomerate (Fig. 6.25a), which presumably provide evidence of

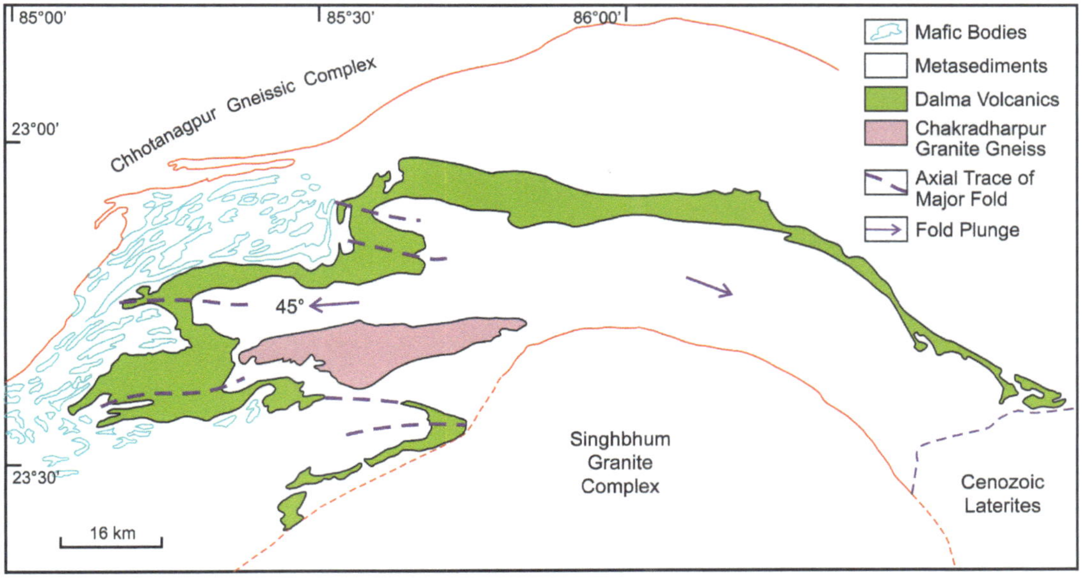

FIG. 6.23

Map showing the outcrop pattern of the Dalma Volcanics and associated lithoformations of the North Singhbhum Fold Belt.

FIG. 6.24

Dalma Volcanics form prominent hill ranges that at places attain precipitous heights.

the close proximity of volcanic foci (Pascoe, 1973). The agglomerates in particular are made almost entirely of ejected boulders some of which are quite large in dimensions (Fig. 6.25b). Locally, thin beds of rhythmically bedded carbonaceous phyllite occur within the lava flows, which are thought as tuff beds (Fig. 6.26) (Pascoe, 1973). There are also some phyllite-like 'enclaves' enclosed in the volcanic flows that have been interpreted as 'inliers' of the underlying rocks (Pascoe, 1973).

FIG. 6.25

(a) A piece of agglomerate with large volcanic 'bombs' of different shapes and sizes embedded in finer ejecta materials. (b) A volcanic bomb (ejecta) with its longest dimension measuring about 10cm across.

FIG. 6.26

A hand specimen of carbon phyllite showing rhythmic alterations of black and white bands representing the beds.

Field studies of the base of the Dalma Lavas over the entire length of its 'strike continuity' in the southern side do not indicate a complete conformable relationship with the underlying phyllites and the associated rocks at the base (Pascoe, 1973). Instead, the evidences of overlap have been recognized at some places, particularly on the northern side of the Dalma outcrops. The discordance between the two successions is also manifested in the occurrences of quite dissimilar lithological formations on the either side of the Dalma Volcanic outcrops (Pascoe, 1973). Regarding the relationship between the volcanics and its underlying rocks, Dunn and Dey (1942) had suggested a time gap between the outpouring of the volcanics and the inversion of the underlying basins in which the pelitic and the associated sediments were deposited.

The Dalma Volcanics (Dalma Group of Ramakrishnan and Vaidyanadhan, 2008) is divided into lower and upper formations. A brief hiatus between the two is indicated by the occurrence of reworked agglomerate and at some place by the presence of weathered rocks like metamorphosed chlorite phyllites on the top of the lower succession. Lithologically, the lower formation consists of tuff (carbon phyllite) and quartzite interlayered with volcanics. Upper formations of the Dalma Volcanics are made up of mafic flows (commonly pillowed, Fig. 6.27) and agglomerates with Mg-rich volcano-clastic grading between picrite and komatiite along with comagmatic intrusions.

FIG. 6.27

Stacks of 'pillow lava' structure in the Dalma Volcanics.

According to Bose et al. (1989), the basaltic flows are comparable in geochemical character with the modern-day Mid-Oceanic Ridge Basalts (MORB). Based on trace elements, REE geochemistry and Sr–Nd isotope systematics, Roy et al. (2002) suggested a Mantle Plume origin of the gabbro-pyroxenite bodies in the Dalma Volcanics.

Structurally, the Dalma Volcanics have been visualized as isoclinally folded fold (Sarkar and Saha, 1962) refolded on a latter westerly plunging antiformal fold with broadly east-west axial trace (Fig. 6.16). However, the geological evidence of isoclinal first folding is truly lacking. On the other hand, the report of diverse character of the lithological types occurring on the either side of the Dalma Volcanics (Pascoe, 1973) provides proof that the Volcanics do not represent coalesced limbs of isoclinal folding. It also seems to be an incongruous proposition that the Dalma Volcanics could have been folded independently without participation of the older lithostratigraphic formations.

The outcrop pattern of the Dalma Volcanics near the western end defines a major westerly plunging antiformally folded closure with east-west axial trace. There are several subsidiary asymmetric folds that have developed on either side of the hinge region (Fig. 6.23). On northern limb, the subsidiary folds define S-shaped sinistral folds (Sarkar, 1982). The fold geometry on the southern side (assuming that Ongabira lavas constitute a part of the Dalma Volcanics) has Z-shaped or dextral fold geometry. The overall fold geometry of this westerly plunging antiform represents a classic example of flexural slip folding on a very large scale. The apparent detachment of the Ongabira part from the main body in the southwestern part is because of the movement along the Shear Zone tearing off the 'Ongabira lava' part of the Dalma Volcanics from the main body.

A rather unusual feature noted in the outcrop pattern of the Dalma Volcanics is the crowding of small bands and lenses of mafic bodies to the northwest of the main fold closure continuing for a short distance up to the hinge zone northern sinistral fold (Fig. 6.23). In absence of any information on the structure and petrology of these lithological formations, irrespective of their sizes, no possible explanation can be offered about the stratigraphic significance of these packs of lenticular outcrops of mafic bodies.

The Dalma Volcanics in the western part show evidence of strong shearing and mylonitization. The sheared and mylonitic rocks have also been observed all along the southern boundary of the Dalma Volcanics at the contact of the Singhbhum Group rocks in the south (Roy, 1965). A similar shear zone has also been noted in the north but is of much lesser intensity. There are some isolated folds developed on the northeastern part of the Dalma outcrops. These NNE-SSW trending folds have also affected the underlying Singhbhum Group rocks. No information is known about the causative forces responsible for the development of these youngest set of folds.

6.4.5 CHAKRADHARPUR GRANITE, ARKASANI GRANITE, SODA-GRANITE

Patches of granitic rocks of various compositions are observed in the southern part of the Singhbhum Shear Zone mainly in the central part. The largest of these bodies is known as the Chakradharpur Granite (also described as Chakradharpur Gneiss). It is an elongate body of granitic gneiss occurring in the southwestern part of the Singhbhum Antiform, close to the western folded outcrops of the Dalma Volcanics (Fig. 6.15). Almost in continuation of the Chakradharpur Granite in the east occur several narrow, linear outcrops of granitic rocks within the Singhbhum Shear Zone. These bodies are known as the Arkasani Granite (also described as the Arkasani Granophyre) or Soda Granite. In fact, the earlier workers (Pascoe, 1973) preferred to use the term Arkasani Granophyre for all granitic rocks occurring in the Singhbhum Shear Zone. In the present discussion, however, we prefer to use the term Chakradharpur Granite for the part of the granitic rock (much of which is gneissic in character) that forms a large outcrop west Rajkharswan (22°44′:85° 49′). The term Arkasani Granite will be used for all the granitic rocks that occur in the Singhbhum Shear Zone east of Rajkharswan. The main body of the Arkasani Granite southwest of Jamshedpur (22°48′:86° 12′) shows a sinuous outcrop pattern with long fingerlike projections and set within extremely sheared mylonitic rocks.

The different types of the Arkasani Granite are compositionally comparable with the western body of Chakradharpur Granite. Petrologically, both the Chakradharpur Granite and the Arkasani Granite include two major types: (i) the trondjhemite-tonalite-granodiorite gneiss that covers the largest aerial extent (Fig. 6.28a) and (ii) the coarse-grained, massive granodiorite grading to granite (Fig. 6.28b). The massive granite in which albite constitutes the dominant feldspar has been described as the Soda Granite by the early workers. In addition to the two different types of granitic rocks described above, the eastern outcrops of the Arkasani Granite include a special textural type called 'granophyric' in which the groundmass shows an intergrowth of alkali feldspar (mostly albite) and vermicular (worm-like) quartz (Fig. 6.28c). Intergrowth of quartz is generally observed near the margins of alkali feldspar crystals in coarse- and medium-grained acid and intermediate igneous rocks. The sheared varieties of both the Chakradharpur and the Arkasani Granite show the development of secondary slip/shear surfaces in the Shear Zone, some of which show contorted forms (Fig. 6.28d).

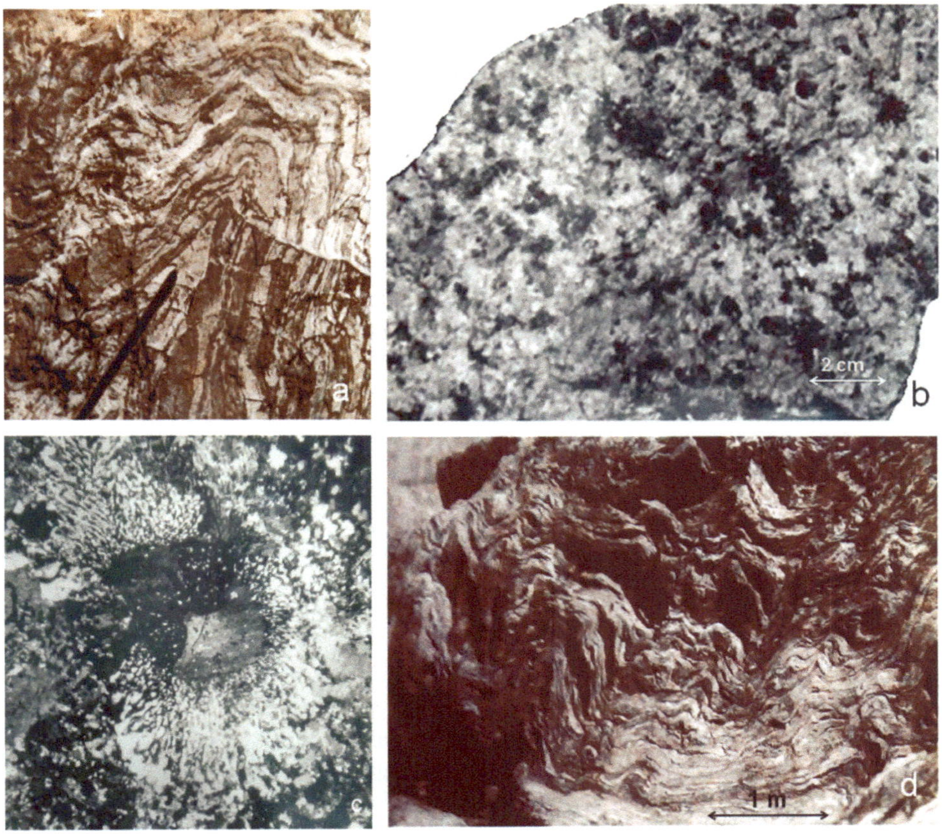

FIG. 6.28

(a) The banded tonalite-trondjhemite-granodiorite occurring in folded in an outcrop southwest of Jamshedpur. (b) The coarse-grained, massive granodiorite grading to granite. (c) Intricate pattern of intergrowth texture (granophyric) formed by quartz and feldspar in massive varieties of the Arkasani Granite. (d) Contorted bands of sheared and foliated granite-gneiss occurring in the Singhbhum Shear Zone.

The Archaean antiquity of the Chakradharpur Granite is hinted by Sarkar and Gupta (2012) based on the physical and geochemical characteristics of the granite-gneisses of tonalite-trondjhemite-granodiorite type. Comparing with the Tamperkola Granite in the western part of the Singhbhum Granite Complex, an age of ~2800 Ma has been suggested for the trondjhemite gneiss bodies of the Chakradharpur Granite (Bandyopadhyay et al., 2001).

The Proterozoic age of the Arkasani Granite/Granophyre rocks from the Singhbhum Shear Zone has been reported by Bhattacharya et al. (2015) based on the SHRIMP U–Pb zircon date of 1861 ± 6 Ma. Quite similar ages have also been reported by Pal et al. (2011) from the Singhbhum Shear Zone rocks that host uranium mineralization. These are U–Pb ICP-MS laser ablation ages suggesting 1882 ± 23 and 1885 ± 31 Ma determined from uraninite and monazite, respectively. However, from the occurrence of typical banded gneissic rocks (Fig. 6.28a), it appears that much of the Arkasani Granite, in all probability, have Archaean antiquity, as indicated by the TD_M model ages as also by the Sm–Nd ages quoted by Ramakrishnan and Vaidyanadhan (2008). The massive granites (Fig. 6.28b) must have evolved postdating the tectonism related to the shearing and thrusting along the Singhbhum Shear Zone. In other words, the 1861 ± 6 Ma age reported by Bhattacharya et al. (2015) constrains the age of the massive variety of the Arkasani Granite occurring in the Shear Zone. The report of ~1.0 Ga Rb/Sr isochron age determined by Sengupta et al. (1994) could possibly be related to the youngest thermal event reported from the Singhbhum Shear Zone. Further, in the occurrence of kyanite-bearing high-aluminous rocks at the contact of the Arkasani Granite (locally named the Chakradharpur Granite), we may logically suggest the Archaean antiquity of this granite-gneiss body.

Summarizing, the isotope data from the different granitic rocks occurring in the Singhbhum Shear Zone that are known by different names like the Chakradharpur Granite in the west, the Arkasani Granite, and the Soda Granite in the east help to confirm the multiphase evolutionary history of the Arkasani Granite and the related rocks that occur in the Singhbhum Shear Zone.

6.4.6 KUILAPAL GRANITE

Several oval-shaped outcrops of granite-gneiss occur in the eastern part of the Chandil region near the northern trail of the Dalma Volcanic outcrop. The largest of these is known as the Kuilapal Granite. Compositionally, it ranges from trondhjemite, through granodiorite to adamalite, to granite proper (Dunn and Dey, 1942; Saha, 1994). The Archaean antiquity of the Kuilapal and other associated granite-gneiss bodies can be presumed from the composition of some of these exposed rocks appearing as grey gneisses, traditionally referred to as tonalite-trondjhemite-granodiorite gneisses showing complex tectonothermal reconstitution. Although originally described as a gneissic body with evidence of 'injection' of gneissic material within the surrounding mica schist along the schistosity (Dunn and Dey, 1942), some later studies indicated that at least a component of the gneissic rocks had formed synchronously during the later phase of deformation of the surrounding metapelites. Sastry et al. (2013) reported Rb–Sr whole isochron age of 1792 ± 98 Ma (MSWD = 13) of the 'granite-gneiss' bodies in the Kuilapal. The Pb–Pb age of the same granite-gneiss body is 1863 ± 80 Ma (MSWD = 11.3). Earlier, Sengupta et al. (1994) had reported Rb–Sr whole-rock isochron age of 1638 ± 38 Ma.

6.5 SUMMARY OF THE TECTONO-STRATIGRAPHIC EVOLUTIONARY HISTORY

The Singhbhum Protocontinent comprises three major lithotectonic blocks: the Singhbhum Granite Complex in the south, the Chhotanagpur Gneissic Complex in the north, and the North Singhbhum Fold Belt in the middle (Fig. 6.2).

Reappraisal of field relationships between the different lithological ensembles supported by available geochronological data, and taking due note of the tectonometamorphic, magmatic, and sedimentation history helped to build up a coherent crustal evolutionary history of the different crustal blocks that make up the Singhbhum Protocontinent. The early Archaean history is well recorded in the southern crustal block known as the Singhbhum Granite Complex. The oldest age of this crustal block comes from the tonalite-trondjhemite-granodiorite gneisses indicating ca. 3700 Ma age. Deposition of the oldest dominantly metasedimentary supracrustals, the Older Metamorphics, was initiated at around

3380 Ma. The cratonization of this Mesoarchaean crustal block was around 3285 Ma, and subsequent to that, no other fabric-forming ductile deformation and metamorphism is noted in the domain of the Singhbhum Granite Complex. The subsequent geological event was marked by the deposition of the thick bands of banded iron formation coevally with the intrusion of mafic and ultramafic bodies and some 'nonorogenic' granitoids at ca.3100 Ma. Following a period of tectonic quiescence, a huge pile of the mafic volcanics with intercalated beds of arenite was formed during the phase of Plume impingements at around 2800 Ma in the oval-shaped belts of Dhanjori and Simplipal belt, in the northeastern and eastern part of the Archaean crustal block. The end-Archaean intrusion of Newer Dolerite dykes in conjugate sets, and the deposition of Kolhan Group marked the culmination of the Archaean Crust-building activity in the Singhbhum Granite Complex.

The Chhotanagpur Gneissic Complex that occupies a huge area in the northern part of the Protocontinent is a poorly known gneissic terrain. However, the Archaean antiquity of the terrane is indicated by the occurrence of the tonalite-trondhjemite-granodiorite gneisses along with the diorite gneisses. Further, the occurrence of isolated bodies of banded iron formation, metacarbonates and some metasedimentary bodies along with the mafic and ultramafic ensembles suggests greenstone-type evolution of these supracrustals. The entire domain of the Chhotanagpur Gneissic Complex shown in the map is not made entirely of the Archaean rock assemblages. There are plenty of evidences that suggest the reconstitution of this Archaean Crust later during the Proterozoic. The confirmation for this comes from the suggestion that much of the southern part of this terrane represents migmatized 'paragneisses' that formed due to the alternation of the pelitic protoliths that could be equivalent to the metasediments of the North Singhbhum Fold Belt in the south. The latter thermal event includes the intrusion of anorthosite magma and the phase of deformation and associated metamorphism leading to the formation of the patches of the high-PT granulites and intrusion/emplacement of the 'grey' and 'pink' granites. Single-zircon studies indicated a concordant U–Pb crystallization age of 1550 ± 2 Ma of the anorthosite (Chatterjee et al., 2008). The postcrystallization thermal reconstitution of the anorthosite body along with associated metapelite was during 947 ± 27 and 950 ± 20 Ma (Chatterjee et al., 2008). The little older age of 995 ± 24 Ma thermal event is recorded in the pelitic granulite located north of the anorthosite body (Chatterjee et al., 2008; Maji et al., 2008). In summary, we may say that unlike the southern Singhbhum Granite Complex, the evolutionary history of the Chhotanagpur Gneissic Complex is much more prolonged extending (in all probability) from Mesoarchaean to the early Neoproterozoic.

The North Singhbhum Fold Belt constitutes a narrow arcuate belt of the Proterozoic metasediments and metavolcanics (Dalma Volcanics) between the two granite-gneiss complexes in the northern and southern parts of the Singhbhum Protocontinent. Tectonically speaking, the dominant components of the North Singhbhum Fold Belt, the metasediments along with volcanics and some reconstituted basement granites were evolved during a single orogenic phase described as the Singhbhum Orogeny by Sarkar and Saha (1962). The tectonometamorphic history is well established in some limited areas mainly in the southeastern part which provide evidence of a 'Barrovian' type of regional metamorphism coevally with the initial folding phase followed by retrogressive metamorphism concomitant with the phase of extensive shearing during the waning phase of the orogeny. The Dalma Volcanics which form 'as median zone' in the eastern part of the fold belt apeears comparable with the present-day Mid-Oceanic Ridge basalts. The Dalma Volcanics were folded coevally with the metasediments during the north-south shortening of the fold belt.

From the sedimentological point of view, the belts of the metasedimentary supracrustals are divisible into two diverse depositional basin types. The metasediments that occur between the Dalma Volcanics in the north of the Singhbhum Granite Complex were deposited in a 'carbonate-free' shale-sand depositing basin. On the other hand, the entire Western Domain and the Chandil Domain together represent a restricted basin depositing chemogenic sediments along with fine-grained clastics with very little arenites.

The age of the supracrustals in the North Singhbhum Fold Belt is debatable and indeed poorly known. However, a Palaeoproterozoic age of these metasediments may be assumed taking into consideration the association of manganese with carbonate in the lower succession of the Gangpur Group. The similar rock association recorded in the Sausar Group in the adjacent Bastar Protocontinent is also considered to be of the same age.

There are hardly any isotope data available for constraining the age of basin opening of the Singhbhum Group. However, assuming the U–Pb zircon date of 1861 ± 6 Ma of the Arkasani Granite (Granophyre) (Bhattacharya et al., 2015) marks the closing of the Singhbhum Orogeny. Somewhat similar ages, Rb–Sr whole isochron age of 1792 ± 98 and 1863 ± 80 Ma Pb–Pb age, have been reported from the Kuilapal area in the Chandil belt (Sastry et al., 2013). Considering these isotope ages, we may presume that the age of the basin opening and deposition of metasediments in different subbasins in the North Singhbhum Fold Belt must have taken place much earlier and possibly during 2200 and 2100 Ma as suggested for the manganese-bearing rocks of the Sausar Group in the adjacent Bastar Protocontinent.

REFERENCES

Bandyopadhyay, P.K., Chakrabarti, A.K., Deomurari, M.P., Misra, S., 2001. 2.8 Ga old anorogenic granite-acid volcanics association from western margin of the Singhbhum Orissa craton, eastern India. Gondwana Res. 4, 465–475.

Banerjee, P.K., 1967. Revision of the stratigraphy, structure and metamorphic history of the Gangpur Series Sundargarh District in Orissa. Rec. Geol. Surv. India 95, 327–346.

Banerjee, P.K., Ghosh, S.K., 1994. Is the Simlipal Complex a product of shallow plume tectonics? J. Geol. Soc. India 43, 353–359.

Banerji, A.K., 1991. Geology of the Chotanagpur region. Indian J. Geol. 63, 275–282.

Basu, A., Ray, S.L., Saha, A.K., Sarkar, S.N., 1981. Eastern Indian 3800-million year old crust and early mantle differentiation. Science 212, 1502–1505.

Bhattacharya, D.S., 1978. Contrast across the shear zone in the Precambrian rocks of western Singhbhum. Geol. Mijnb. 57, 59–63.

Bhattacharya, H.N., 1991. A reappraisal of the depositional environment of the Precambrian metasediments around Ghatshila-Galudih, Eastern Singhbhum. J. Geol. Soc. India 37, 47–54.

Bhattacharya, H.N., Bandyopadhyay, S., 1998. Seismites in a Proterozoic tidal succession, Singhbhum, Bihar, India. Sediment. Geol. 119, 239–252.

Bhattacharya, H.N., Mahapatra, S., 2008. Evolution of the Proterozoic rift margin sediments—North Singhbhum Mobile Belt, Jharkhand-Orissa, India. Precambrian Res. 162, 302–316.

Bhattacharya, P.K., Mukherjee, S., 1987. Granulites in and around the Bengal anorthosite, eastern India; genesis of coronal garnet, and evolution of the granulite-anorthosite complex. Geol. Mag. 12, 21–32.

Bhattacharya, H.N., Nelson, D.R., Thern, E.R., Altermann, W., 2015. Petrogenesis and geochronology of the Arkasani Granophyre and felsic Dalma volcanic rocks: implications for the evolution of the Proterozoic North Singhbhum Mobile Belt, east India. Geol. Mag. 152, 492–503.

Bose, M.K., Chakrabarti, M.K., Saunders, A.D., 1989. Petrochemistry of the lavas from Proterozoic Dalma volcanic belt, Singhbhum, eastern India. Geol. Rundsch. 78, 633–648.

Chakraborty, K.L., Chakraborty, T.L., 1984. Geological features and origin of the chromite deposits of Sukinda valley, Orissa, India. Mineral. Deposita 19, 256–265.

Chatterjee, N., Crowley, J.L., Ghose, N.C., 2008. Geochronology of the 1.55 Ga Bengal anorthosite and Grenvillian metamorphism in the Chhotanagpur gneissic complex, eastern India. Precambrian Res. 161, 303–316.

Chaudhuri, A.K., Pal, A.B., 1983. Structural history as an aid in Precambrian Stratigraphic correlation: an example from Gangpur Group in Eastern India. J. Geol. Soc. India 24, 522–532.

Dunn, J.A., Dey, A.K., 1942. The Geology and Petrology of eastern Singhbhum and surrounding area. Mem. Geol. Surv. India 69 (2), 281–450.

Ghosh, D.K., Sarkar, S.N., Saha, A.K., Ray, S.L., 1996. New insights on the early Archaean crustal evolution in Eastern India: re-evaluation of lead-lead, samarium-neodymium and rubidium-strontium geochronology. Indian Minerals 50, 175–188.

Ghosh, N.C., Chatterjee, N., Mukherjee, D., Kent, R.W., Saunders, A.D., 2008. Mineralogy and geochemistry of the Bengal Anorthosite Massif in the Chhotanagpur Gneissic Complex at the Eastern Indian Shield Margin. J. Geol. Soc. India 72, 263–277.

Goswami, J.N., Mishra, S., Wiedenbeck, M., Ray, S.L., Saha, A.K., 1995. 3.55 Ga old zircon from Singhbhum-Orissa iron ore Craton, Eastern India. Curr. Sci. 69, 1008–1011.

Haldar, S.K., 2017. Platinum-Nickel-Chromium Deposits: Geology, Exploration and Reserve Base. Elsevier. 322p.

Jones, H.C., 1934. The iron ore deposits of Bihar and Orissa. Mem. Geol. Surv. India 71, 105–120.

Kanungo, D.N., Mahalik, N.K., 1967. In: Structure and stratigraphic position of Gangpur Series in the Archaeans of Peninsular India. Proceeding of Symposium on Upper Mantle Project, Hyderabad, India, pp. 458–476.

Knopf, E.B., 1931. Retrogressive metamorphism and phyllonitization. Am. J. Sci. 21, 1–37.

Krishnan, M.S., 1937. Geology of Gangpur State, Eastern India. Mem Geol. Surv. India 71. 181p.

Lal, R.K., Singh, J.B., 1978. Prograde polyphase regional metamorphism and metamorphic reactions in the pelitic schists in Sini, district Singhbhum India. Neus Jahrb. Miner. Abh. 124, 294–325.

Mahadevan, T.M. 1992. Geological evolution of the Chotanagpur Gneiss Complex in a part of Purulia district, West Bengal, Indian J. Geol., v. 64, pp. 1–22.

Mahadevan, T.M., 2002. Geology of Bihar and Jharkhand. Geological Society of India, Bangalore. 563p.

Maji, A.K., Goon, S., Bhattacharya, A., Mishra, B., Mahato, S., Bernhardt, H.J., 2008. Proterozoic polyphase metamorphism in the Chotanagpur Gneiss Complex (India), and implications for trans-continental Gondwana correlation. Precambrian Res. 162, 385–402.

Mazumdar, S.K., 1996. Precambrian geology of Peninsular eastern India: a selective review. Indian Minerals 50, 139–174.

Misra, S., 2006. Precambrian chronostratigraphic growth of Singhbhum-Orissa craton, Eastern Indian Shield: an alternative model. J. Geol. Soc. India 67, 356–378.

Misra, S., Johnson, P.T., 2005. Geochronological constraints on the evolution of the Singhbhum Mobile Belt and associated basic volcanics of eastern Indian shield. Gondwana Res. 8, 129–142.

Misra, S., Deomurari, M.P., Wiedenback, M., Gosswami, J.N., Ray, A., Saha, A.K., 1999. 207Pb/206Pb zircon ages and the evolution of the Singhbhum craton, Eastern India: an ion-microprobe study. Precambrian Res. 93, 139–151.

Mondal, S.K., Ripley, E.M., Li, C., Frei, R., 2006. The genesis of Archaean chromitites from the Nuasahi and Sukinda massifs in the Singhbhum Craton, India. Precambrian Res. 148, 45–66.

Naha, K., 1962. Precambrian sedimentation around Ghatshilain east Singhbhum, eastern India. Proc. National Inst. of Sci. India 27A, 261–272.

Naha, K., 1965. Metamorphism in relation to stratigraphy, structure and movements in part of East Singhbhum, Eastern India. Q. J. Geol. Min. Meta. Soc. India 37, 41–85.

Nelson, D.R., Bhattacharya, H.N., Misra, S., Dasgupta, N., Altermann, W., 2007. In: Banerjee, S. (Ed.), New SHRIMP U/Pb zircon dates from the Singhbhum craton, Jharkhand-Orissa region, India. Abstract. International Conference on Precambrian Sedimentation and Tectonics and the Second GPSS Meeting. Indian Institute of Technology, Bombay.

Nelson, D.R., Bhattacharya, H.N., Thern, E.R., Altermann, W., 2014. Geochemical and ion-microprobe U–Pb zircon constraints on the Archaean evolution of Singhbhum Craton, eastern India. Precambrian Res. 255, 412–432.

Pal, D.C., Chaudhury, T., McFarlane, C.H., Mukherjee, A., Sarangi, A.K., 2011. Mineral chemistry and in situ dating of allanite, and geochemistry of its host rocks in the Bagjata Uranium Mine, Singhbhum Shear Zone, India: implications for the chemical evolution of REE mineralization and mobilization. Econ. Geol. 106, 1155–1171.

Pascoe, E.D., 1973. A Manual of the Geology of India and Burma. vol. 1. Geological Survey of India, Calcutta. 485p.

Ramakrishnan, M., Vaidyanadhan, R., 2008. Geology of India. vol. 1. Geological Society of India, Bangalore. 556p.

Rao, P., et al., 1964. In: Associated sedimentary sequences in parts of Keonjhar, Cuttack, Dhenkanal Sundargarh districts, Orissa, India. Internatinal Geological Congress, New Delhi. Pt. 10, pp. 72–87.

Ray, K.K., Ghosh Roy, A.K., Sengupta, S., 1996. Acid volcanic rocks between the Dalma volcanic belt and Chhotanagpur Gneissic Complex, East Singhbhum and Purulia districts of Bihar and West Bengal. Indian Minerals 50, 1–8.

Roy, A.B., 1965. A note on the Dalma thrust along the southern boundary of the Dalma "Lavas" in central of Singhbhum, Bihar. Geol. Soc. India Bull. 2, 81–83.

Roy, A.B., 1966. Interrelation of metamorphism and deformation in Central Singhbhum, Eastern India. Geol. En Mijnb 45, 365–374.

Roy, A.B., 2012. Indian Shield: insight into the pristine size, shape and tectonic framework. Ind. J. Geosci. 66, 181–192.

Roy, A.B., Bhattacharya, H.N., 2012. Tectonostratigraphic and geochronologic reappraisal constraining the growth and evolution of Singhbhum Archaean Craton, Eastern India. J. Geol. Soc. India 80, 455–469.

Roy, A., Sarkar, A., Jeyakumar, S., Aggrawal, S.K., Ebihara, M., 2002. Mid-Proterozoic Plume-related thermal event in Eastern Indian Craton: evidence from trace elements, REE geochemistry and Sr-Nd isotope systematics of basic-ultrabasic intrusives from Dalma Volcanic Belt. Gondwana Res. 5, 133–146.

Roy, A.B., Dutt, K., Rathore, S., 2016. Development of ductile shear zones during diapiric magmatism of nepheline syenite and exhumation of granulites, examples from central Rajasthan, India. Curr. Sci. 110, 1094–1101.

Saha, A.K., 1994. Crustal Evolution of Singhbhum, North Orissa, Eastern India. Geological Society of India, Bangalore. 341p.

Saha, A.K., Ray, S.L., Sarkar, S.N., 1988. Early history of the Earth: evidence from eastern Indian Shield. In: Mukhopadhyay, D. (Ed.), Precambrian of the Eastern Indian Shield. Memoirs of the Geological Society of India, 8, pp. 13–37.

Sarkar, A.N., 1982. Structural and Petrological Evolution of the Precambrian rocks in western Singhbhum, Bihar. Mem. Geol. Surv. India 113. 97p.

Sarkar, S.C., Gupta, A., 2012. Crustal Evolution and Metallogeny in India. Cambridge University Press (840 p).

Sarkar, S.N., Saha, A.K., 1962. A revision of Precambrian Stratigraphy and tectonics of Singhbhum and adjacent region. Q. J. Geol. Min. Meta. Soc. India 34, 97–136.

Sarkar, A., Pati, U.C., Panda, P.K., Patra, P.C., Kundu, H.K., Ghosh, S., 2000. Late Archaean Charnockite rocks from northern marginal zone of Eastern Ghats. Geol. Surv. India Spec. Publ. 57, 177–179.

Sastry, D.V.L.N., Pandey, U.K., Pandey, B.K., 2013. Rb-Sr and Pb-Pb Geochronological studies on the granite gneisses of Kuilapal, Purulia-Bankura-Midnapore Districts, West Bengal. J. Appl. Geochem. 15, 19–24.

Saxena, V.P., Krishnamurthy, P., Murugan, C., Sabot, H.K., 1992. Geochemistry of the granitoids from the central Surguja shear zone, India: geological evolution and implication on uranium mineralization and exploration. Expl. Res. Atom. Minerals 5, 27–40.

Sengupta, S., Paul, D.K., Bishui, P.K., Gupta, S.N., Chakravarty, R., Sen, P., 1994. Geochemical and isotopic study of Kuilapal and Arkasani Granophyre from eastern Indian craton. Indian Minerals 48, 77–88.

Sharma, R.S., 2009. Cratons and Mobile Belts of India. Springer. 304p.

Singh, Y., Krishna, V., 2009. Rb-Sr Geochronology and Petrogenesis of Granitoids from the Chotanagpur Granite Gneiss Complex of Raikera-Kunkuri Region, Central India. J. Geol. Soc. India 74, 200–208.

Valdiya, K.S., 2010. The Making of India: Geodynamic Evolution. Macmillan. 816p.

CHAPTER 7

RAJMAHAL PROTOCONTINENT

7.1 INTRODUCTION: DEFINING RAJMAHAL PROTOCONTINENT

The proposed scheme of dividing the Indian Shield by 'Joins' marked by the Gondwana Rift Basins helps to delimit the Singhbhum Protocontinent in the north. This leaves a small region between the Damodar Valley Gondwana Basins in the south and the Narmada-Son Lineament in the northwest for the newly defined Protocontinent, which has been named the Rajmahal Protocontinent. Objectively speaking, the precise tectonostratigraphic status and the eastward continuation of the Rajmahal Protocontinent is little known except the fact that it is carved out of the huge outcrop of dominantly granite-gneiss terrane used to be known by the pristine name the Chhotanagpur Gneissic Complex. The eastern boundary of this newly defined Protocontinent is clogged by elongate outcrops of the north-south running outcrops of the Cretaceous Rajmahal Trap, and beyond that by the huge pile of the sediments of the Bengal Basin.

Lithocomponents of the Rajmahal Protocontinent include diverse rock ensembles starting with dominantly granite-gneiss in the south to the zone riddled with mica-rich pegmatites in the north. Earlier, considering the poor availability of geological information including reliable geochronological data, Mahadevan (2002) had divided the pristine Chhotanagpur Gneissic Complex into five major divisions or belts, based mainly on the broad lithological ensembles (Fig. 7.1a). These east-west trending metamorphic belts from south to north are (1) South Palamau-Gumla-Ranchi-Purulia Belt; (2) Daltonganj-(North Palamau)-Dumka

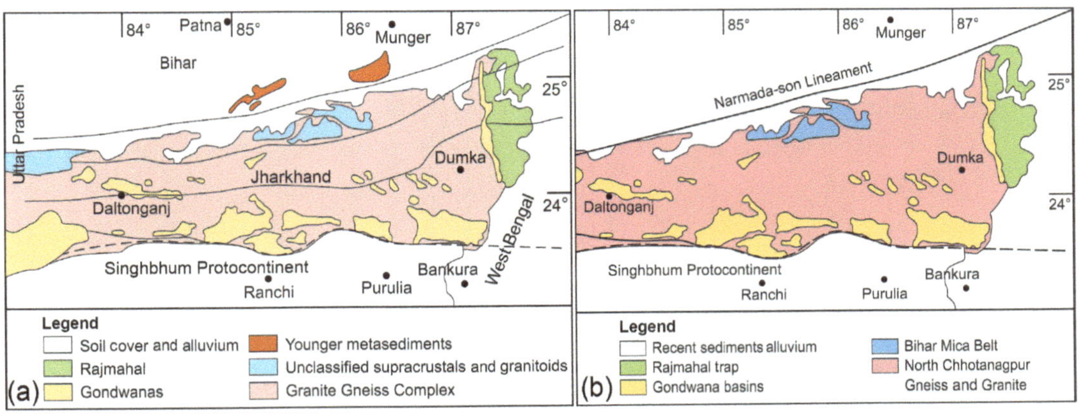

FIG. 7.1

(a) Tectonic division of the 'pristine' Chhotanagpur Gneissic Complex showing different belts. (b) The extent of the Rajmahal Protocontinent between Singhbhum Protocontinent in the south and Narmada-Son Lineament in the north. The eastern boundary is marked by the outcrops of the Rajmahal Trap.

Reproduced with modification from Mahadevan (2002) with permission.

Indian Shield. https://doi.org/10.1016/B978-0-12-809839-4.00007-2

Belt; (3) North Garhwa-Chatra-Girdih-Deogarh-Madhupur Belt; (4) unclassified supracrustals and granitoid, including the Bihar Mica Belt; and (5) Rajgir-Kharagpur Belt.

Out of the above stated tectonic blocks, the southernmost Belt 1 is now considered a part of the Singhbhum Protocontinent, while the Belt 5 in the north is included in the domain of the Vindhyan Basin in the north and west, separated from the southern belt by the Narmada-Son Lineament. The Belts 2, 3, and 4 constitute the Rajmahal Protocontinent. However, for the sake of convenience, the discussion on the Rajmahal Protocontinent would be divided into two belts: (1) North Chhotanagpur Gneissic Complex and (2) Bihar Mica Belt (Fig. 7.1b).

7.2 NORTH CHHOTANAGPUR GNEISSIC COMPLEX

The most dominant lithology of the North Chhotanagpur Gneissic Complex is granitic gneisses of different petrochemistry, geological age and structure, which along with granites of different ages constitutes the oldest basement in the Rajmahal Protocontinent. The basement rocks also host several supracrustal ensembles like mica schists, quartzite, calc-silicate rocks, amphibolite, and metamorphosed intrusive bodies like metadolerite-amphibolite and ultramafic rocks. The outcrops of gneiss-granite-bearing rocks show evidence of extensive partial melting, resulting in the development of 'migmatite' gneisses of different types. A greater part of the basement complex is covered by a wide variety of low-lying 'dome-shaped' gneissic outcrops generally showing northeast-southwest to east-west oriented foliation surfaces (Chatterjee and Sengupta, 1980).

Petrologically, the granite-gneiss bodies in the North Chhotanagpur Gneissic Complex show wide range of variation in their petrochemical character from normal granitic to quartz diorite (Mahadevan, 2002). The gneissic rocks are generally described by the different field terms like biotite gneiss, hornblende-biotite gneiss, hornblende gneiss, and banded gneiss (Chatterjee and Sengupta, 1980). In spite of extensive reconstitution, some of the gneissic rocks look very similar to the tonalite-trondjhemite-granodiorite gneiss described from the Archaean basement rocks in different parts of the Indian Shield (Fig. 7.2a). Granitic rocks in the North Chhotanagpur Gneissic Complex vary in texture

FIG. 7.2

(a) A typical banded tonalite-trondjhemite-granodiorite-type granite-gneiss from North Chhotanagpur Gneissic Complex. (b) Massive, coarse-grained granite revealing faint traces of relict foliation, which is indicative of its gneissic prehistory.

Reproduced from Chatterjee and Ghose (2011) with permission.

from fine- to coarse-grained porphyritic types. There is also a wide variation in the petrochemistry of these rocks from normal granite to granodiorite, tonalite, and quartz diorite. Some of nonfoliated leucogranite has reportedly formed from the anatexis of granulite (Mahadevan, 2002). In most other cases, however, faint trace relict foliation is detectable in the massive granites, which are indicative of their gneissic pre-history (Fig. 7.2b). Late intrusive nonfoliated granites invariably cut across the basement gneisses and the supracrustals (Rode, 1948).

Several bodies of supracrustal rocks occur within or skirting the granite-gneiss outcrops. Most of these bodies, mainly the metasediments, are in highly dismembered state because of successive deformation and plutonic intrusions. The most important components of the supracrustal ensembles include mica schist with linear bands of quartzite, along with calc-silicate rocks and carbonates, occurring in widely dispersed outcrops in the gneiss-granite terrane. Chatterjee and Sengupta (1980) have mapped some disjointed, complexly folded bodies of metasedimentary bodies within the gneiss-granite terrane in the northeastern part of the North Chhotanagpur Gneissic Complex (Fig. 7.3).

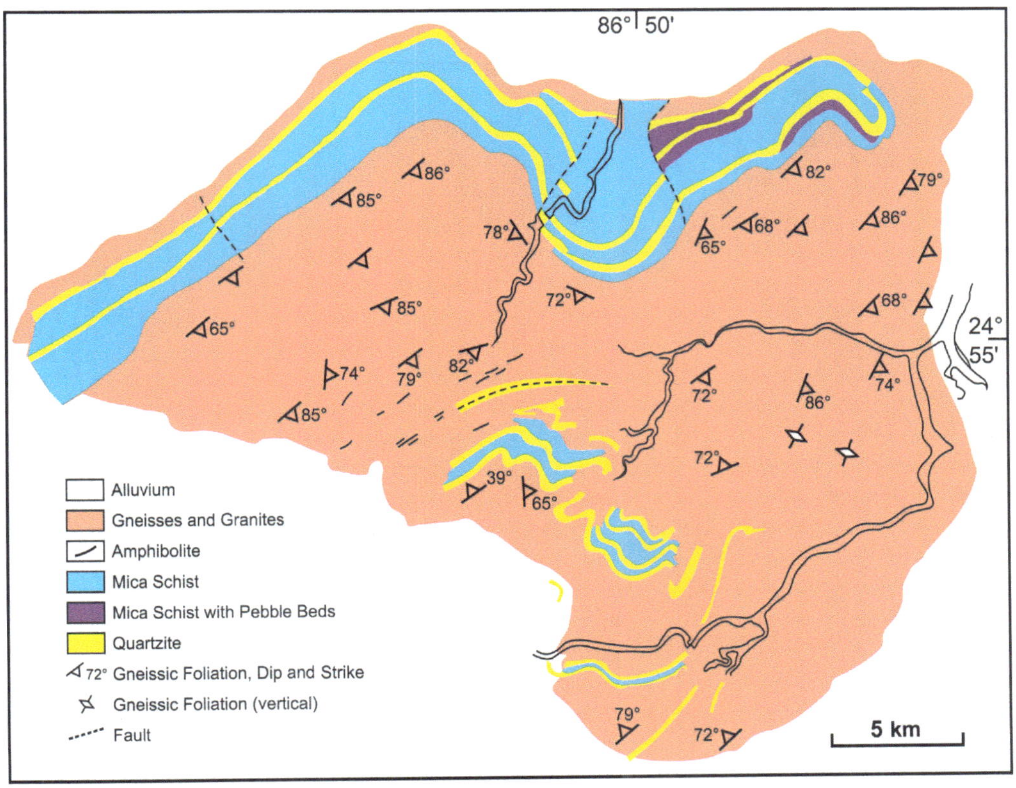

FIG. 7.3

Occurrence of deformed and disjointed bodies metasediments (mainly quartzite and mica schist) within granite-gneiss of different compositions from the Bhagalpur region in the northeastern part of the North Chhotanagpur Gneissic Complex. The supracrustal outliers have been repeatedly folded with the host gneissic rocks.

Reproduced from Chatterjee and Sengupta (1980) with permission.

In addition to the mica schists showing amphibolite facies metamorphism, there are sizeable outcrops of orthopyroxene-clinopyroxene-bearing pelitic granulites like Khondalite and charnockite that occur along with calc-silicate granulites of different types and composition in different parts of the North Chhotanagpur Gneissic Complex. The best known outcrops are reported from the Daltonganj (24°01′ and 84°06′) area west and Dumka (24°23′ and 87°18′).

In the Daltonganj and surrounding areas, thin bands of metasedimentary enclaves occur within granite-gneiss outcrops. These include graphite-sillimanite-bearing mica schists (khondalite), crystalline limestone/marble, and quartzite. Development of cordierite and andalusite in some pelitic granulites indicates their development due to the thermal metamorphism of the precursor rocks at the contact of intrusive granitic bodies.

The high-grade metasediments along with the concordant mafic intrusives commonly occur within the granite-gneisses in the Dumka and adjacent areas (Fig. 7.4). The metasediments in this belt include sillimanite-graphite-bearing pelitic schist, commonly described as khondalite, calc-silicate rocks, quartzite, and charnockite. The metasedimentary protoliths of charnockites is proved by the close association of sillimanite and cordierite-andalusite-bearing mica schists with the charnockitic rocks in the region.

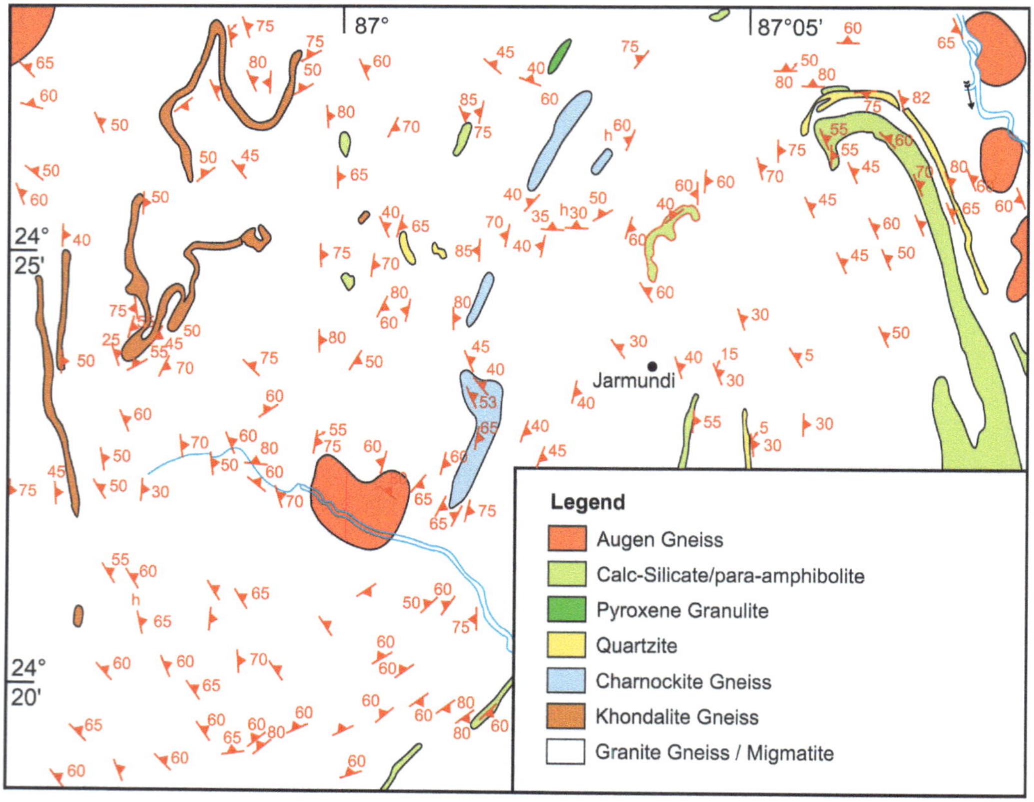

FIG. 7.4

Geological map of a part of the Santhal Parganas area, northwest of Dumka, Jharkhand.

The metabasic rocks represented by amphibolite, metadolerite, and metanorite occur as concordant bodies in the granite-granite terrane. The less deformed and metamorphosed bodies have retained their original texture and mineralogy. These metabasics show evidence of migmatization along their margins. Like the mica schists and quartzites, these metabasics have also shared much of the deformation history of the gneiss-granite hosts. Based on this evidence, it is suggested that these metabasics formed prior to the earliest deformation of the supracrustal rocks.

Anorthosite forms a large crescent-shaped outcrop near Dumka. There are also some minor bands of anorthosite bodies which occur as concordant bands within the gneiss-granite terrane. These bodies of anorthosite occur concordantly with the pyroxene granulite bodies of the area (Mahadevan, 2002).

An interesting feature is the occurrence of 'low-grade' metamorphic rocks having considerably different lithologic association than the known metasediments observed in the terrane (Mahadevan, 2002). Ball (1881), however, had earlier reported several such deposits in the northeastern part of the part of the Rajmahal Protocontinent. The most significant of these are the outcrops that occur close to Gore (23°58′ and 83°58′) and Biwabathan (23°55′ and 83°03′) described by Krishnan (1954). The Gore metasedimentary deposits have north northwest-south southwest trend and are traced for over 1 km in length (Fig. 7.5). The rock sequence includes calc-silicate and limestone overlying gneiss-granite-bearing basement rocks. The sedimentary sequence is in turn overlain by tremolite schist with or without magnetite. These rocks are intruded by bodies of later dolerite. The map pattern and the

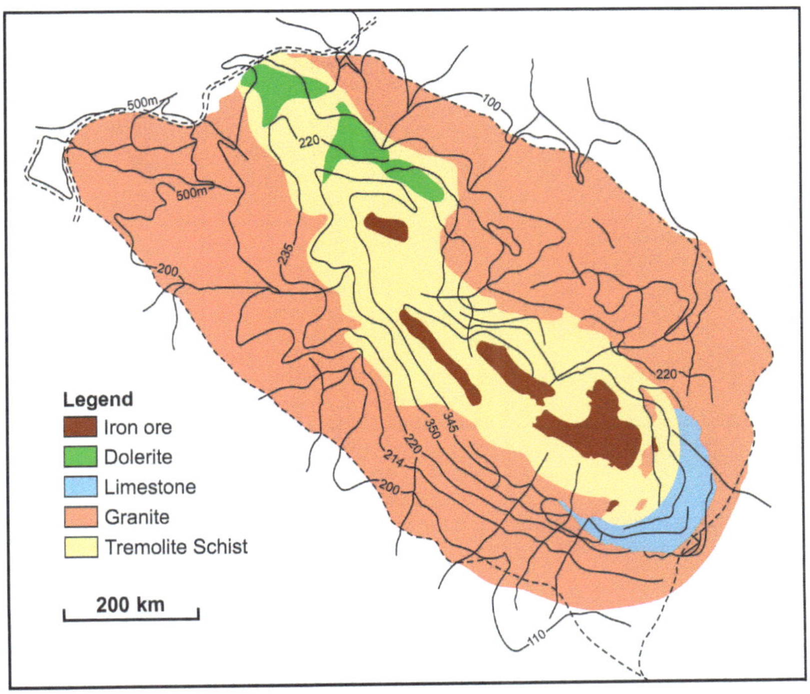

FIG. 7.5

Geological map of the Gore iron-ore deposit in Daltonganj.

Reproduced from Mahadevan (2002) with permission.

description of lithologies suggest that the total sequence represents a typical greenstone belt association comparable with those reported from the different Archaean greenstone terranes in the Indian Shield.

The structural complexity of the supracrustal assemblages and basement gneiss-granite sequence is clearly indicated by the map patterns of the Bhagalpur and Dumka areas (Figs. 7.3 and 7.4). Three phases of folding is reported from these belts that affected both the gneiss-granite basement and supracrustal assemblages. The early two phases of folds in the supracrustals are isoclinal in geometry, producing hook-shaped outcrop pattern. The youngest deformation is represented by open folds having vertical to steeply inclined axial surfaces. The complexity of deformation in the basement gneissic rocks is illustrated in the development of polyphased folding in these rocks in several parts of the terrane (Fig. 7.7a and b).

7.3 BIHAR MICA BELT

The Bihar Mica Belt includes a sequence of arenaceous and pelitic rocks, interbanded with amphibolites. The entire ensemble is intruded by granites of different types, followed by younger mafic dykes. The geological map prepared by Ramachandran and Sinha (1992) shows an extremely complex outcrop pattern produced by multiphase folding deformation of cover succession in the basement gneiss (Fig. 7.6).

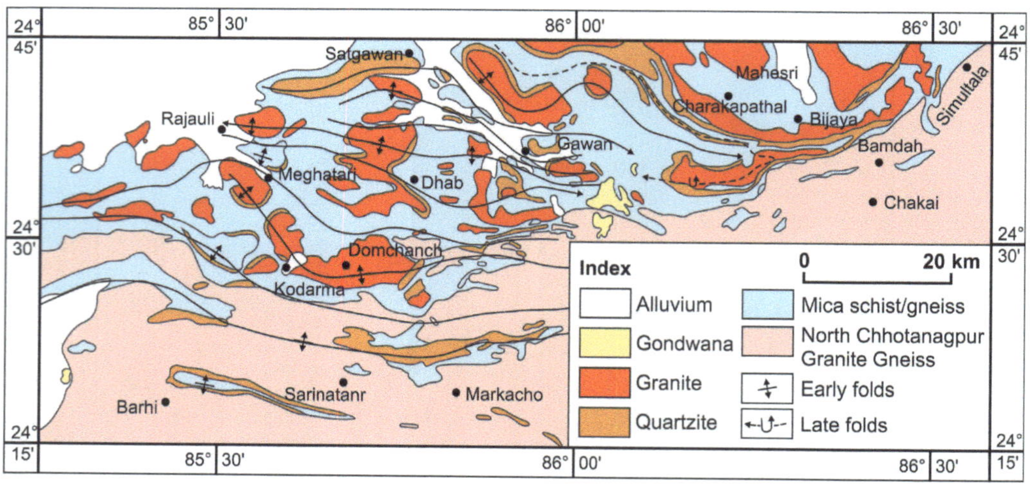

FIG. 7.6

Illustrations of multiphase folding deformation in the basement gneiss in the Deogarh-Gomo region Bihar.

Modified from Ramachandran and Sinha (1992).

The dominant lithology in the well-known mica-producing belt in the Rajmahal Protocontinent includes mica schist, quartzite, calc-silicate rocks, amphibolite, and subordinate carbonate rocks. The entire lithological ensemble is grouped under the Kodarma Group (Mahadevan, 2002). The sedimentary depositional character of the different lithologies is indicated in the preservation of primary sedimentary structures, especially in quartzite. Small localized bodies of conglomerate have been reported at different places in spite of considerable dismemberment due to the intense and repeated deformation and metamorphism. The calc-silicate rocks containing zoisite, diopside, and hornblende are subordinate to the pelitic assemblages. The pelitic schists show upper amphibolite facies metamorphism and have undergone partial melting yielding a variety of migmatitic gneisses. Mahadevan (2002) reported occurrence of granitic rocks in the mica belt that occur as intrusive plutons of different shapes and sizes (Fig. 7.6). Some of the dome-shaped bodies are described as the dome gneisses. The core of the granite domes comprises peraluminous monzonite containing enclaves of tonalities and other supracrustal rocks (Mahadevan, 2002). Pegmatites intruding into pelitic schists host important mica deposits. Some of the late-REE-enriched pegmatites are potential source of radioactive minerals.

The Bihar Mica Belt rocks show complex deformation pattern involving three phases of ductile deformation and hosts of minor faulting. In small scale Sarangi and Mohanty (1998) has reported occurrence of a series of sheath folds in the metasedimentary rocks of the Bihar mica belt (Fig. 7.7a and b). According to Misra and Dey (2002), the pegmatites intruding the metasedimentary rocks of the Bihar Mica Belt were emplaced after the folding deformation of the host metasediments and are not strictly time-correlative with the emplacement of the granite plutons. The necessary fluids for the formation of pegmatites were generated by muscovite dehydration reaction of the muscovite-bearing metasedimentary country rocks during metamorphism.

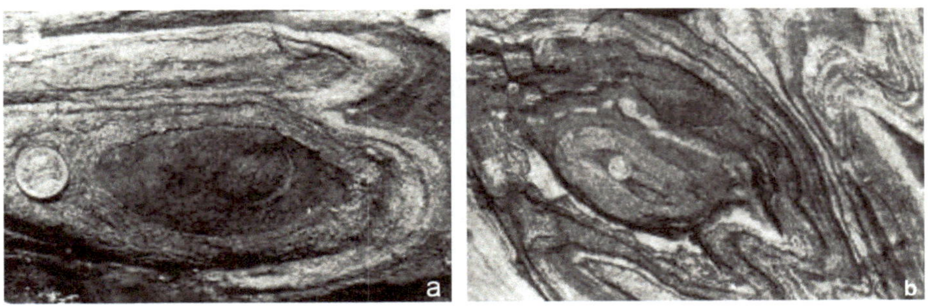

FIG. 7.7

(a and b) Examples of sheath folds in zones of ductile deformation in the Bihar Mica Belt.

Reproduced from Sarangi and Mohanty (1998).

7.4 SUMMARY OF GEOLOGICAL EVOLUTION OF THE RAJMAHAL PROTOCONTINENT

The basement granite-gneiss in the North Chhotanagpur Gneissic Complex includes scattered bodies of tonalite-trondjhemite-granodiorite gneiss, traditionally recognized as 'TTG gneiss' or 'grey gneiss'. Such a rock association is common in different Palaeo-/Mesoarchaean basement assemblage in different parts of the Shield areas. The field description of the occurrence of the tonalite-trondjhemite-granodiorite gneiss by different authors (Chatterjee and Ghosh, 2011; Mahadevan, 2002; Rode, 1948) seems to confirm the presence of Archaean basement rocks in the Northern Chhotanagpur Gneissic Complex, in spite of the absence of any precise isotope/geochronological data. Further confirmation about the Archaean status of the gneiss-granite basement rocks is the occurrence of typical greenstone belt association in the North Chhotanagpur Gneissic Complex. No less important is the description of the basement rocks as biotite gneiss, hornblende-biotite gneiss, and hornblende gneiss (possibly representing diorite gneiss or quartz diorite) by different authors (Chatterjee and Sengupta, 1980; Mahadevan, 2002).

Another feature that missed the attention of different workers is the reference to the localized presence of 'sillimanite patches' amidst the biotite granite-gneiss in the terrane (Mahadevan, 2002). Such local occurrences of sillimanite dominant rocks in the gneiss-granite terrane have been recognized as the Precambrian palaeosol (Golani, 1989). In the Aravalli Protocontinent, the occurrence of palaeosol has been recorded as the marker of the Archaean-Proterozoic boundary (Roy and Jakhar, 2002).

The geological map prepared by Chatterjee and Sengupta (1980) from the area around Jamua-Kakwara-Bhitia belt in Bhagalpur district, Bihar, is very educative in understanding the geological evolution of the region. The complexity of deformed and polymetamorphosed detached bodies of metasediments comprising the shelf facies sedimentary sequence represented by metamorphosed sand-shale-carbonate association suggests its evolution much before the emplacement of massive granites that intruded all the pre-existing rocks.

Chatterjee and Ghosh (2011), based on studies of monazite and xenotime studies suggested crystalline age of 1694 and 1704 Ma age for the undeformed and unmetamorphosed porphyritic granite from the Gaya region. This age matches well with the Rb/Sr whole rock age of 1741 ± 65 Ma suggested by Ray Barman et al. (1994). It may, therefore, be logically assumed that the deformed and metamorphosed metasediments of the North Chhotanagpur Gneissic are much older than 1700 Ma. Considering this age data, an early Palaeoproterozoic age of the metasedimentary cover rocks becomes a natural conclusion.

The metamorphic state of the granite-gneiss complex that constitutes the basement for the overlying Palaeoproterozoic succession varies between lower and upper amphibolite facies (Mahadevan, 2002). The granulite facies metamorphism noted locally both in the basement and in the cover succession must have resulted due to the localized thermal rise in the region. The timing for this thermal rise and the consequent granulite facies metamorphism is unknown but must have post-dated the deposition, deformation, and low-grade metamorphism of the Palaeoproterozoic supracrustal rocks. There is hardly any field data that indicate the timing of the events like granulite facies metamorphism, emplacement of granitoids, and the formation of the mica-bearing pegmatites.

The three commonly suggested ages from the Northern Chhotanagpur Gneissic Complex are (i) c.1.7, (ii) 1.55, and (iii) 1.1–0.96 Ga (Pandey et al., 1986; Mallik et al., 1991; Ray Barman et al., 1994; Maji et al., 2008; Chatterjee et al., 2008; Chatterjee and Ghosh, 2011). These ages mark the three successive thermal events in the entire Rajmahal Protocontinent as indicated below:

~1.7 Ga—granulite facies metamorphism
~1.55 Ga—emplacement of massive porphyritic granite
~1.1–0.96 Ga—emplacement of the youngest granite and development of mica pegmatite

REFERENCES

Ball, V., 1881. Geology of the districts of Manbhum and Singhbhum. Mem. Geol. Surv. India, XVIII, 150 p.

Chatterjee, N.J., Ghosh, N., 2011. Extensive Early Neoproterozoic high-grade metamorphism in North Chhotanagpur Gneissic Complex of the Central Indian Tectonic Zone. Gondwana Res. 20, 362–379.

Chatterjee, S.R., Sengupta, D.K., 1980. Structural and petrological evolution of the rocks around Jamua-Kakwara-Bhitia, Satpura Orogeny, Bhagalpur district, Bihar. J. Geol. Soc. India 21, 171–183.

Chatterjee, N., Crowley, J.L., Ghose, N., 2008. Geochronology of the 1.55 Ga Bengal anorthosite and Grenvillian metamorphism in the Chotanagpur gneissic complex, eastern India. Precambrian Res. 161, 303–316.

Golani, P.R., 1989. Sillimanite-corundum deposits of Sonapahar, Meghalaya, India: a metamorphosed Precambrian Palaeosol. Precambrian Res. 43, 175–189.

Krishnan, M.S., 1954. Iron ore, iron and steel. Series A. Econ. Geol. Bull Geol. Surv. India, 116–120.

Mahadevan, T.M., 2002. Geology of Bihar and Jharkhand. Geol. Soc. India, Bangalore. 563 p.

Maji, A.K., Goon, A., Bhattacharya, A., Mishra, B., Mahato, S., Bernhardt, H.-J., 2008. Proterozoic polyphase metamorphism in the Chhotanagpur Gneissic Complex (India), and implication for trans-continental Gondwanaland correlation. Precambrian Res. 162, 385–402.

Mallik, A.K., Gupta, S.N., Ray Barman, T., 1991. Dating of early Precambrian granite-greenstone complex of the Eastern Indian Precambrian shield with special reference to the Chotanagpur granite gneiss complex. Rec. Geol. Surv. India 124, 20–21.

Misra, S., Dey, S., 2002. Bihar Mica Belt Plutons—an example of post-orogenic granite from eastern Indian Shield. J. Geol. Soc. India 59, 363–377.

Pandey, B.K., Gupta, J.N., Lall, Y., 1986. Whole rock and Rb–Sr isochron ages for the granites from Bihar mica belt of Hazaribagh, Bihar, India. Indian J. Earth Sci. 12, 157–162.

Ramachandran, S., Sinha, R.P., 1992. Pegmatites of the Bihar mica belt in relation to the Chhotonagpur granite gneiss and their columbite-tantalite potential. Indian J. Geol. 64, 276–283.

Ray Barman, T., Bishui, P.K., Mukhopadhyay, K., Ray, J.N., 1994. Rb–Sr geochronology of the high grade rocks from Purulia, West Bengal, and Jamua-Dumka sector, Bihar. Indian Miner. 48, 45–46.

Rode, K.P., 1948. On the charnockite rocks of Palamau, Bihar, India. Schweiz. Mineral. Petrogr. Mitt. 28, 288–302.

Roy, A.B., Jakhar, S.R., 2002. Geology of Rajasthan: Precambrian to Recent. Scientific Publishers (India), Jodhpur. 421 p.

Sarangi, S., Mohanty, S., 1998. Structural studies in the Chhotanagpur Gneissic complex near Gomo, Dhanbad, Bihar. Indian J. Geol. 70, 73–80.

CHAPTER 8

BUNDELKHAND PROTOCONTINENT

8.1 INTRODUCTION: EXTENT AND GEOLOGICAL CONSTITUENTS

The Bundelkhand Protocontinent occurring between the Aravalli Protocontinent in the west and southwest and the Narmada-Son Lineament in the south and southeast constitutes a distinctive Precambrian crustal block in the northern part of the Peninsular Indian Shield. The major Precambrian constituents of the Bundelkhand Protocontinent comprise the semi-circular granitoid-dominated terrane (described as the Bundelkhand Granite) and the Vindhyans (the Vindhyan Supergroup) that virtually encircle it except in the alluvial covered northern part. Much of the Precambrian geology of the Protocontinent is also obscured because of the cover of Deccan Volcanics in the southern and southwestern part (Fig. 8.1). Some isolated patches of the Vindhyan rocks occur as 'inliers', mostly in the southeastern parts, which help in defining the extent of the Vindhyan Basin in the south.

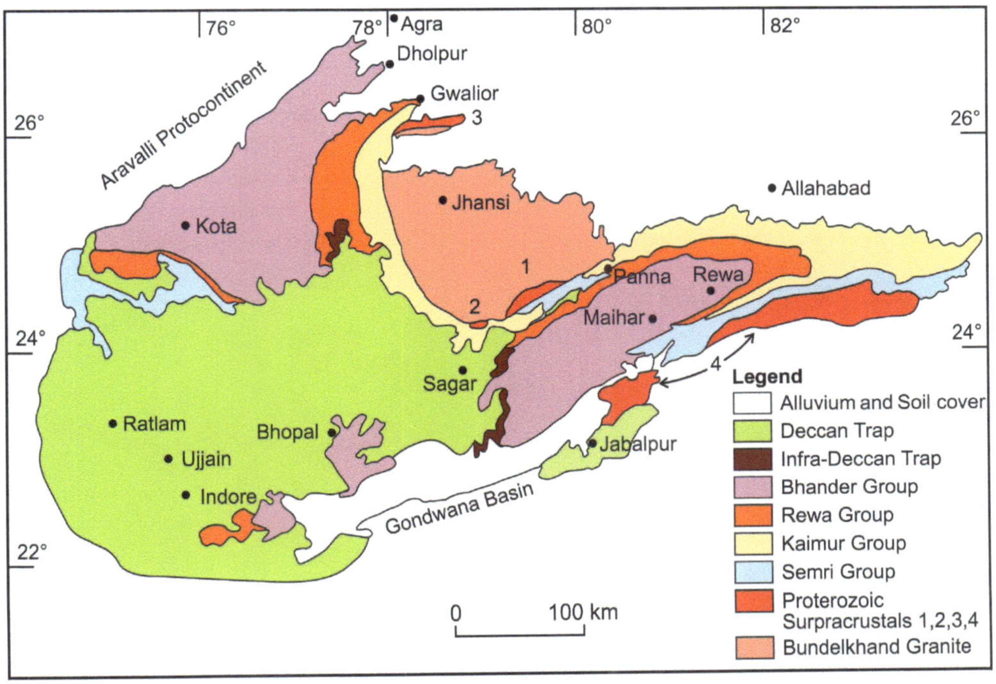

FIG. 8.1

Geological map of the Bundelkhand Protocontinent: (1) Bijawar Basin, (2) Sonrai Basin, (3) Gwalior Basin, and (4) Pre-Vindhyans.

Indian Shield. https://doi.org/10.1016/B978-0-12-809839-4.00008-4

BOX 8.1

Geologically, an 'inlier' is defined as an area where older rocks are found to be surrounded by younger rocks. Inliers are typically formed by the erosion the of the overlying younger rocks uncovering a limited exposure of the older underlying rocks. Conversely, an 'outlier' is an area of younger rock completely surrounded by older rocks.

8.2 GEOLOGY OF BUNDELKHAND GRANITE

The dominant lithological component of the 'Bundelkhand Granite' is the granitoids of diverse composition and textural characters. Apart from that, the terrane includes outcrops of some volcano-sedimentary rocks representing dismembered greenstone belts along with the isolated outcrops of granite-gneiss that constitutes the oldest Archaean basement (Singh, 2012; Singh and Slabunov, 2016) (Fig. 8.2). The dominantly granitoid terrane is bounded both in the northwest and in the southeast by two major sedimentary basins named the Gwalior and Bijawar Basins, respectively. The Sonrai Basin occurring south of Girar (Fig. 8.2) in all likelihood occurs in continuation of the larger Bijawar Basin in the northeast. The geological depiction of the Bundelkhand Granite would be incomplete without the mention of two other rather 'unique' features that also characterize the terrane. These are (i) the northeast-southwest trending Quartz reefs and (ii) the WNW-ESE trending mafic dykes cross-cutting the reefs described latter.

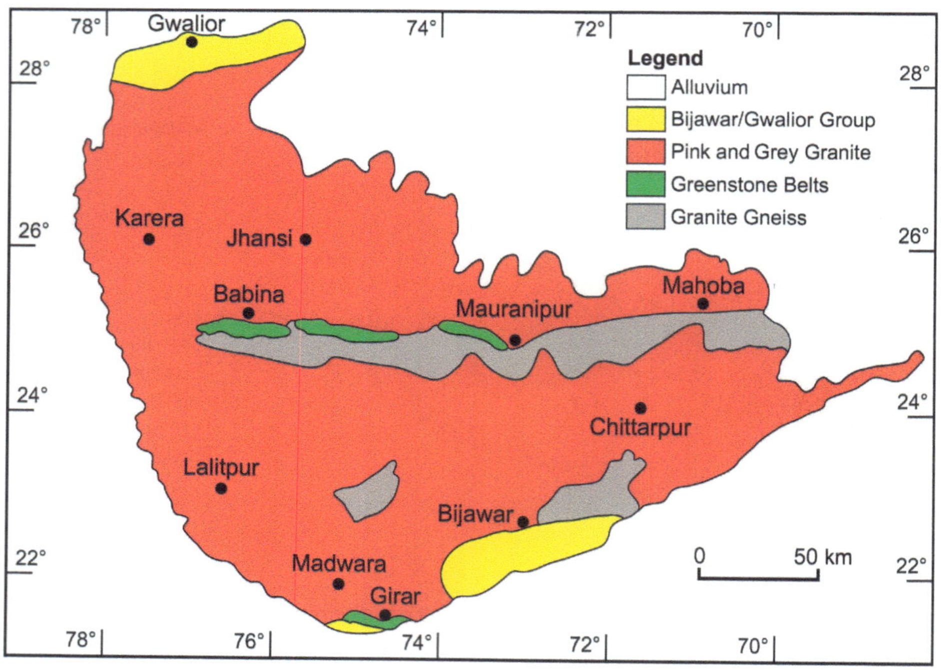

FIG. 8.2

Geological map of the Bundelkhand Craton showing distribution of different lithotectonic assemblages.

BOX 8.2

In the present discussion, the term 'Bundelkhand Granite' is used for the dominantly Archaean semi-circular crustal block in preference to the frequently used misnomer terms like 'Bundelkhand massif' or 'Bundelkhand granite/gneiss'.

8.2.1 BANDED GRANITE GNEISS: THE BASEMENT COMPLEX

The oldest rocks recorded in the Bundelkhand Granite also include a group of granite gneisses of diverse composition and character, which occur as slivers, rafts, and enclaves within it's outcrops. The most characteristic feature in these rocks is the gneissic banding, which in certain instances show complex deformation features. Large, linear, E-W trending outcrops of banded granite-gneisses with s series of mafic enclaves occur in different parts of the massif around Mahoba, Kuraicha, Babina, and Karera. The enclaves of the gneiss rocks of various dimensions also occur within the younger bodies of granitoids (Mondal et al., 2002). Though occur mainly along east-west running bands within the younger granitoids, the foliation trend in the gneissic rocks is generally in the WNW-ESE direction. Apart from occurring in the east-west running bands in the central part of the Bundelkhand Granite, the banded granite-gneisses also occur as small irregular-shaped enclaves in the younger granitoids.

Very little information is available on the petrology and the petrochemical character of the Bundelkhand granite-gneisses. However, the general impression one gets from the description of the rocks is that these are petrologically similar to the trondjhemite-tonalite-granodiorite gneiss that commonly occur in the Archaean basement complexes world over. Some of the dark-coloured varieties represent quartz diorite (Fig. 8.3a and b). In general, the typically banded Bundelkhand granite-gneiss shows highly fractionated REE patterns with HREE and Yb depletion showing no significant Eu anomaly. These petrochemical characters compare well with the rocks described from the different Archaean terranes (Mondal, 2010).

FIG. 8.3

(a and b) Outcrops of the typical banded granitic gneiss occurring within the Bundelkhand Granite.

Image courtesy: (a) Asima Saikia and (b) Surya P. Singh.

There are reported occurrences of some high-aluminous rocks along with clinopyroxene-bearing amphibolites in the basement gneisses. The common mineral assemblage of the high-aluminous metasediments includes K-feldspar, plagioclase, biotite, cordierite, sillimanite, ±garnet, and quartz (Singh and Dwivedi, 2015). The presence of pockets of such rocks in a milieu of tonalite-trondjhemite-granodiorite-bearing gneisses is a common feature in the different Archaean basement rocks of the Indian Shield. It may be suggested that the occurrence of the high-aluminous sediments in the basement association represents the localized zones of palaeosols that formed on top of the exposed basement rocks, and formed and metamorphosed at different times along with the associated rocks. The close association of the high-aluminous metasediments with the 'quartz-free Fe, and Na-poor high-Mg schists' has been reported from the Bundelkhand Granite by Saha et al. (2016). The association of such unusual metasedimentary rock association may be explained by assuming these as palaeosols drawing sediments from two different sources, the granitic and the Mg-rich ultramafic rocks.

The $^{207}Pb/^{206}Pb$ zircon ages of the banded gneisses from the Kuraicha region yielded 3297 ± 8 and 3270 ± 8 Ma ages (Mondal et al., 2002). A mafic enclave within the banded gneiss from the adjacent Babina area yielded 3249 ± 5 (Mondal et al., 2002). Kaur et al. (2014) based on LA-ICP-MS study using the combined U–Pb and Lu–Hf isotope of zircon from the similar gneissic rocks suggested 3551 ± 6 and 3551 ± 30 Ma ages. A little older U–Pb and Lu–Hf isotope age of 3.59 Ga is reported by Saha et al. (2016) from the gneisses in the nearby area.

8.2.2 GREENSTONE BELT ASSOCIATIONS

The Bundelkhand Granite terrane includes some dismembered greenstone belts in different parts. The important ones occur at (i) the Babina-Kuraicha-Mauranipur-Mahoba belt in the central part, and (ii) the Madawra to Baraitha belt through Girar (Fig. 8.2) in the southern part of the Bundelkhand Granite. The dominant metasedimentary constituent of these greenstone belts is the banded iron formation (Fig. 8.4a) that occurs in association with amphibolite and ultramafic schist along with

FIG. 8.4

(a) Outcrop of banded iron formation from the Babina region. (b) Fuchsite-bearing quartzite from the greenstone belt near Jhansi.

(a) Image courtesy: Surya P. Singh.

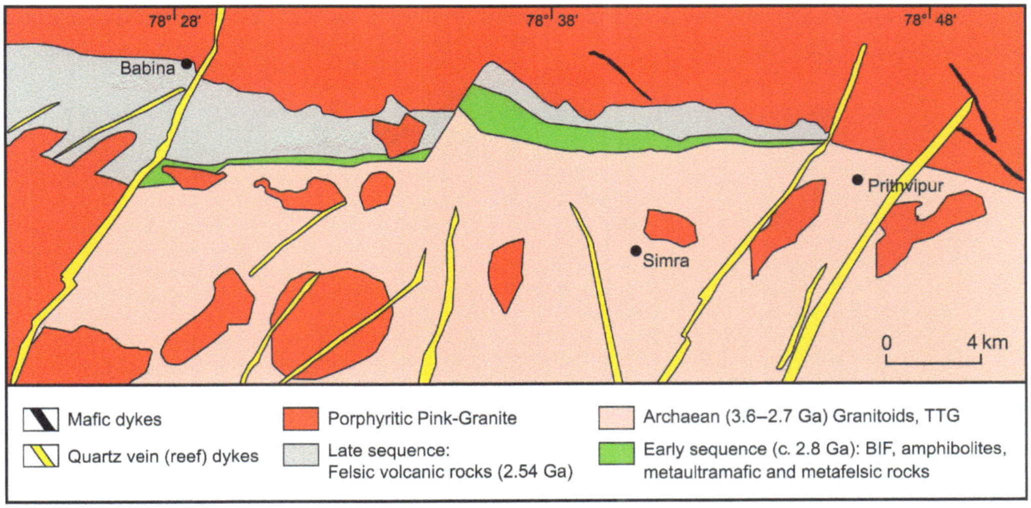

FIG. 8.5

Geological map between Babina and Prithvipur, southeast of Jhansi showing disposition of rock assemblages.

Reproduced from Singh and Slabunav (2015) with permission.

quartzite (generally fuchsite bearing) (Fig. 8.4b), metapelite, and marble (Singh and Slabunov, 2016). The ultramafic rocks include peridotite, dunite, pyroxenite, and gabbro in close association with metabasics. Veinlets of chrysotile asbestos are noted in the ultramafic rocks. At places, smaller bodies of felsic volcanics of dacite-rhyolite composition have also been reported (Fig. 8.5) (Singh and Slabunov, 2016).

The greenstone belts show a complex pattern of deformation. The most notable ones are observed in the banded iron formation that generally shows the presence of tight to isoclinal folds (Fig. 8.4a). The banded iron formation shows tectonic contacts with the sheared mafic, ultrabasic rocks, and the felsic volcanics (Singh and Slabunav, 2015).

The Babina-Kuraicha-Mauranipur-Mahoba greenstone belt generally shows an east-west linear trend (Fig. 8.2) bounded between two parallel shear zones. Even the contact between different components of the greenstone belts appears tectonic, marked by Ductile Shear Zones.

The U–Pb zircon age of early felsic volcanics in the Babina area is 2813 ± 20 Ma (Singh and Slabunav, 2015). Similarly, ~2.8 Ga age is also reported for the banded iron formation and amphibolite of the Babina-Kuraicha-Mauranipur-Mahoba Belt (Singh and Slabunav, 2015). On the other hand, there are some monazite and zircon ages that vary in age between 2.73 and 2.54 Ga in the Babina area (Saha et al., 2011; Singh and Slabunav, 2015). The younger ages could have resulted from the overprinting in the younger thermal events.

8.2.3 YOUNGER MASSIVE GRANITOIDS

The major component of the Bundelkhand Granite is the granitoids of different composition, which have been studied in detail by several workers in recent years (Basu, 1986; Mondal and Zainuddin, 1996; Singh et al., 2007; Singh and Slabunov, 2016). Described as a vast plutonic complex of 'batholithic dimension'

(Ramakrishnan and Vaidhyanadhan, 2008), the region is better known as a 'granite country' marked by rounded outcrops, ranging from small mounds and tors of granite to large dome-shaped hillocks (Fig. 8.6a).

FIG. 8.6

(a) The granite country is marked by rounded outcrops, ranging from small mounds and tors of granite.
(b) The polished slabs of pink biotite granite from the Bundelkhand Granite.

(a) Image courtesy: Erfan Mondal.

Petrologically, a variety of granitic rock is observed in the region, which varies widely in terms of colour, grain size, and physical character. Based mainly on the dominant petrological character, all the components of the massive granitoids have been classified into four different petrologic types: (i) hornblende granite, (ii) biotite granite, (iii) hornblende-biotite granite, and (iv) leucogranite. The leucogranite, apart from occurring in diverse shades between grey and pink, shows wide variation in grain size from fine to very course. Some of these show coarse porphyritic character in which large grains of feldspar crystals (usually pink-coloured k-feldspar) occur within a fine-grained matrix. The uniform-textured, medium-grained pink varieties of biotite granite are used as building stones (Fig. 8.6b).

Based on the cross-cutting field relationship, the hornblende-bearing granitoids have been classified as the oldest granitic bodies (Mondal and Zainuddin, 1996). The youngest ones are represented by the coarse porphyritic leucogranites. Compositionally, the granitoids range between quartz diorite and 'normal' k-feldspar-bearing granite. The petrochemistry of these granitoids indicates a calc-alkaline trend showing meta-aluminous character (Mondal, 2010).

Barring local shear-zone-related deformation and some fluid-induced hydrothermal changes, all the granitic bodies are undeformed and unmetamorphosed in character, often containing small, angular enclaves of older rocks (Fig. 8.7a). The presence of faint traces of planar fabric defined by the traces of dark and lighter shades even within the massive varieties. Such a feature, in all probability, points to their formation from gneissic protoliths (Fig. 8.7b). The confirmation for this comes from the field evidence of transformation of granitic gneisses into coarse porphyritic granite (Basu, 1986).

Studies on $^{207}Pb/^{206}Pb$ zircon geochronology (Mondal et al., 2002) indicated 2516 ± 4 Ma age of the hornblende-bearing quartz diorite. Virtually similar 2521 ± 7 Ma has been ascribed to the age of the biotite-bearing granitoids (Mondal et al., 1998). The leucogranitoids from the Bundelkhand Granite yielded 2492 ± 10 Ma age. The emplacement of these end-Archaean, undeformed, and unmetamorphosed granitoids marks the cratonization of the Archaean Crust in the region (Mondal, 2010).

FIG. 8.7

(a) Coarse, massive 'pink granitoid' enclosing angular blocks of foliated amphibolite. (b) Massive granitoid showing relic traces of gneissic banding marked by pink- and dark-coloured parallel bands.

(a) Image courtesy: Erfan Mondal.

8.2.4 QUARTZ REEFS AND DOLERITE DYKES

The quartz reefs and the dolerite dykes are the two key features that need special mention in understanding postcratonization tectonics of the Bundelkhand Granite (Fig. 8.8). The two features show cross-cutting relationship making an obtuse angle of about 105° in the easterly direction. Such a strict geometric relationship needs explanation though there is definite age difference between the two. In view of that, the two features are described together.

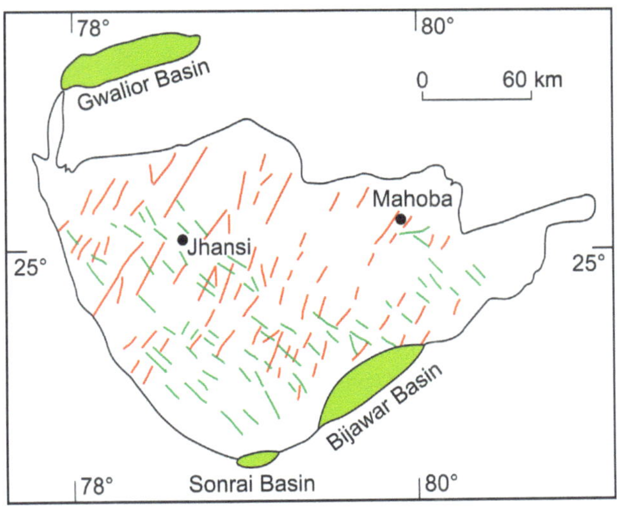

FIG. 8.8

Generalized geological map of the Bundelkhand Craton showing disposition of quartz reefs and the dolerite dykes.

The quartz reefs, according to Pascoe (1973), form a series of wall-like topographic feature abruptly rising to the formidable heights, at places about 175 m above the ground level. Basu (1986) has recorded about

11 major reefs in the entire region, spaced at 12.5–19 km apart. The longest reef runs for about 100 km. There are numerous smaller bodies of quartz reefs that are spaced at an average of 6 km. The average trend of the reefs is N35°E–S35°W. The variation in trend barring a few isolated instances of coalescing and bifurcation is hardly significant. The veinlike reefs are composed almost essentially of coarse-grained quartz having diffused grain boundaries. Less than 5% of the composition is made of impurities like hematite and pyrite. The veins show virtually monotonous grey-white colour. Though generally strain-free, evidence of shearing is noted along the margin of the veins with the host bodies at some places. Field observations indicated some sinistral displacement (Basu, 1986; Singh et al., 2007) of the cross-cutting host lithologies.

The dolerite dykes like the quartz reefs represent another common feature in the Bundelkhand Granite. Locally, these are so numerous that some authors described these intrusives as the 'mafic dyke swarms'. The dykes, unlike the quartz reefs, do not form local topographic highs but generally occur in levelled ground (Fig. 8.9a).

Like the quartz reefs, most of the dykes are discontinuous and run for variable lengths. Many dykes instead being a single band occur in a number of parallel bands having different widths (Fig. 8.9b). The maximum width of the dykes is around 45 m in southwestern corner of the Bundelkhand Granite. The dolerite dykes generally show dark, greyish green colour in the exposed surfaces. Like quartz reefs, the dykes also maintain a very persistent trend. The average trend of these dykes is N40°W–S40°E. At some localities, the dolerite dykes are seen to cut across the quartz reefs. The reverse relationship has not been observed in the field. The information indicates that the quartz reefs were formed earlier than the intrusion of the dolerite dykes.

FIG. 8.9

(a) Scattered boulders of dolerite occurring almost at the ground level near Lalitpur-Mehrauli road near Chapru. (b) Several set of thin, parallel dolerite dykes intruding into Bundelkhand Granite.

(a) Reproduced from Kaur et al. (2016) with permission. (b) Image courtesy: Erfan Mondal.

Petrologically, most of the dolerite dykes are typically tholeiites, and quartz normative types (Mallikharjuna Rao et al., 2005). The ultramafic types are komatiite or basaltic komatiite in composition and show olivine normative character. $^{40}Ar/^{39}Ar$ determinations of the dolerite suggest two phases of dyke activity at c.2150 and 2000 Ma. A polyphase evolution of the mafic-ultramafic dykes is also supported by the Palaeomagnetic study of these dykes (Mallikharjuna Rao et al., 2005).

Some dykes, rarely though, show different orientations like the prominent Mahoba dyke that runs for about 20 km in the ENE-WSW direction. Petrochemically, the ENE-WSW-oriented Mahoba dyke is said to be similar to the WNW-ESE trending dykes. Detailed field studies, however, indicated occurrence of dykes of diverse petrochemical character irrespective of their orientation. Palaeomagnetic studies indicated occurrences of number of phases of mafic-ultramafic dyke intrusion in the Bundelkhand Craton (Mallikharjuna Rao et al., 2005).

8.2.5 YOUNGER SUPRACRUSTALS

8.2.5.1 Bijawar Group

The Bijawar Basin is a ENE-WSW trending, about 100km long, narrow basin in the southeastern part of the Bundelkhand Granite (Fig. 8.2). The Bijawar sediments with some volcanics were deposited on the eroded surfaces of the gneiss-granite complex of the Bundelkhand Granite. At Girar, in the southern part of the Granite body, however, the Bijawar sediments lie over the banded iron formation of the greenstone assemblage (Basu, 1986). The Bijawar rocks that generally have gently dipping beds locally show folding and faulting. The basal succession starts with quartz pebble conglomerate overlain by thick pile of mafic lava. The next overlying unit includes stromatolite-bearing dolomitic limestone. The youngest succession consists of ferruginous sandstone and shale. Some lenses of conglomerate occur in the upper section. Kumar et al. (1990) have suggested the following stratigraphic succession of the Bijawar Group:

Karri Sandstone
Hirapur Phosphorite
Malehra Chert Breccia
Pukhra Sandstone
Dargawan Sill
Bajno Dolomite
Bhusar Basalt
Kawar Conglomerate

The suggested stratigraphic framework may need revision in view of the possibility that the repetition of beds shown in succession might be fault controlled, and not due to repetitive depositional events (Fig. 8.10).

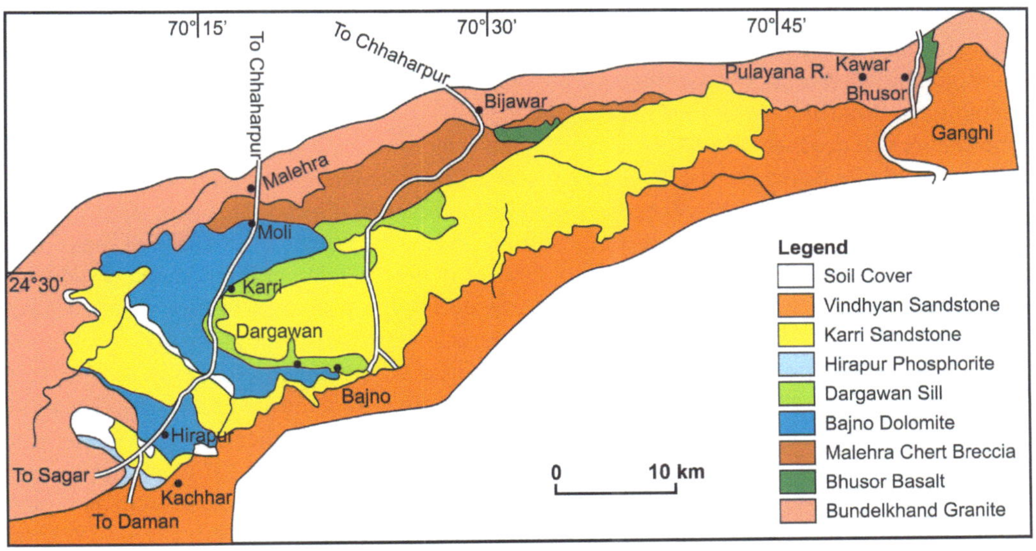

FIG. 8.10

Geological map of the Bijawar Basin.

Compiled from different sources.

No well-constrained data are available for the Bijawar rocks except the age data for the Bijawar basalt (Dargawan Sill) that indicate 1989 ± 71 Ma. The Kurat lava indicates an age of 1691 ± 16 Ma. However, in spite of high error factor, the age of the Bijawar Group is presumed to be between 1700 and 1800 Ma (Haldar and Ghosh, 2000).

8.2.5.2 Gwalior Group

The Gwalior Basin, occurring in the northwestern part of the Bundelkhand Craton (Fig. 8.2), is also a narrow basin unconformably overlying the Bundelkhand Granite. The Gwalior Group consists of a basal sequence (Par Formation) comprising grit and conglomerate which grade upward into glauconitic sandstone and layers of stromatolitic limestone. Overlying the Morar Formation occurs siliceous and ferruginous shale, chert, banded red jasper, limestone, and sandstone, the latter having some volcanic contributions. In addition, several layers of tholeiite basalt and grabboic sills are present in the Morar Formation.

There are some ill-constrained Rb–Sr isotope ages for the Gwalior Group (Geological Survey of India, 2004). The Rb–Sr isochron age of the Gwalior Lava is 1830 ± 200 Ma. The combined Rb–Sr and K–Ar studies indicate ~1800 Ma ages of the mafic dykes (Geological Survey of India, 2004). Age wise, therefore, the Bijawar and the Gwalior Groups appear to be broadly contemporaneous.

8.2.6 SUMMARY OF LITHO-TECTONIC HISTORY AND TECTONOSTRAIGRAPHIC FRAMEWORK

Traditionally known as the 'Bundelkhand massif' of granitoid rocks, the Precambrian history of this crustal block extends over a span of over 2000 million years. However, in a region where the outcrops are scanty barring some localized parts, the building of tectonostratigraphic relationship is truly an arduous task. However, some important field clues have been recorded by Singh et al. (2007) that helped in building up the mutual relationships between different lithotectonic units of the Bundelkhand Granite (Fig. 8.11).

Using the information obtained from the field data and the available isotope data, the following tectonostratigraphic framework of the different components could be built up for the different components constituting the Bundelkhand Granite (Table 8.1).

The feature that needs some additional explanation is the geometric relationship between the quartz reefs and the dolerite dykes and the involved tectonics. The most interesting fact about the occurrence of the two features is the geometric (angular) relationship between the two. This is possible only if the two features have formed along two sets of cross joints during the regime of extensional stress. A very similar feature is reported from the Singhbhum Granite in the Singhbhum Protocontinent, where two set of dykes were emplaced following cross factures during the post-cratonization phase. However, the difference is that, in the case of Singhbhum, the intrusion that took place along two sets was of the same petrologic rock type. However, unlike in the conjugate 'Newer Dolerite' dykes in the Singhbhum Granite terrane, in the case of Bundelkhand, the quartz reefs were emplaced earlier followed by the intrusion of the dolerite dykes at later date. The reefs, which formed earlier due to low-temperature fluid-induced melting of granitic rocks during a phase of shear stress; the mafic and the ultramafic dykes (described as dolerite) formed later along zones of crustal extension. Such a tectonic regime is thought to be a precursor of the basin formations in the Gwalior and Bijawar Basins.

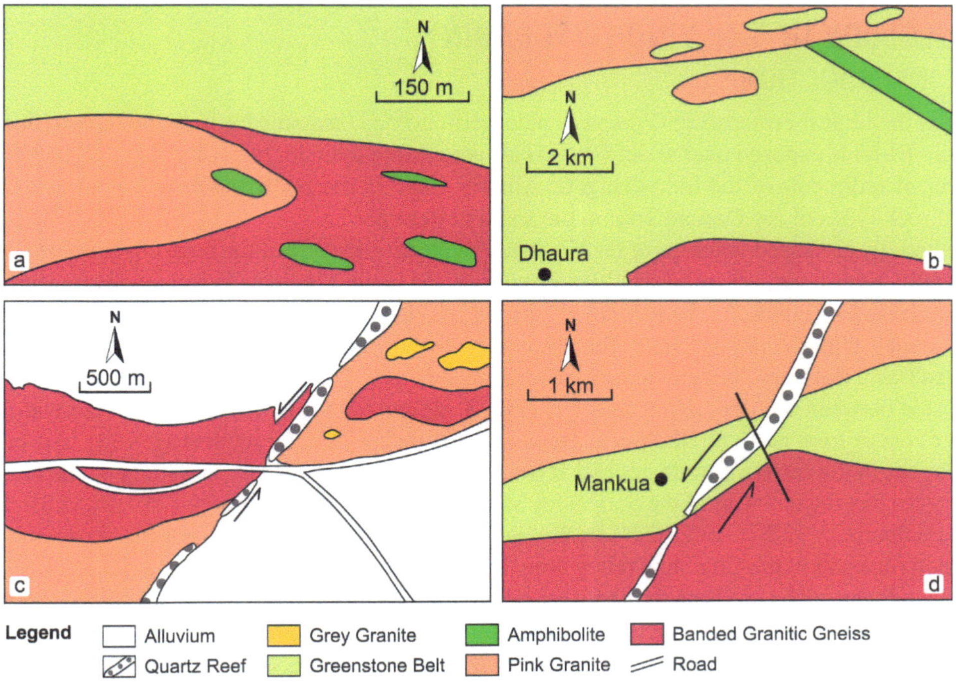

FIG. 8.11

Field relationships between lithotectonic units as illustrated in some key localities in the Bundelkhand Granite. (a) The relationship between the banded granite-gneiss, the greenstone belt rocks, and the massive granitoid (pink granite), about 3 km southwest of Babina. (b) Enclaves of mafic volcanics (greenstone belt rocks) in the massive granitoids at Dhaura. (c) Isolated bodies of banded granite-gneiss in massive granitoids (pink granite). Evidence of sinistral displacement of the banded gneiss by the northeast trending quartz reefs near Mauranipur. (d) A northeast-southwest quartz reef sinistrally displacing bands of the host rocks, Mankuan.

Images courtesy: Surya P. Singh.

Table 8.1 The Table Showing Tectonostratigraphic Evolution of the Bundelkhand Granite

1850–1800 Ma 2150–2000 Ma	Deposition of Bijawar and Gwalior Groups Intrusion of dolerite and other mafic dykes Formation of Quartz reefs
Development of postcratonization fractures	
2500 Ma 2800 Ma 3200–3550 Ma 3800 Ma	Emplacement of massive granitoids, and Cratonization Evolution of the Greenstone belts Formation of Bundelkhand granite-gneiss (the basement) Growth of Early Crust

8.3 GEOLOGY OF THE VINDHYAN BASIN

8.3.1 INTRODUCTION

Known as the largest Proterozoic (Purana in older terminology) cratonic basin in the Indian Shield, the Vindhyan Basin is exposed over about 60,000 km^2 area. More than double of the exposed area is under the cover of sediments of the Indo-Gangetic Alluvial Plain in the north and the volcanics and the associated rock types of the Deccan Trap in the southwestern part.

The broadly U-shaped outcrops of the Vindhyan sediments, in spite of the huge cover of Deccan Trap in the southwestern part, are traditionally referred to as the Great Vindhyan Basin mainly because of the huge area it covers (Fig. 8.1). The half-moon-shaped Bundelkhand Craton occurs as an inlier occupying the core region of the Vindhyan Basin. Purely for the convenience of description, the exposed area of the Vindhyan Basin has been divided into two geographical parts: (i) the Son Valley Vindhyans in the southeastern side between Sasaram and Kuteshwar (Fig. 8.12) and (ii) the Chambal Valley Vindhyans where exposures occur from north of Dholpur to Neemach and Jhalawar in the south (Fig. 8.13). The outcrops of Vindhyans in the eastern part do not end at Sasaram as normally inferred. The study of satellite imagery indicates that the outcrops of the Son Valley Vindhyans extend beyond Rajgir (25°01′05″:85°24′58″) up to Jamalpur (25°32′57″:86°23′27″) in the northeast. The broad lithostratigraphic types in the Son Valley outcrops and in the Chambal Valley domain appear quite similar.

The Vindhyan rocks are broadly undeformed and unmetamorphosed. However, some deformation structures occur in the western part, east of Chittaurgarh and in the southeastern and southern part of the Bundelkhand Protocontinent. In the western part, a series of north-south trending alternate antiforms and synforms have developed in the Vindhyan rocks (Fig. 8.14), which gradually die out in the easterly direction from the east of Dhaneshwar (24°41′17″:74°58′57″).

Strongly deformed Vindhyan rocks are also observed in the eastern side close to the Narmada-Son Lineament zone in the Son Valley region (Fig. 8.15). A series of parallel folds having ENE-WSW trend have developed close to the eastern margin between Sasaram and Hoshangabad (Auden, 1933). The zone

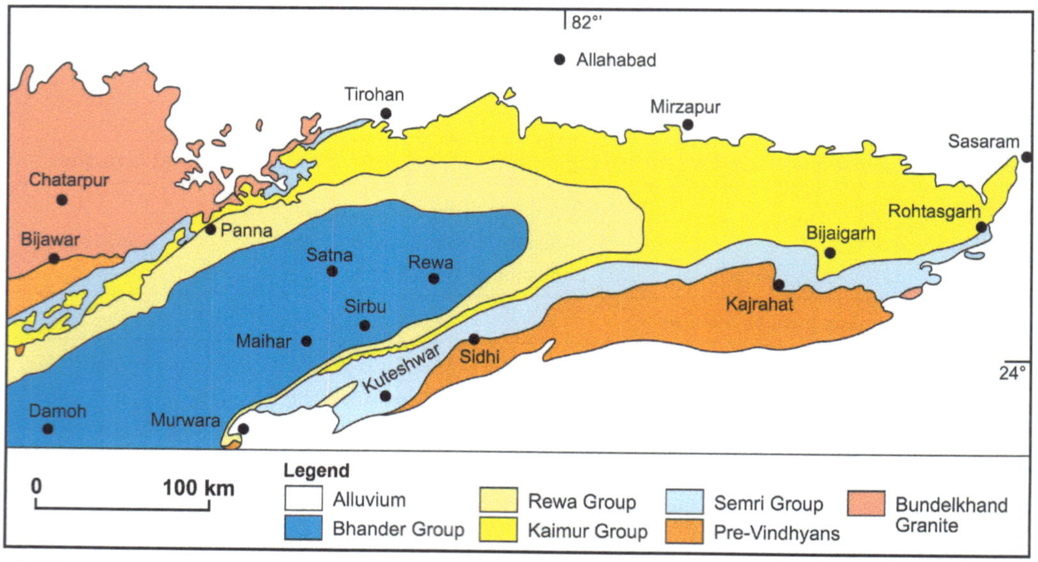

FIG. 8.12

Generalized geological map of the Son Valley Vindhyans.

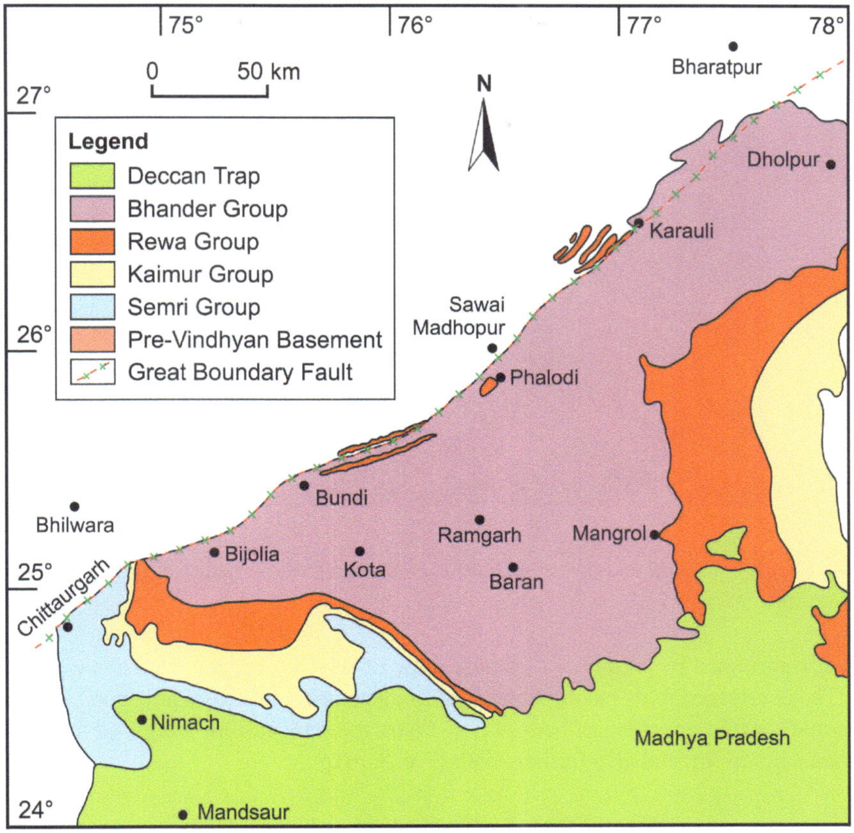

FIG. 8.13

Geological map showing distribution of the Vindhyan rocks in the Chambal Valley and extension into Chittaurgarh-Jhalawar belt in the south.

Reproduced from Roy and Jakhar (2002) with permission.

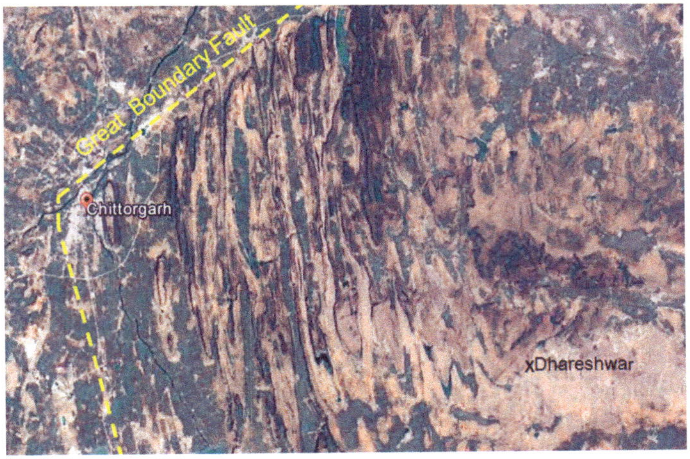

FIG. 8.14

Google Satellite Imagery showing occurrence of a series of north-south-oriented folds in the westernmost part of the Vindhyan Basin. The folds gradually die out eastward from Dhaneshwar where the Vindhyans assume subhorizontal dip of the beds.

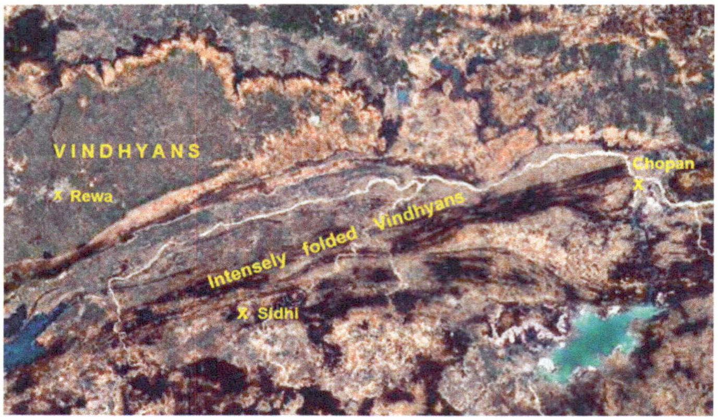

FIG. 8.15

Goggle Satellite Imagery showing an intensely folded zone along the eastern strip of the Vindhyan Basin between Chopan and Sidhi.

of folding is also traversed by a number of faults, some of which represent high-angle overthrusts (Ram et al., 1996). The overturned folds having tight to isoclinal geometry are also quite common. The intensely folded belt is marked by steepening of beds as noted near the southeastern margin of the Vindhyan Basin between Chopan and Sidhi, and also further south (Fig. 8.16).

FIG. 8.16

Vertical to subvertical dip of the Vindhyan rocks marked by a scarp face near the southeastern margin of the Vindhyan Basin.

Reproduced from Chakraborty and Bhattacharjee (1996) with permission.

The deformation of the Vindhyans cannot be described as Precambrian features. These are reflections of very young, latest Phanerozoic tectonism that have been witnessed in different parts of the Peninsular India (Roy, 2004).

8.3.2 LITHOSTRATIGRAPHY OF THE VINDHYANS

The earliest study of the Vindhyan Basin was by Thomas Oldham (1856) who worked on these rocks about one-and-half century ago. Significant contributions have been made by Mallet (1869) and Auden (1933) during later periods, which formed the basis for all the latter works on the Vindhyans.

Oldham (1856) was the first to propose a lithostratigraphic succession subdividing the Vindhyan rocks into three 'series', namely, the Kaimur, Rewa, and Bhander, presuming the Semri as a part of the pre-Vindhyan basement. Medlicott (1859) modified the Vindhyan succession by including the Semri as the basal part of the Vindhyan.

Conventionally, the stratigraphic scheme adopted for the rocks developed in the Son Valley region (Mallet, 1869; Auden, 1933) is used for the entire Vindhyan Basin. Following the 'code of the stratigraphic nomenclature', the term Group has been introduced in place of 'series' while elevating the stratigraphic status of the Vindhyans to the level of Supergroup. The understanding of Vindhyan stratigraphy up to the formational level is greatly hindered because of a variety of reasons. The most significant one is the plethora of nomenclature suggested by different authors working in the different parts of the Vindhyan Basin (Ramakrishnan and Vaidhyanadhan, 2008). However, for the benefit of the readers, a generalized, conventionally used lithostratigraphy of the Vindhyan Supergroup is provided in Table 8.2.

Table 8.2 Generalized Lithostratigraphic Succession of the Vindhyan Supergroup

	Son Valley	Chambal Valley
Bhander Group	Bhavpura Formation	
	Balwan Limestone	
	Silkaoda Formation	Upper Bhander Sandstone
	Sirbu Shale	Sirbu Shale
	Bundi Hill Sandstone	Bundi Hill Sandstone
	Lower Bhander Limestone	Lower Bhander Limestone
	Ganurgarh Shale	Ganurgarh Shale
Rewa Group	Upper Rewa Sandstone	Govindgarh Sandstone
	Jhiri Shale	Jhiri Shale
	Lower Rewa Sandstone	Lower Rewa Sandstone
	Panna Shale	Panna Shale
Kaimur Group	Bijaigarh Shale	Upper Kaimur Sandstone
	Markundi Quartzite	Ghaghar Sandstone
	Gurma Shale	Susnai Breccia
	Ghagar Quartzite	Lower Kaimur Sandstone

Continued

Table 8.2 Generalized Lithostratigraphic Succession of the Vindhyan Supergroup—cont'd

	Son Valley	Chambal Valley
Semri Group	Rotash Limestone	
	Basuhar Sandstone	Suket Shale
	Bargawan Limestone	Nimbahera Limestone
	Kheinjua Shale	Bari Shale
		Jiran Sandstone
		Cohri Malan Conglomerate
		Binota Shales
		Parli Shale
	Chopan Porcellanite	Porcellanite
		Sawa Sandstone
	Kajrahat Limestone	Bhagwanpura Limestone
	Arangi Conglomerate	Khardeola Sandstone

8.3.2.1 Semri Group

The Semri Group in the Son Valley developed over the pre-Vindhyan rocks like laterite-capped Bijawar Group and at places over the granite-gneisses of the Bundelkhand Granite (Fig. 8.18a). The basal beds include patches of conglomerate that at places appear as mass of debris-flow-type oriented clasts (Fig. 8.18b). A diamictite-type conglomerate is reported from the Gangau region (24°16′:79°31′) that has been described as 'glacial tilloids' formed by debris-flow-type deposits (Ahmad, 1955) (Figs. 8.17 and 8.18b).

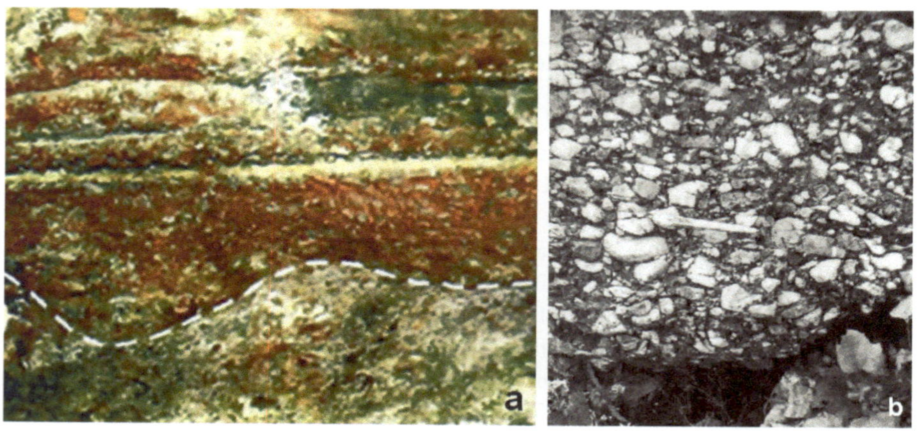

FIG. 8.17

(a) An outcrop in the Pandav Fall Forest Park showing the basal beds of Semri Group overlying the Archaean granite-gneiss. Angular fragments of the basement rocks are noted in the cover sediments. (b) Assorted, oriented clasts in the basal (Semri) conglomerate representing the debris-flow-type deposit in the Son Valley region.

(a) Image courtesy: S. Kumar and M. Sharma. (b) Reproduced from Chakraborty and Bhattacharjee (1996) with permission.

The Semri Group is dominantly calc-argillaceous in nature, intercalated with shale and porcellanite beds. The carbonate formations are the more dominant component (Fig. 8.18a and b) compared with the other lithologies. The carbonate sequences are reported to have been deposited below the wave base in the marine environment. Towards the later part of carbonate succession, extensive development of dolomitic bioherms like stromatolites, other microbial structures have developed in the shallow shelf region of the basin.

FIG. 8.18

(a) Finely laminated Kajrahat Limestone (basal beds) from Maihar area, Son Valley. (b) Nimbahera Limestone (upper section) characteristically showing the development of thin, uniformly planar beds from the Chittaurgarh area.

(a) Image courtesy: S. Kumar and M. Sharma.

The carbonate depositional condition was interrupted by the volcanic formations when the porcellanite beds were deposited. In the Son Valley, volcanism occurred after the deposition of the Kajrahat Limestone. It was an explosive volcanism, penecontemporaneous with the sedimentation on land and in water in tidal flats. The acidic volcanism in the Chopan area consists of rhyolitic tuff, vesiculated and accretionary lapilli, ignimbrite and agglomerate with fragments of pumice, and volcanic bombs (Srivastava et al., 2001; Bose et al., 2001). Petrologically, much of the porcellanite beds are constituted of the rhyolite tuff.

The Semri Group shows variable stratigraphic thickness in different parts of the Vindhyan Basin. It is much thicker in the Son Valley compared to that in the Chambal Valley region. The estimated thickness of the Semri sediments ranges between 20 m near Chandola (24°17′:79°05′) and nearly 4345 m near Deoland (24°14′:81°10′). This suggests that the depositional floor depths of the Vindhyan Basin were uneven at least during the initial stages.

8.3.2.2 Kaimur Group

The Kaimur Group overlies the carbonate dominated Semri Group with an unconformity in between. The nature of contact between the two seems to vary from region to region. A conformable relationship between the two is observed in the Chambal Valley outcrops in Rajasthan, especially near Chittaurgarh. At Bundi, lenses of conglomerate occur at the base of the Kaimur Group overlying the Semri. In the eastern belts, the conformable relationship is observed in the Chitrakoot-Kohari-Panna area in Madhya Pradesh. However, an angular unconformity is reported near the Kalinjar Fort (25°28′20″:79°06′13″) where the subhorizontal Kaimur sediments lie on mildly folded Semri Group.

The Kaimur is dominantly arenitic in composition with some shale beds. The sequence locally shows interbeds of the porcellanite and silty shales. The sandstones are thick bedded, grey and pink coloured. These are commonly crossbedded and ripple marked. Mud cracks are noted in the fine-grained rocks. The Bijaigarh Shale containing pyrite is interpreted as a quiet water swamp or lagoonal deposit. The Kaimur Group of sandstones are spread almost all over the Vindhyan Basin, at places, transgressing into the boundaries of the Semri Group. The rocks have their maximum thickness (about 400 m) near Rotashgarh. The Kaimur sandstones are extensively used as building stones from the historical times.

8.3.2.3 Rewa Group

The Rewa Group is characterized by lateral Facies variations and is best developed along the northern part of the Vindhyan Basin, particularly in the Panna and Satna districts in Madhya Pradesh. In the western part of the Son Valley, the thickness of the Rewa Group varies from about 180 to about 300 m. In the Panna area, its thickness is about 109 m. The Rewa is well-known for having some diamond-bearing conglomerate horizons in the Panna area. The Panna shales and the Lower Rewa Sandstones occurring in the lower part of the Rewa Group have developed as a regressional sequence on the sloping floor of the Kaimur. The Rewa Group rocks show depositional features in the arenitic beds (Fig. 8.19a and b).

FIG. 8.19

(a) Ripple marks in Rewa Sandstones. (b) Deformed cross laminations in Govindgarh Sandstone, Rewa Group.

Image courtesy: S. Kumar and M. Sharma.

One of the most fascinating aspects of this group of the Vindhyan Basin is the fact that the exquisite statues, figures, and panels that decorate the Khajuraho Temples near Panna were carved out from greyish- to cream-coloured Rewa Sandstones (Figs. 8.20 and 8.21).

8.3.2.4 Bhander Group

The Bhander Group represents the youngest lithostratigraphic unit in the Vindhyan succession. Lithologically, it forms the typical shale-sand-carbonate succession deposited in stable continental shelf zone. The Bhander sediments show their deposition under a shallow water depositional condition in basins like lagoons, tidal flats with channels, and beaches (Chakraborty, 2001). The sand-shale-dominated Bhander rocks show alternate thicker sandstone alternating with shale layers, preserving hosts of primary depositional features (Fig. 8.21a). Finely laminated algal mats and cyanobacterial stromatolites formed bioherms in the supratidal to subtidal environments, especially in the western Chambal Valley sector (Akhtar, 1996). The Bhander limestones are grey-coloured stromatolitic type having high concentration

FIG. 8.20

The exquisitely carved statues and murals in the Khajuraho Temples near Panna are made of Rewa Sandstones.

of CaO and low Mg. The chemical character of the carbonates makes it useful for the development of cement factory. The red-coloured variety of sandstones is being widely used as building stones from historical times like the Qutub Minar at Delhi and the Red Forts at Delhi and Agra (Fig. 8.21b).

FIG. 8.21

(a) A vertical section showing well-bedded shale-sandstone alternation in the Bhander succession. (b) Qutub Minar is one of the many beautifully carved structures made of Bhander sandstone.

(a) Image courtesy: S. Kumar and M. Sharma.

8.3.3 DEPOSITIONAL ENVIRONMENT, SEDIMENTATION PATTERN AND LIFE IN VINDHYANS

The Vindhyan sedimentation is traditionally attributed to near-shore marine depositional environment (Valdiya, 1982). Recent studies, however, indicated highly varied depositional environments ranging from fluvial to deep marine environment during the Vindhyan sedimentation (Bhattacharya and Morad, 1993; Bose and Chakraborty, 1994; Akhtar, 1996; Chakraborty et al., 1996,b). The basal succession (Semri Group) consists dominantly of carbonate association, which shows profuse development of stromatolitic structures and laminated algal mat (Fig. 8.22a and b) indicating deposition under subtidal to supratidal conditions. The rare presence of gypsum layers indicated conditions favourable for the formation of evaporates (Akhtar, 1996).

FIG. 8.22

Images of different forms of stromatolites in the Semri Group. (a) Columnar Stromatolite in Kajrahat Limestone, Semri Group. (b) Algal laminites in the Salkhan Limestone of Semri Group, Maihar.

(a) Reproduced from Soni et al. (1987) with permission. (b) Image courtesy: S. Kumar and M. Sharma.

The Kaimur Group, consisting dominantly of siliciclastic rocks (shale, siltstone, and sandstone), shows a depositional setting that included tidal-flat lagoons, barrier-island complex, shoal-sand bar complex, and other tidally influenced environments. Intermittent emergent basin condition is indicated by the occurrence of desiccation cracks, raindrop imprints, and rill marks (Akhtar, 1996). The main sandstone facies rocks of the Rewa Group have been described as the deltaic facies (Bhardwaj, 1977). On the other hand, the shale-sandstone association shows a coarsening upward regressive sequence generated by rapid build-up of a series of deltas along the coastal plains. The youngest Vindhyan succession, the Bhander Group, includes an important limestone formation (the Lakheri Limestone) that represents a transgressive sequence deposited in a protected, low-energy marine to marginal intermittently exposed tidal flats and shallow subtidal lagoons (Akhtar, 1996).

Mention has already been made about occurrence of stromatolitic limestone in the Semri sequence (Fig. 8.21a). Apart from the microbialites, there are references of micro- and macrofossils in the Vindhyan Supergroup. Several distinctive morphological forms have been recognized by different workers. A number of planktonic and benthic microfossil assemblages have been recorded from the chemogenic and argillaceous sediments of the Vindhyan Supergroup. Azmi (1998) reported occurrences of small shelly fossils and brachiopods from the Rotashgarh Limestone (Semri Group). There are also several reports of enigmatic forms supposed to represent macrofossil from the Vindhyan rocks. However, no undoubted marine fossil, which is typical of Lower Palaeozoic, has so far been reported from the Vindhyan Supergroup.

Seminal studies made by Bose et al. (2015) indicated occurrences of distinctive facies assemblages and diverse palaeocurrents in different exposures of the Semri Group that play a pivotal role in the interpretation. Analysis of outcrop and subsurface data from under the Gangetic alluvium in the north of the Vindhyan outcrops supports the hypothesis that an east-west-elongated basement ridge initially separated the 'master' Vindhyan Basin from smaller contemporary basins to the north. According to Bose et al. (2015), the deposition initially took place in isolated lacustrine and fluvial basins north of the divide, while a marine realm prevailed in the south. The uniform character of the Upper Vindhyans (the Kaimur, Rewa, and Bhander Groups) throughout the Vindhyan Basin is because of the subsequent drowning of the dividing ridge causing unification of all of the smaller subbasins, primarily as the consequence of regional northward tilt of the basin (Bose et al., 2015).

Description on the sedimentation and depositional pattern would be incomplete without the information on the contributions from volcanogenic sources at different times. Traditionally, the occurrences of porcellanite beds have been described from the basal Semri Group rocks in different parts of the Vindhyan deposits as the effusive-enriched bodies. Apart from that, very little was known about the volcanic contributions in other lithostratigraphic formations.

Three distinct volcanic eruptions have been detected in the Son Valley porcellanite-bearing formation at Chopan (Srivastava et al., 2003) in the form of (i) ashfall (volcanic ash/dust), (ii) pyroclastic breccia and ignimbrite, and (iii) finer pyroclastics and silicified tuff (often described as chert). In some other instances, white, massive porcellanite beds have formed. Contrary to the earlier perceptions, the detailed studies indicated occurrences of pyroclastic deposits from the dominantly arenitic Kaimur and Rewa Groups (Chakraborty et al., 1996). In the Kaimur Group, lenses of pyroclastic layers have been recorded that occur as discontinuous layers varying from about 50 m to a few millimetres in thickness. Microscopic studies revealed occurrences of flow banding and stretched pumice fragments and locally concentric and radial tensional cracks with the development of peripheral devitrification (Fig. 8.23a). Scanning electron microscope studies indicated occurrence of frozen droplets of viscous fluid with wrinkle marks and such other features (Fig. 8.23b). Volcaniclastics are reported in different forms in the shale units of the Rewa Group. Petrologically, three different types are noted: (i) pyroclastic falls, (ii) flow breccia, and (iii) epiclastics (Chakraborty, 1996).

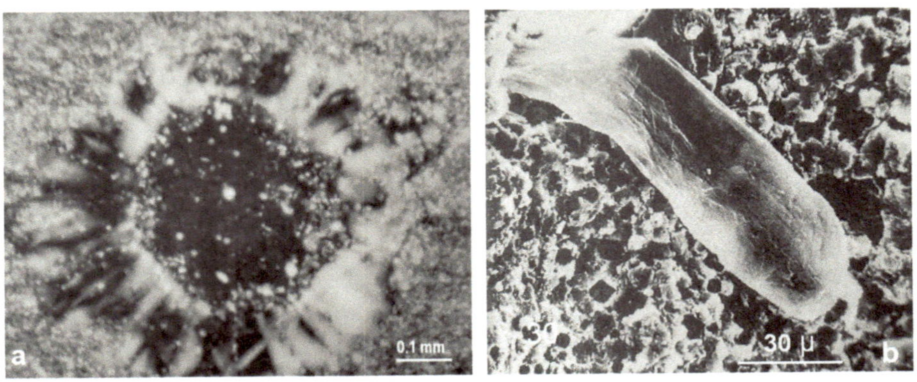

FIG. 8.23

(a) Photomicrograph showing the development of radial quenching feature within ash flow, Kaimur Group.
(b) Scanning electron microphotograph showing a frozen droplet of a viscous liquid with wrinkle marks within an ash-flow deposit, Kaimur Group.

Reproduced from Charkaborty et al. (1996) with permission.

An important feature reported by Chakraborty et al. (1996) is that all the volcano-sedimentary deposits occurring at different stratigraphic levels are succeeded by major events of regression indicating regional uplift of the source area followed by erosion. The information greatly helps in the understanding the basin evolution and sedimentation tectonics of the Vindhyan rocks.

Though smaller in dimension, the kimberlite and lamproite pipe-shaped plugs that occur at two places, Hinota and Majhgawan near Panna (24°43′05″:80°10′54″) (Fig. 8.24a and b), deserve special mention. The Majhgawan pipe in particular is characterized by diamond and phlogopite. Covering about 500 m ground surface, the cone-shaped Majhgawan plug is made of olivine-lamproite-type lapilli tuffs of the crater face. The diamond distribution is concentric within the body. The eroded materials from the plugs occur in the conglomerate horizons of the Rewa succession.

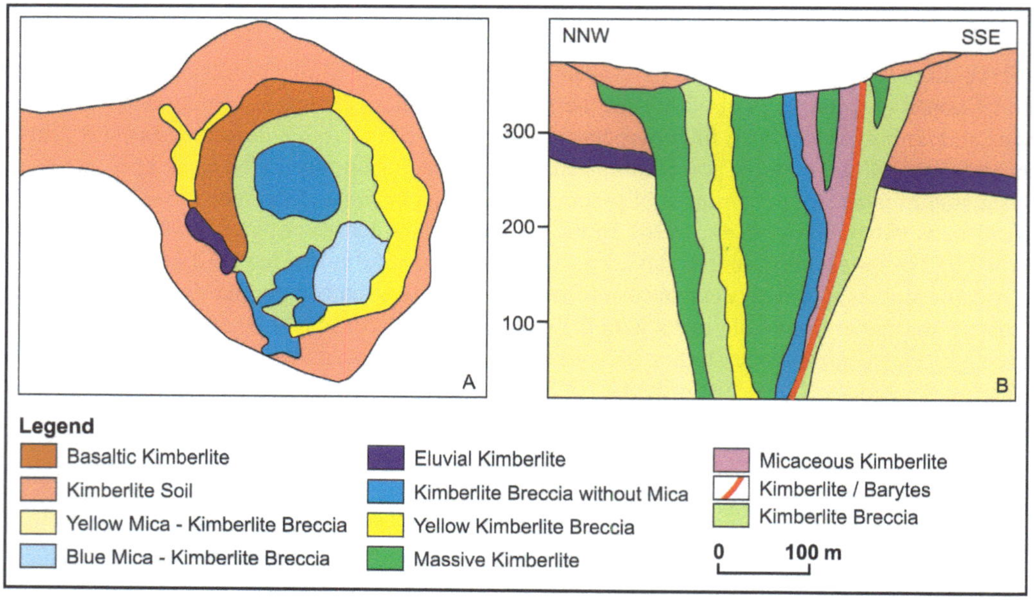

FIG. 8.24

Surface outcrop (a) and a section through the Majhgawan Plug (b) showing structure and compositional characteristics of the Majhgawan 'Pipe'.

Reproduced from Chatterjee and Rao (1995) with permission.

8.3.4 BASIN TECTONICS: DEPOSITIONAL AND POST-DEPOSITIONAL

During recent years, there has been a tremendous boost in the understanding of different aspects of basin evolution, inversion-related tectonics, and post-depositional tectonics of the Vindhyan Basin, mainly because of the accumulation of data on remote sensing, gravity, magnetic, magnetotelluric, and seismic reflection studies. All these data suggest a rift model for the Vindhyan sedimentation. Vindhyan Basin is constrained by two major fault-lineaments, the Great Boundary Fault virtually marking the northwestern and the Narmada-Son Lineament bounding the southern and the southeastern margins. Evidences suggest that these two basin margin faults-lineaments actually acted as rift shoulders during different periods of deposition of the Vindhyan sediments. Available regional gravity and magnetic data coupled with the analysis of the seismic reflection data coupled with the analysis of the seismic reflection profiles indicate southward deepening of the basin, the deepest part occurring close to the southern basin margin (Ram et al., 1996).

The Narmada-Son Lineament constitutes the southern basin margin fault during the Vindhyan sedimentation. Although no detailed subsurface data are available from the western Vindhyan Basin, it has been argued that the Great Boundary Fault, representing northwestern margin of the western Vindhyan Basin, also behaved in a similar fashion (Verma, 1996).

Although the nature of movements observed along the intrabasinal and the major basin margin faults is dominantly of the strike-slip type, the interpretation of thickness variation and the abrupt changes in lithofacies across the faults indicate that these are post-depositional faults.

The opening of the Vindhyan Basin was due to transtensional rifting (Ram et al., 1996) during the extension of the stabilized parts of the surrounding Crust. The north-south trending open folds east of Chittaurgarh resulted due to the easterly push of the Aravalli Protocontinental block. On the other hand, the tight to isoclinal folds along the southeastern margin were formed due to the transpressional movement of southeastern crustal block. It is possible that the differential movements of the different crustal blocks had also resulted during the late Quaternary. The present seismicity along the Narmada-Son Lineament is a proof of such pattern of the Quaternary and the recent crustal movements (Roy, 2006).

8.3.5 AGE OF THE VINDHYANS

Based on the study of the multicellular palaeobiological evidence, Venkatachala et al. (1996) suggested that the 'Purana' sediments of the Vindhyan antiquity range in age from the Lower Riphean to Upper Riphean (i.e. between ~1650 and ~600 Ma). These authors also inferred Crygenian age (850–650 Ma) of the Vindhyans. There is, however, no record of any incontrovertible evidence to mark the Precambrian-Cambrian boundary in the Vindhyan sediments.

Several geochronological dates are now available from different parts of the Vindhyan succession. The Rb–Sr study of glauconites from the basal Semri Group yielded ages of 1449 ± 15, 1461 ± 15, 1504 ± 15, and 1531 ± 15 Ma (Anil Kumar et al., 2001). Ray et al. (2002) reported 1631 ± 5 Ma zircon ages using U–Pb zircon dates and Sr isotope systematics from the porcellanite beds in the Semri Group. Further, 1628 ± 8 and 1599 ± 8 Ma U–Pb zircon ages are reported by Rasmussen et al. (2002) from the Chorhat Sandstone-bearing evidence of burrows of the triploblastic animals in the Lower Vindhyan sequence. Records of much older ages come from 1729 ± 110 Ma Pb–Pb age from the Kajarhat Limestone in the Semri Group (Sarangi et al., 2004). Similar older age of ~1721 Ma comes from the study of trace element and Nd isotope data by Chakraborty et al. (2007), who indicated that these older ages are from the source rocks of the Vindhyan sediments.

The K–Ar age of the phlogopite of the Majhgawan pipe is 1140 ± 112 Ma (Paul et al., 1975). The Rb–Sr isochron age of lamproite body is 1067 ± 31 Ma (Kumar et al., 1993). Since the kimberlite pipes are intrusive in the Kaimur Group rocks, it may be assumed that the depositional age of the Kaimur rocks must be older.

Collating the available geochronological data, it may be suggested that the Vindhyan sedimentation was initiated with the deposition of Semri Group at around 1600 ± 50 Ma. The kimberlite ages on the other hand suggest that the deposition of the arenite dominant Kaimur Group must have taken place before the intrusion of the diamond-bearing plugs around 1050 Ma. There is no reliable isotope data to fix the age of deposition of the youngest units, the Rewa and the Bhander. Inferences have been made about the Crygenian age (850–650 Ma) of the younger Vindhyan sequences based on the occurrence of multicellular palaeobiological evidence. In short, it may be suggested that the geological evolutionary history of the Vindhyans spanned over about 1000 million years during the Meso- and Neoproterozoic.

REFERENCES

Ahmad, F., 1955. Glaciation in the Vindhyan System. Curr. Sci. 24, 231.

Akhtar, K., 1996. Facies sedimentation process, and environment in the Proterozoic Vindhyan Basin. In: Bhattacharya, A. (Ed.), Recent Advances in the Vindhyan Geology. Mem. Geol. Soc. India, 36, pp. 127–136.

Auden, J.B., 1933. Vindhyan sedimentation in the Son Valley, Mirzapur District. Mem. Geol. Surv. India 62 (2), 141–250.

Azmi, R.J., 1998. Discovery of Lower Cambrian small shelly fossils and brachiopods from the Lower Vindhyan of Son Valley, central India. J. Geol. Soc. India 52, 381–389.

Basu, A.K., 1986. Geology of parts of Bundelkhand granite massif, Central India. Rec. Geol. Surv. India 117, 61–124.

Bhardwaj, B.D., 1977. Stratigraphy and depositional environments of the Rewa Group (Upper Vindhyan) Bhainsrorgarh, Rajasthan. J. Indian Assoc. Sediment. 1, 52–61.

Bhattacharya, A., Morad, S., 1993. Proterozoic braided ephemeral fluvial deposits; an example Dhandraul sandstone formation of the Kaimur Group, Son Valley, central India. Sed. Geol. 84, 101–114.

Bose, P.K., Chakraborty, P., 1994. Marine to fluvial transition: Proterozoic Upper Rewa Sandstone, Maihar, India. Sed. Geol. 89, 285–301.

Bose, P.K., Sarkar, S., Chakraborty, S., Banerjee, S., 2001. Overview of the Meso- to Neoproterozoic evolution of the Vindhyan basin, Central India. Sed. Geol. 141–142, 395–419.

Bose, P.K., Sarkar, S., Das, N.G., Banerjee, S., Mandal, S., Chakraborty, N., 2015. Proterozoic Vindhyan Basin: configuration and evolution. In: Mazumder, R., Eriksson, P.G. (Eds.), Precambrian Basins of India: Stratigraphic and Tectonic Context. Geol. Soc. London, Mem. 43, pp. 85–102.

Chakraborty, C., 1996. Sedimentary Records of Erg development over Braid plain: Proterozoic Dhandraul Sandstone, Vindhyan Supergroup, Son Valley. In: Bhattacharya, A. (Ed.), Recent Advances in the Vindhyan Geology. Mem. Geol. Soc. India, 36, pp. 77–100.

Chakraborty, C., 2001. Lagoonal-tidal flat sediments in a foreland moat: Deoland Formation, Vindhyan Supergroup, India. Geol. J. 36, 125–141.

Chakraborty, P.P., Banerjee, S., Das, N.G., Sarkar, S., Bose, P.K., 1996. Volcaniclastics and their sedimentological bearing in Proterozoic Kaimur and Rewa Groups in Central India. In: Bhattacharya, A. (Ed.), Recent Advances in the Vindhyan Geology. Mem. Geol. Soc. India, 36, pp. 59–75.

Chakraborty, R., Basu, A.R., Chakraborty, A., 2007. Trace element and Nd-isotope evidence for sediment sources in the mid-Proterozoic Vindhyan Basin, central India. Precambrian Res 153, 260–274.

Chakraborty, C., Bhattacharya, A., 1996. Fan-Delta sedimentation in the foreland moat: Deoland Formation, Vindhyan Supergroup, Son Valley. In: Bhattacharya, A. (Ed.), Recent Advances in the Vindhyan Geology. Mem. Geol. Soc. India, 36, pp. 27–48.

Chakraborty, T., Sarkar, S., Chaudhuri, A.K., Dasgupta, S., 1996. Depositional environment of Vindhyan and other Purana basins: a reappraisal in the light of recent findings. In: Bhattacharya, A. (Ed.), Recent Advances in the Vindhyan Geology. Mem. Geol. Soc. India, 36, pp. 101–126.

Chatterjee, A.K., Rao, K.S., 1995. Majhgawan diamondiferous pipe Madhya Pradesh, India: a review.. J. Geol. Soc. India 45, 175–189.

Geological Survey of India, 2004. Manual Geology of India, v. 1, Precambrian, Pt IV (N & NW part of Peninsula), fourth ed. Geol. Surv. India, Spec. Publ. 77. 257 p.

Haldar, D., Ghosh, R.N., 2000. Eruption of Bijawar lava: an example of Precambrian volcanicity under stable cratonic condition. Geol. Surv. India, Spec. Publ. 57, 151–170.

Kaur, P., Zeh, A., Chaudhri, N., 2014. Characterisation and U–Pb–Hf isotope record of the 3.55 Ga felsic crust from the Bundelkhand Craton, northern India. Precambrian Res. 255, 236–244.

Kaur, P., Zeh, A., Chaudhri, N., Eliyas, N., 2016. Unravelling the record of Archaean crustal evolution of the Bundelkhand Craton, northern India using U–Pb zircon–monazite ages, Lu–Hf isotope systematics, and whole-rock geochemistryof granitoids. Precambrian Res 281, 384–413.

Kumar, A., Gopalan, K., Rajagopalan, G., 2001. Age of Lower Vindhyan sediments, Central India. Curr. Sci. 81, 806–808.

Kumar, A., Kumari, P., Dayal, A.M., Murthy, D.S.N., Gopalan, K., 1993. Rb–Sr ages of Proterozoic kimberlites of India: evidence for contemporaneous emplacements. Precambrian Res. 62, 227–237.

Kumar, A.V., Srivastava, R.K., Jha, D.K., Pant, N.C., Bhandari, B.K., 1990. A revised stratigraphy of the type area of Bijawar Group in central India. Indian Minerals 44, 303–314.

Mallet, F.R., 1869. On the Vindhyan Series as exhibited in the Northeastern and Central Provinces of India. Mem. Geol. Surv. India 7 (1), 1–129.

Mallikharjuna Rao, J., Poornachandra Rao, G.V.S., Widdowson, M., Kelley, S.P., 2005. Evolution of Proterozoic mafic dyke swarms of Bundelkhand granite massif, central India. Curr. Sci. 88, 502–506.

Medlicott, H.B., 1859. Vindhyan rocks and their associates in Bundelkhand. Mem. Geol. Surv. India 2 (1).

Mondal, M.E.A., 2010. Geochemical evolution of the Archaean-Proterozoic Bundelkhand Craton, Central Indian Shield: revisited. J. Econ. Geol. Georesour. Manage. 7, 69–80.

Mondal, M.E.A., Goswami, J.N., Deomurari, M.P., Sharma, K.K., 2002. Ion microprobe for crustal evolution of the Bundelkhand–Aravalli Protocontinent. Precambrian Res. 117, 85–100.

Mondal, M.E.A., Sharma, K.K., Rahman, A., Goswami, J.N., 1998. Ion microprobe 207Pb/206Pb zircon ages for the gneiss-granitoid rocks from Bundelkhand massif: evidence for the Archean components. Curr. Sci. 74, 70–75.

Mondal, M.E.A., Zainuddin, S.M., 1996. Evolution of the Archean-Palaeoproterozoic Bundelkhand Massif, central Inda—evidence from granitoid geochemistry. Terra Nova 8, 532–539.

Oldham, T., 1856. Remarks on the classification of rocks of Central India resulting from the investigation of the Geological Survey. J. Asiat. Soc. Bengal 25, 224–256.

Pascoe, E.D., 1973. A Manual of Geology of India and Burma. vol. 1. Geological Survey of India, Calcutta. 485 p.

Paul, D.K., Rex, D.C., Harris, P.G., 1975. Chemical characteristics and K-Ar ages of Indian Kimberlites. Geol. Soc. Am. Bull. 86, 364–366.

Ram, J., Shukla, S.N., Pramanik, A.G., Varma, B.K., 1996. Recent Investigations in the Vindhyan Basin: implications for the basin tectonics. In: Bhattacharya, A. (Ed.), Recent Advances in the Vindhyan Geology. Mem. Geol. Soc. India, 36, pp. 267–286.

Ramakrishnan, M., Vaidhyanadhan, R., 2008. Geology of India. Geological Society of India, Bangalore. 556p.

Rasmussen, B., Bose, P.K., Sarkar, S., Banerjee, S., Fletcher, I.R., McNaughton, N.J., 2002. 1.6 Ga U-Pb zircon age for the Chorhat Sandstone, Lower Vindhyan, India: Possible implications for early evolution of animals. Geology 30, 103–106.

Ray, J.S., Martin, M., Veizer, J., Bowring, S.A., 2002. U-Pb zircon dating and Sr isotope systematics of the Vindhyan Supergroup India. Geology 30, 131–134.

Roy, A.B., 2004. Phanerozoic reconstitution of Indian Shield as the aftermath of break-up of the Gondwanaland. Gond. Res. 7, 387–406.

Roy, A.B., 2006. Seismicity in the Peninsular Indian Shield: Some Geological considerations. Curr. Sci. 91, 456–463.

Saha, L., Frei, D., Gerdes, A.J.K., Pati, J.K., Sarkar, S., Patole, V., Bhandari, A., Nasipuri, P., 2016. Crustal geodynamics from the Archaean Bundelkhand Craton, India: constraints from zircon U–Pb–Hf isotope studies. Geol. Mag. 153, 179–192.

Saha, L., Pant, N.C., Pati, J.K., Bernd, J., Upadhyay, D., Bhattacharya, A., Satyanarayanan, M., 2011. Neoarchaean high-pressure margarite-phengitic muscovite-chlorite corona mantled corundum in quartz-free high-Mg, Al phlogopite- chlorite schists from the Bundelkhand craton, north central India. Contrib. Mineral. Petrol. 161, 511–530.

Sarangi, S., Gopalan, K., Kumar, S., 2004. Pb-Pb age of the earliest megascopic, eukaryotic algal-bearing Rotash Formation, Vindhyan Supergroup, India: implication for Precambrian atmospheric oxygen evolution. Precambrian Res. 132, 107–121.

Singh, S.P., 2012. Archaean Geology of Bundelkhand Craton, Central India: an overview. Gondwana Geol. Mag. 13, 125–140.

Singh, S.P., Dwivedi, S.B., 2015. High grade metamorphism in the Bundelkhand massif and its implications on Mesoarchean crustal evolution in central India. J. Earth Syst. Sci. 124, 197–211.

Singh, S.P., Singh, M.M., Srivastava, G.S., Basu, A.K., 2007. Crustal Evolution in Bundelkhand area, Central India. Himal. Geol. 28, 79–101.

Singh, V.K., Slabunav, A., 2015. The Central Bundelkhand Archaean greenstone complex, Bundelkhand craton, central India: geology, composition, and geochronology of supracrustal rocks. Int. Geol. Rev. 57, 1349–1364.

Singh, V.K., Slabunov, A., 2016. Two types of Archaean Supracrustal Belts in the Bundelkhand Craton, India: geology, geochemistry, age and implication for Craton Crustal Evolution. J. Geol. Soc. India 88, 539–548.

Soni, M.K., Chakraborty, S., Jain, V.K., 1987. Vindhyan Supergroup: a review. In: Purana Basins of Peninsular India. Geol. Soc. India. Mem.6. pp. 87–138.

Srivastava, R.N., Srivastava, A.K., Singh, K.N., Radcliff, R.P., 2001. Precambrian acid volcanism in the Lower Vindhyan basin of the Sonbhadra district and adjoining areas in Uttar Pradesh—a review. Spl. Publ. Geol. Surv. India 55, 205–218.

Srivastava, R.N., Srivastava, A.K., Singh, K.N., Redcliff, R.P., 2003. Sedimentation and depositional environment of the Chopan Porcellanite Formation, Semri group, Vindhyan Supergroup in parts of Sonbhadra District, Uttar Pradesh. J. Palaeontological Soc. India 48, 167–179.

Valdiya, K.S., 1982. Tectonic perspective of the Vindhyachal region. In: Valdiya, K.S., Bhatia, S.B., Gaur, V.K. (Eds.), Geology of Vindhyachal. Hindustan Publishing Corporation, New Delhi.

Venkatachala, B.S., Sharma, M., Shukla, M., 1996. Age and Life in the Vindhyans-Facts and Conjectures. In: Bhattacharya, A. (Ed.), Recent Advances in the Vindhyan Geology. Mem. Geol. Soc. India36. pp. 137–165.

Verma, P.K., 1996. Evolution and the age of the Great Boundary fault of Rajasthan. In: Bhattacharya, A. (Ed.), Recent Advances in the Vindhyan Geology. Mem. Geol. Soc. India, 36, pp. 197–212.

CHAPTER

ARAVALLI PROTOCONTINENT

9

9.1 INTRODUCTION

The Aravalli Protocontinent comprises the region of Aravalli Mountains and adjacent areas in Rajasthan and parts of neighbouring States of Gujarat and Madhya Pradesh. This crustal block in the northwestern part of the Indian Shield is endowed with well-preserved records of a long and protracted history of Precambrian geological evolution that spans over 3000 million years, unknown in any other part in the world. The foundation for the geological studies in this terrain was laid by the early workers, like Coulson (1933), Gupta (1934), Gupta and Mukherjee (1938), and Heron (1917a,b, 1936). Excellent geological maps prepared by these workers helped Heron (1953) to erect a stratigraphic succession shown in Table 9.1.

Table 9.1 Geological Formations of Rajasthan (Former Rajputana) and Adjoining Areas

Algonkian	Vindhyan system	
	Malani series	
	Delhi system	Ajabgarh series
		Alwar series
Eparchaean interval	Raialo series	
Archaean	Aravalli system	
	Banded Gneissic Complex and Bundelkhand Gneiss	

Studies made during the last seven decades or so on aspects of tectonics, stratigraphy, metamorphism, migmatization, metallogeny, and geochronology (Roy and Jakhar, 2002) helped to highlight some deficiencies in the Heron's (1953) summary of geological formations necessitating wide-ranging revisions in the lithostratigraphic framework of the Precambrian rocks of the Aravalli Mountains and adjoining areas (Fig. 9.1). The most significant revisions proposed by different workers are as follows:

1. Geochronological data mainly from the southeastern part around Udaipur and neighbouring region helped to identify the granite-gneiss complexes including the enclosed greenstone belts as undoubted Archaean basement rocks.
2. In addition to the major bodies, isolated granite-gneiss bodies that occur around Udaipur in the south, earlier classed as intrusive into cover (Aravalli) succession, have been identified as 'inliers' within younger cover metasediments.
3. Field data based on the discovery of the oldest palaeosols, aided by the geochronological information, helped in delimiting an erosion-related unconformity globally recognized as the

Indian Shield. https://doi.org/10.1016/B978-0-12-809839-4.00009-6

Archaean-Proterozoic Boundary at the contact zone of the Archaean basement with the oldest Proterozoic metasedimentary-metavolcanic cover succession, the Aravalli Supergroup.

4. Because of the strong tectonic reconstitution along zones of high strain, the basement-cover relationship is erased at many places. This has made the understanding of the relative stratigraphy of different litho-units very difficult.
5. Intense basement reconstitution also been noted in the region mapped earlier as the 'BGC II' (Gupta, 1934) around Sandmata Hills due to emplacement of bodies of granulite during late Palaeoproterozoic time. A new nomenclature, Sandmata Complex, is proposed for the tectonothermally reconstituted Archaean basement east of the Main Delhi Basin in the central part (Fig. 9.1).
6. In southeastern Rajasthan, a single continuous lithostratigraphic succession has been recognized, named the Aravalli Supergroup, in place of the earlier known divisions into the Aravalli system, Raialo series and the 'outliers' of the Delhi system as the oldest Proterozoic ensemble overlying the Archaean basement (Fig. 9.1).
7. No separate stratigraphic status is recognized for the 'Raialo series' that Heron (1953) recognized from different parts of the Aravalli Mountains and thought to have been deposited during the 'Eparchaean Interval' (meaning Archaean-Proterozoic Break).
8. On the other hand, a separate Raialo Group is recognized as a lithostratigraphic unit for those rocks that occur at the base of the Delhi succession in the Alwar and Bayana Subbasins in the northeastern part of the Delhi Basin.
9. A diachronous evolutionary history of the Delhi Supergroup, popularly described as the Delhi Fold Belt, has been suggested with an older North Delhi Fold Belt and a younger South Delhi Fold Belt based on poorly constrained isotope (Rb/Sr) dates of some of the granites.
10. A newly defined Sirohi Group is recognized as a distinct lithostratigraphic formation overlying the Delhi Supergroup (Roy and Sharma, 1999). Occurring on the western side of the Aravalli Mountains, this lithostratigraphic formation was earlier considered time equivalent of the 'Aravalli system' by Coulson (1933) and Heron (1953).
11. Identification of the Malani Group ('Group' replacing 'series' of early workers) as a volcano-sedimentary-plutonic ensemble considered as geological manifestation of the Plume impingement during late Neoproterozoic (Roy, 1998).
12. Recognition of Marwar Supergroup as a typical cratonic platform-type sequence unconformably overlying the Malani Group (Roy and Jakhar, 2002). The newly defined lithostratigraphic ensemble was earlier considered equivalent to the Vindhyans occurring east of the Aravalli Mountains and was described as the 'Trans-Aravalli Vindhyans'.

The following discussion is based on the critical evaluation of the apparently contentious issues with ultimate aim of building up the relative stratigraphy of different formations.

BOX 9.1

The tectonostratigraphic status of the earliest formed basement was subjected to serious questioning with the publication of a new map by Gupta et al. (1980), followed by detailed description (Gupta et al., 1997). The most remarkable feature in the new map is the recognition of the metasedimentary ensemble of the Bhilwara-Chittaurgarh-Hindoli belt as a pre-Aravalli Archaean basement (termed Bhilwara Supergroup), contradicting the correlation of the 'Raialo series' and the 'Aravalli system' by the earlier workers Gupta (1934) and Heron (1953). In short, the 'revised' new geological map has been proposed without providing any stratigraphic reasoning and without following the 'Standard Code Stratigraphic Nomenclature' while introducing hosts of new names for the assumed geological formations. In view of this, we would prefer to avoid any detailed discussion on the new map by Gupta (SN) et al. (1980).

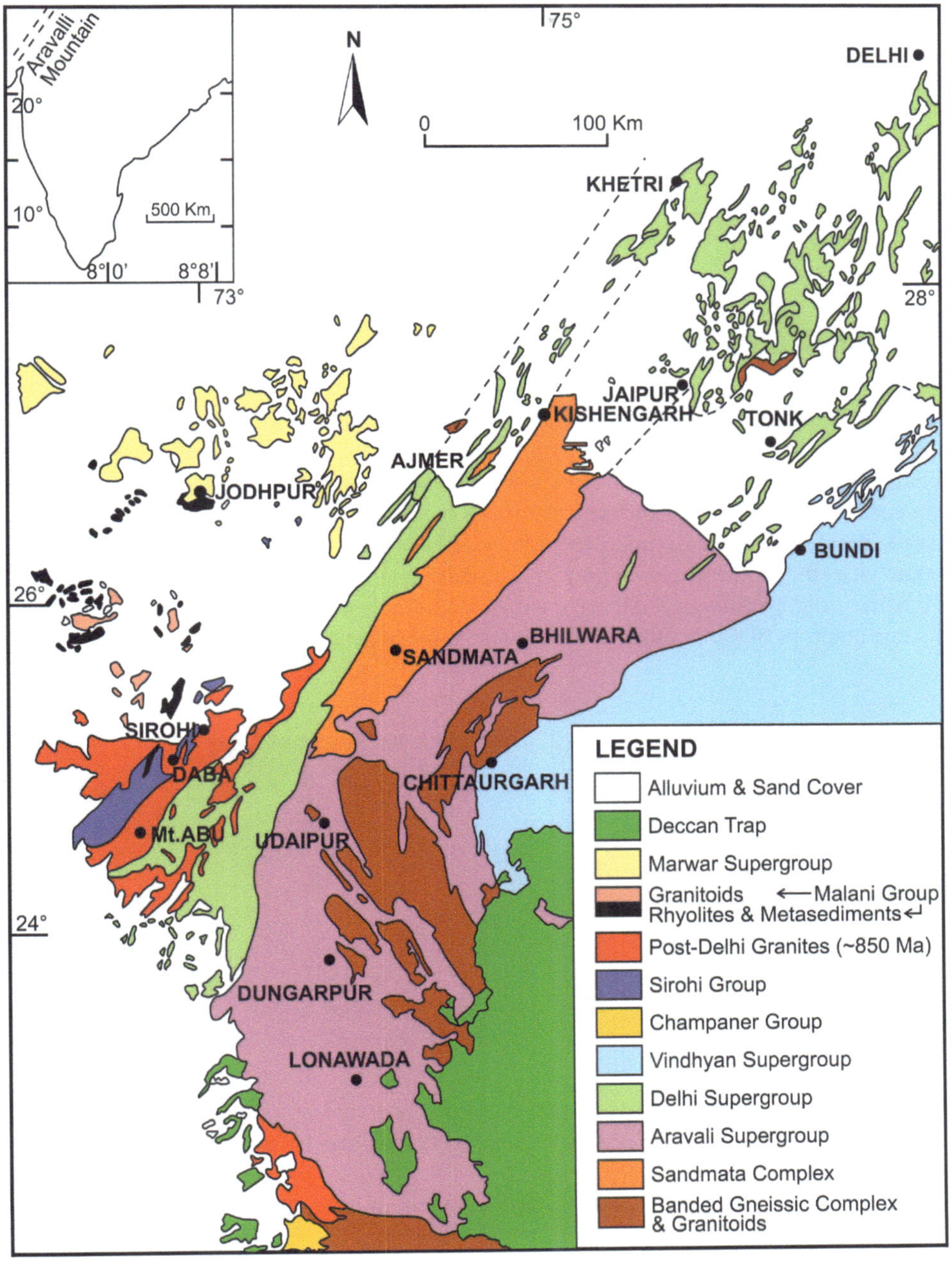

FIG. 9.1

Generalized geological map of the Aravalli Mountains and neighbouring areas showing distribution of the major lithostratigraphic units.

Reproduced from Roy and Jakhar (2002) with permission.

Purely for the descriptive purpose, we propose to follow the tentative scheme of stratigraphic order (Fig. 9.1), initially suggested by Gupta (1934) and Heron (1953):

Marwar Supergroup.
Malani Group.
Sirohi Group.
Delhi Supergroup.
Aravalli Supergroup.
Banded Gneissic Complex.

9.2 RECOGNITION AND AGE OF THE ARCHAEAN BASEMENT

The recognition of oldest basement and the understanding of the basement-cover relationship are the two most burning questions of the Precambrian geology of Rajasthan that hindered erection of a coherent lithostratigraphic framework of the region. The early geologists who mapped the different parts of the Aravalli Mountains had identified a group of banded gneisses along with granitoids as the oldest Archaean basement. Considering the typical banded character of the gneissic rocks, the ensemble was named as the Banded Gneissic Complex (popularly known by the abbreviated name BGC) (Fig. 9.2a and b). Also known as the biotite gneiss or grey gneiss, these gneissic rocks petrochemically fall in the category of tonalite-trondhjemite-granodiorite gneiss (also described as grey gneiss or biotite gneiss). There are also some outcrops of dark-coloured banded gneiss having chemical characteristics of diorite gneiss.

FIG. 9.2

(a) The outcrops of the Banded Gneissic Complex showing banded character that typifies the Archaean basement rocks. (b) Highly tectonized basement gneiss *(grey gneiss)* in the Sandmata Complex, the reconstituted Archaean basement.

(a) Image courtesy: N.K. Chauhan; (b) Image courtesy: Azimuddin Sirwani.

BOX 9.2

There were proposals to change the nomenclature of the Archaean basement rocks to Mangalwar Complex (Gupta et al., 1980), Mewar Gneiss or Mewar Gneiss Complex (Roy, 1988; Roy and Kröner, 1996), etc. We propose to retain the traditionally accepted name 'the Banded Gneissic Complex' for the Archaean basement in the present description.

The Banded Gneissic Complex covers a wide region in the southeastern and south-central part of the Aravalli Mountains. This, broadly gneissic terrane includes bands and enclaves of amphibolite showing a variety of shapes and sizes. Apart from the boudin-like bodies (Fig. 9.3a and b), there are also a few larger bodies of amphibolite that show schistose character and mineralogically quite homogeneous. Petrochemistry of the majority of the amphibolite bodies indicates their tholeiite basalt character. A very few amphibolites have komatiitic composition (Upadhyaya et al., 1992).

FIG. 9.3

(a) Series of rhomb-shaped boundinaged bodies of amphibolite within the banded gneisses. (b) Small, elongate patches of amphibolite occur as irregular boudin-like structures, cofolded with the intricately deformed banded gneiss.

In addition to the banded gneisses and amphibolites, several bodies of granites of different sizes and shapes, like the Bundelkhand Granite (rechristened as Berach Granite; Crawford, 1970), the Gingla Granite, and the Untala Granite, occur within and outside the domains of the Banded Gneissic Complex. All these granitoid bodies differ considerably in their physical character (varying from banded gneiss to massive granitoid) and in their petrochemistry. Detailed field studies in and around Udaipur (Roy and Jakhar, 2002) also helped to identify several other granite-gneiss bodies, like the Ahar River Granite, the Udaisagar Granite, and the Jaisamand Road Granite as the basement 'inliers' surrounded by the cover metasediments, the Aravalli Supergroup (Gopalan et al., 1990; Roy and Kröner, 1996; Roy and Paliwal, 1981; Roy et al., 1988; Wiedenbeck and Goswami, 1994). The best-known 'inlier' of the basement ensemble is known as the Sarara Inlier, which occurs south of the Zawar lead-zinc mines area south of Udaipur. All these granite-gneiss bodies, except the 'Sarara Inlier', were earlier considered parts of the 'post-Aravalli, pre-Delhi' granites described by Heron (1953).

Some sporadic outcrops of dismembered greenstone belts have been recorded within the Archaean basement. These greenstones include large disjointed bodies of quartzite and marble along with mica schist, amphibolite, and ultramafics at several places in southeastern Rajasthan (Upadhyaya et al., 1992; Roy et al., 2000; Roy and Jakhar, 2002). One such greenstone belt occurs at Rakhiawal (24°47′:73°59′) (Fig. 9.4), which include large disjointed bodies of quartzite and marble (not exposed in the map area) along with amphibolite and ultramafic rock (Roy and Jakhar, 2002).

Among the other greenstone belts, an important one occurs near Jagat, southeast of Udaipur (Upadhyaya et al., 1992). The constituents of this highly dismembered greenstone belt include quartzite along with

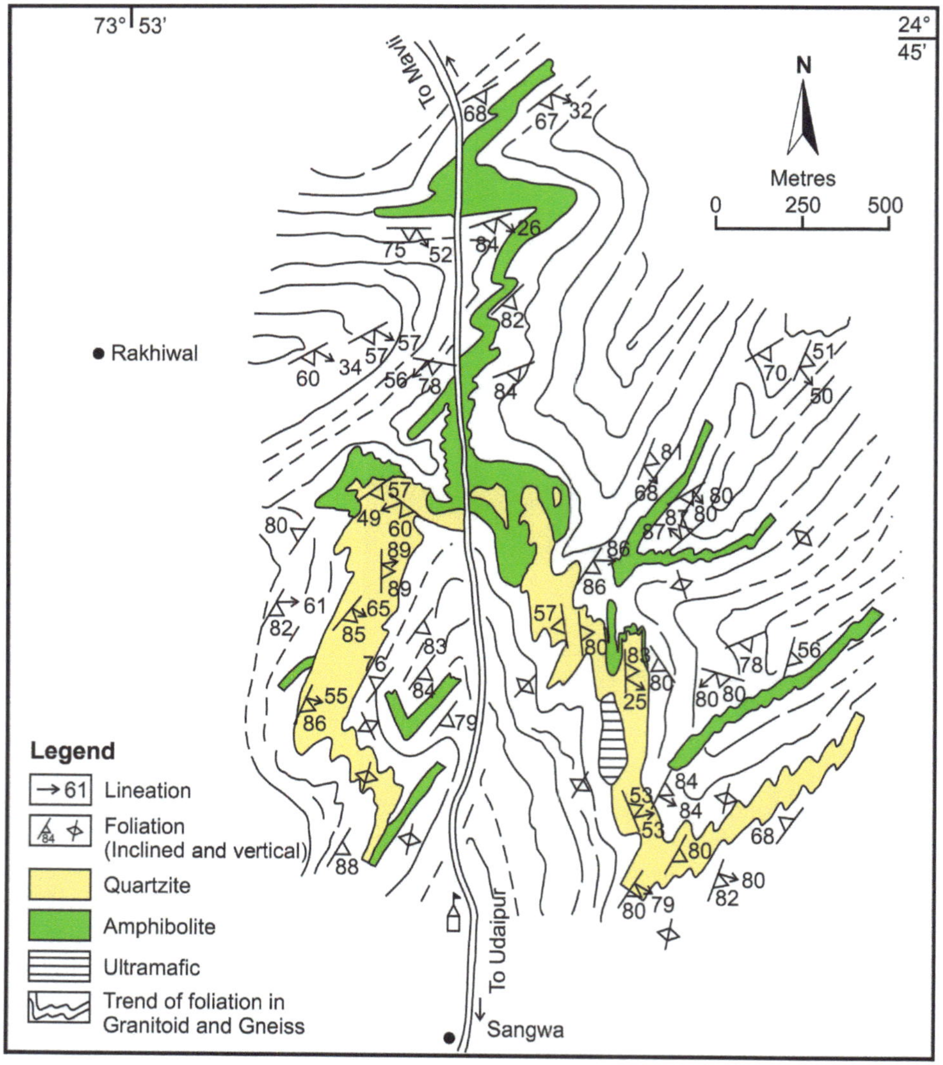

FIG. 9.4

Geological map of the Rakhiawal Greenstone Belt, east of Udaipur.

Reproduced from Roy and Jakhar (2002) with permission.

marble, calc-silicate rocks, banded iron formation, and mica schist. The quartzite is usually greenish in colour containing chrome-bearing mica minerals. The trace and REE compositions of the quartzites show abundance of Cr and Ni. The carbonate rocks in the greenstone belts are represented by low-Mg coarse crystalline marble with thin bands of disseminated graphite. The silicate ironstones contain both banded magnetite quartzite and quartz-magnetite-grunerite rocks. The significant features noted in the greenstone belt rocks are (i) paucity of the ultramafic rocks and (ii) the total absence of conglomerate and coarse clastic rocks (Upadhyaya et al., 1992). There is no known occurrence of metal mineralization in the Archaean basement rocks except some local bodies of barite and talc deposits. The occurrence of barite could be linked with the mafic volcanics formed during the opening of the Aravalli Basins.

The available isotope data not only confirm the existence of an Archaean basement in the Aravalli Mountains and adjacent areas but also helped in erecting a billion-year evolutionary history spanning between 3500 and 2500 Ma (Crawford, 1970; Gopalan and Chaudhary, 1984; Gopalan et al., 1990; Roy and Kröner, 1996; Wiedenbeck and Goswami, 1994; Wiedenbeck et al., 1996; Vinogradov et al., 1964). Kaur et al. (2013) recorded existence of 3.7 Ga old sialic Crust in the Aravalli Mountain region, based on Lu–Hf isotope data from the detrital zircons associated with the younger supracrustal (Delhi Supergroup) assemblages. There are several isotope studies that suggested 3.3–3.28 Ga age of the Archaean Banded Gneissic Complex (Gopalan et al., 1990; Roy and Kröner, 1996; Wiedenbeck and Goswami, 1994; Vinogradov et al., 1964). Gopalan et al. (1990) have recorded 2828 ± 46 Ma age based on the Sm/Nd isotope studies from the amphibolite (meta-basalt) intruding into cofolded quartzite in the Rakhiawal Greenstone Belt, a few kilometre west of Mavli region (Fig. 9.4). Similar age, 2887 ± 5 Ma, was recorded from the reconstituted tonalite-trondhjemite-granodiorite gneiss of the Jagat area by Roy and Kröner (1996). Other than these, there are records of isotope data on the granite bodies that range in age between ca. 2666 and 2450 Ma (Roy and Kröner, 1996; Wiedenbeck et al., 1996).

The geochronological framework of the Archaean basement of the Aravalli Mountains is shown in Table 9.2.

Table 9.2 Geochronologic framework of the Archaean Basement of the Aravalli Mountains

Age (Ma)	Analytical Method Used	Tectonomagmatic Event	Reference
2450 ± 8	Single zircon ion-microprobe U–Pb	Berach Granite	Wiedenbeck et al. (1996)
2506 ± 4	Single zircon ion-microprobe U-Pb	Vali River Granite, Jagat	Wiedenbeck et al. (1996)
2505 ± 3	Single zircon ion-microprobe U-Pb	Pink Granite, Untala	Wiedenbeck et al. (1996)
2532 ± 5	Single zircon ion-microprobe U-Pb	Jhamarkotra Granite	Wiedenbeck et al. (1996)
2562 ± 6	Single zircon ion-microprobe U-Pb	Ahar River Granite	Wiedenbeck et al. (1996)
Shearing and low temperature metamorphism			
2620 ± 5	Single zircon evaporation age	Gingla Granite	Roy and Kröner (1996)
2658 ± 5	Single zircon evaporation age	Gagat Granitoid	Roy and Kröner (1996)
2666 ± 6	Single zircon evaporation age	Untala Trondhjemite (enclave in pink granite)	Roy and Kröner (1996)
Ductile deformation and repeated folding and metamorphism			
2828 ± 46	Sm/Nd isochron age	Mafic dykes, Mavli	Gopalan et al. (1990)
2887 ± 5	Single zircon Evaporation age	TTG Gneiss, Jagat	Roy and Kröner (1996)
~3230	Single zircon Evaporation age	Trondjhemite intrusion into BGC, Jhamarkotra	Roy and Kröner (1996)
~3232	Single zircon evaporation age	Detrital zircon in Quartzite in Rakhiawal Greenstone Belt	Roy et al. (2001)
3281 ± 3	Single zircon ion-microprobe U-Pb	BGC, Jhamarkotra	Wiedenbeck and Goswami (1994)
3307 ± 65	Sm/Nd isochron age	BGC, Jhamarkotra	Gopalan et al. (1990)

Note: TTG Gneiss, tonalite-trondhjemite-granodiorite gneiss; BGC, Banded Gneissic Complex.

After Roy, A.B., Jakhar, S.R., 2002. Geology of Rajasthan: Precambrian to Recent. Scientific Publishers (India), Jodhpur, 421p.

9.2.1 THE BASEMENT-COVER RELATIONSHIPS: RECOGNITION OF UNCONFORMITY

Tectonostratigraphic studies based on continuous mapping in the southern Nathdwara-Udaipur-Lonavada sector in the southeastern part of the Aravalli Mountains (Fig. 9.1) helped to confirm the presence of unconformity that Heron (1953) identified as 'erosion unconformity' between the Archaean gneissic basement, the Banded Gneissic Complex, and the overlying cover metasediments, the Aravalli Supergroup (Roy and Jakhar, 2002). A number of features developed at the contact of the basement gneisses and the cover Aravalli succession helped to recognize the surface of stratigraphic break between the two major formations (Roy et al., 1988; Roy, 1990). These are (i) the occurrence of pockets of high aluminous deposits recognized as palaeosols along the surface of separation between the Banded Gneissic Complex and the overlying (stratigraphically) the Aravalli Supergroup (Roy and Paliwal, 1981; Sreenivas et al., 2001; de Wall et al., 2012) (Fig. 9.5a) and (ii) the local occurrences of basal conglomerate (Fig. 9.5b).

FIG. 9.5

Features indicating unconformity between the basement gneisses and basal beds of the Aravalli Supergroup. (a) Occurrence of isolated pockets of metamorphosed high aluminous deposits (palaeosols) at the contact of the basement gneisses (Banded Gneissic Complex) (flat lying left-hand side) and the basal quartzite (locally overturned). (b) Basal conglomerate showing occurrence of large boulders and pebbles of quartzite, granite, and pieces of vein quartz in arkosic matrix.

Detail mapping at different places also helped to recognize the surface of unconformity between the Archaean basement and the overlying cover succession marked by the major structural and metamorphic breaks in the lithostratigraphic sequence. At Nathuwas southeast of Nathdwara, there are straight running thin basal beds of quartzite that overlie the Banded Gneissic Complex with pockets of high aluminous palaeosols along the surface of unconformity (Fig. 9.6). In addition to the highly tectonized basement rocks, there are several detached and shredded bands of quartzite that show extremely complex pattern of folding. Some of these disjointed bands of quartzite show high angular relationship with the overlying crossbedded quartzite (Fig. 9.6). Additional field evidence comes from some outcrops where the mafic volcanics are noted as having intruded into complexly deformed banded gneiss, sending irregular veins into it (Fig. 9.7).

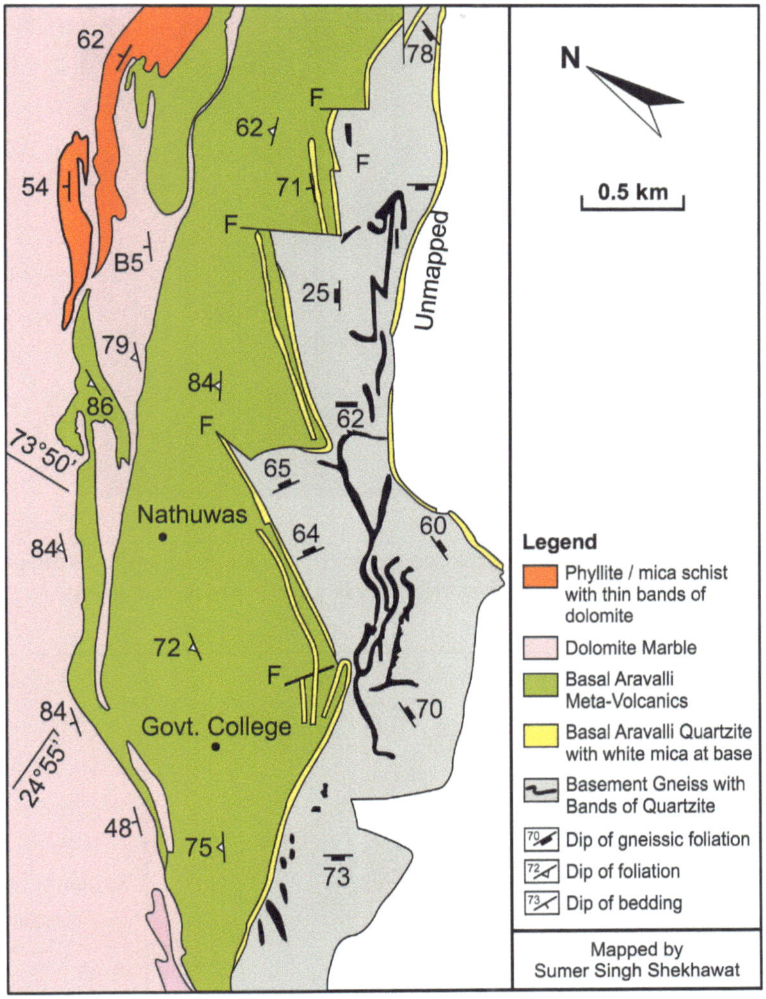

FIG. 9.6

Geological map of the Nathuwas area, southeast of Nathdwara showing the unconformable relationship between the Banded Gneissic Complex and the Aravalli Supergroup.

Reproduced from Roy et al. (1988) with permission.

The strong unconformable relationship is also indicated clearly in the map pattern of the Jhamarkotra region (Fig. 9.8) (Roy et al., 1980). Here, a prominent angular relationship is observed between the bedding surfaces in the overlying straight running basal quartzite (showing the direction of stratigraphic younging away from the Archaean basemnt rocks) and the gneissic foliation in the basement represented by the Banded Gneissic Complex. There also exists a sharp metamorphic break between the two rock sequences.

FIG. 9.7

An outcrop showing the extremely deformed banded gneiss is intruded by the mafic volcanics (in *dark shades*) of the basal Aravalli formation, sending irregular veins into the basement gneiss.

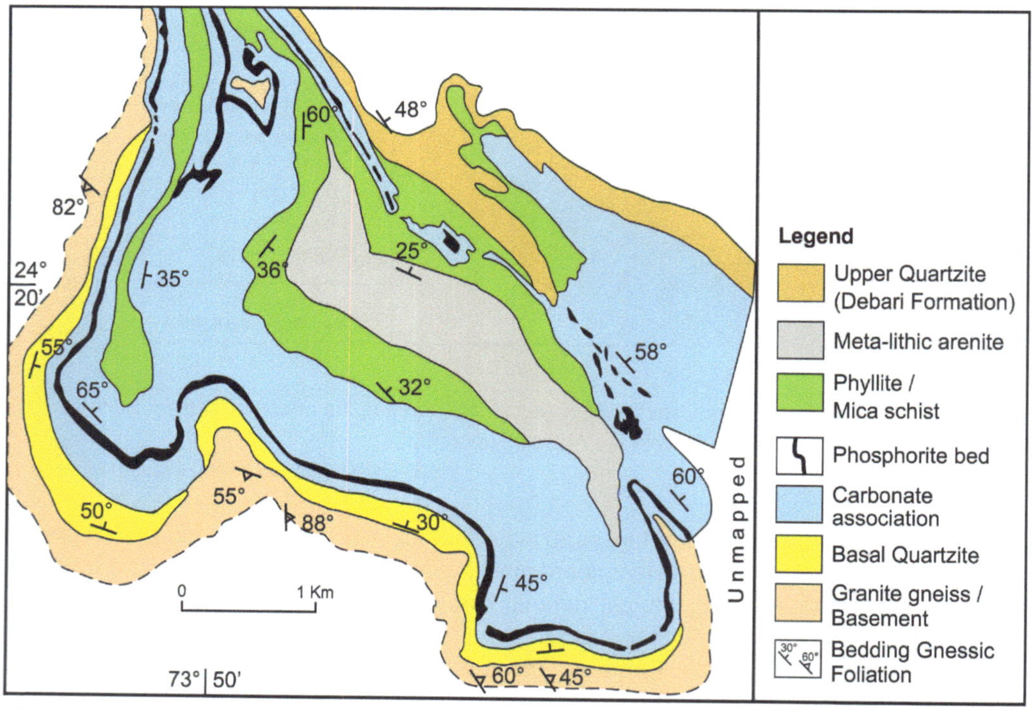

FIG. 9.8

Geological Map of the Jhamarkotra Phosphorite Mine area, southeast of Udaipur.

Reproduced from Roy and Jakhar (2002) with permission.

9.2.2 TECTONO-THERMAL RECONSTITUTION OF THE ARCHAEAN BASEMENT ROCKS

The detailed mapping postdating the publication of the monumental work of Heron (1953) highlighted on several unequivocal evidence of tectonothermal reconstitution of the Archaean basement in parts of central Rajasthan (Naha and Roy, 1983). Most of these reconstitutions took place along the zone of high strain concentrated along the interface of the basement and the cover succession. As a result of that, the earlier developed metamorphic foliation (in the gneisses) has been transposed into uniformly planar foliation that can be described as the mylonitic foliation. Critical analysis of some of the outcrop-level surfaces reveals the presence of tightly pressed isoclinal folds in between the planar (mylonitic) foliation (Fig. 9.9a) (Roy et al., 2016). Gradual changes from the broadly open folds to folds having isoclinal geometry and finally to virtually planar transposed foliation, close to the surface of unconformity between the basement and the cover sequence (Fig. 9.9b).

FIG. 9.9

(a) Uniformly planar mylonitic foliation formed due to transposition of earlier foliation that at times bound hinges of tightly pressed isoclinal folds. (b) An outcrop near the basement-cover interface illustrating gradual change over from broadly open folds to folds having isoclinal geometry finally to transforming into planar transposed foliation at the zone of contact between the basement and the cover.

Apart from the reconstitution along the basement-cover interface, there are instances where widespread transformation of the gneissic rocks took place during the emplacement of late-Palaeoproterozoic (1725 and 1621 Ma) (Roy et al., 2005a; Roy et al., 2012) granulites in the Banded Gneissic Complex occurring north of Nathdwara (mapped as the BGC II by Gupta, 1934). The emplacement of the hot granulite bodies transformed the adjacent host rocks (the Archaean gneisses) into migmatite gneiss (Roy et al. 2005a, 2012).

Field evidence and isotope data from this part of the Banded Gneissic Complex suggest a polycyclic and time-transgressive evolution of the Banded Gneissic Complex in the east-central part of the Aravalli Mountains. In view of this, a new nomenclature, Sandmata Complex, has been introduced for the reconstituted Archaean basement (Roy et al., 2005a, 2005b, 2012).

9.3 ARAVALLI SUPERGROUP: THE OLDEST COVER SUCCESSION

The oldest cover succession unconformably overlying the Archaean gneiss-granite basement is termed the Aravalli Supergroup (Roy, 1988, 1990; Roy et al., 1988). The outcrops of this major lithostratigraphic unit occur over wide areas in the east-central and southeastern part of the Aravalli Mountains, between Jaipur in the north and much beyond Lonavada in the south, Fig. 9.1). For the sake of convenience, we would divide the entire belt into two sectors: the eastern Bhilwara-Chittaurgarh belt in the northeast and the Udaipur-Lonavada belt in the southeast.

A complete stratigraphic succession of the Aravalli Supergroup has been reconstructed in the Udaipur-Lonavada belt that extends between Nathdwara in the north to much beyond Lonavada in the south (Roy and Jakhar, 2002) (Fig. 9.10).

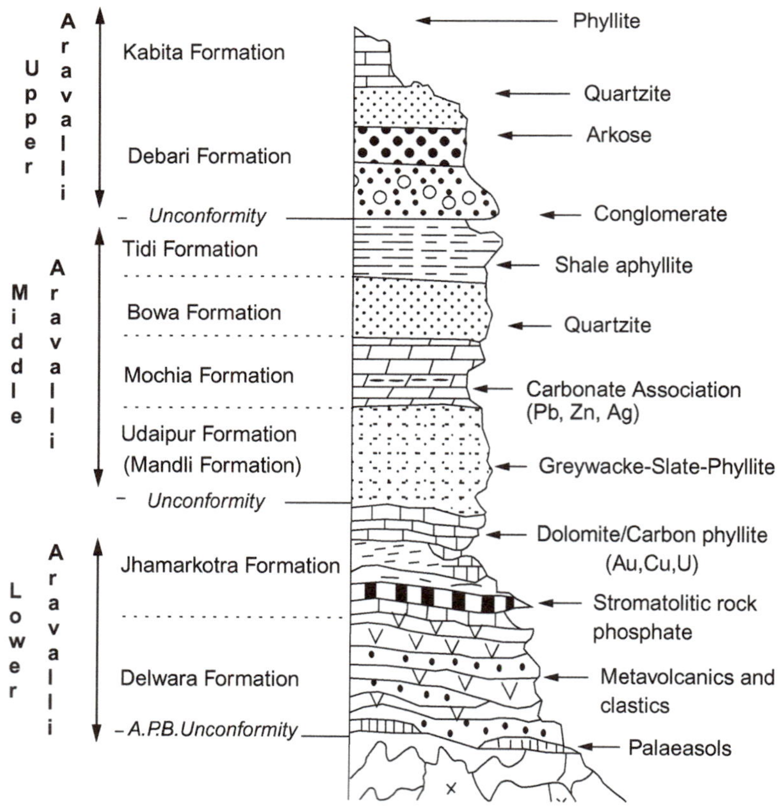

FIG. 9.10

Lithostratigraphic column of the Aravalli Supergroup in the shelf facies zone of the Udaipur sector. APB, Archaean-Proterozoic Boundary.

Reproduced from Roy and Jakhar (2002) with permission.

A modification in the lithostratigraphic column is suggested by Purohit et al. (2012b), who introduced a new carbonate sequence, the Babarmal Formation (Fig. 9.11) between the Jhamarkotra Formation above and the Delwara Formation below. The newly defined litho-unit comprises dominantly of pink calcitic marble, which is devoid of any organic matter and phosphate content unlike the younger Jhamarkotra dolomite.

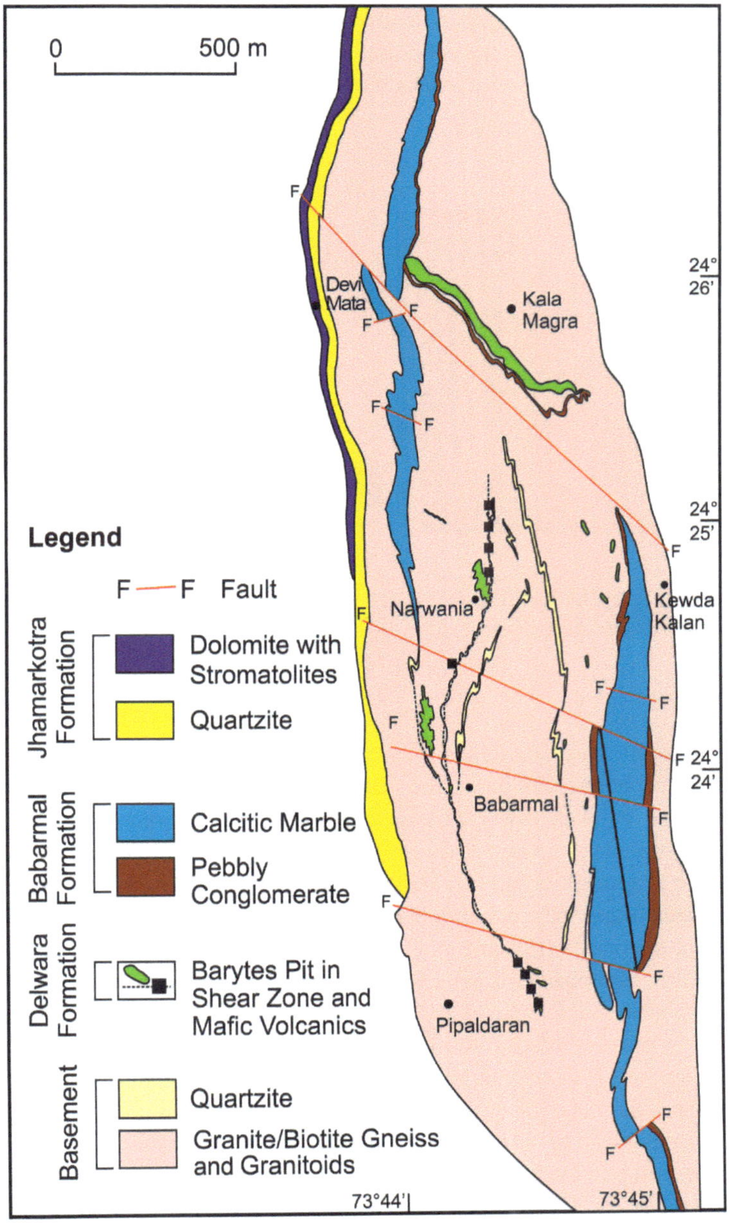

FIG. 9.11

Geological Map of the Babarmal area showing occurrence of pink, calcitic marble overlying basal quartz-pebble conglomerate.

Reproduced from Purohit et al. (2012a) with permission.

The deposition in this part of the Aravalli basin was initiated with the outpouring of mafic volcanics interlayered with thin bands of quartzite mainly in the basal part. The distribution of the Aravalli metasediments in the Udaipur-Lonavada belt shows two distinctive lithologic associations (Roy and Paliwal, 1981; Roy, 2000; Roy and Jakhar, 2002). The eastern belt shows occurrence of arenites, psammopelites, and carbonates hosting stromatolitic phosphorite (Fig. 9.12a). The younger carbonates show *syn*-sedimentary lead-zinc deposits (Fig. 9.12b) and uranium and gold mineralization. The western belt, on the other hand, is represented by a carbonate-free association of dominantly pelitic and psammopelitic rocks with thin bands of arenite. This belt is devoid of any mineralization.

FIG. 9.12

(a) Columnar stromatolite *(grey)* in dolomitic matrix *(dark)*. (b) pattern of lead-zinc mineralization in black-shale facies rocks. Zawar mineralized belt.

The stromatolitic rock-phosphate deposits in the eastern part are related to extensive cyanobacterial activities in the carbonate rocks along the tidal and intertidal flats of the earliest developed epicontinental sea (Roy and Paliwal, 1981). These Paleoproterozoic carbonates not only host the world's oldest sedimentary phosphate deposit but also exhibit near-zero $\delta^{13}C_{carb}$ values that are indicative of a complex carbon cycle (Papineau et al., 2013). The event of ^{13}C excursion was not essentially a time-specific process but was found to be dependent on the variation in the depositional palaeoenvironment prevailing in different subbasins (Purohit et al., 2010). The original microbial biomass is inferred to have been dominated by cyanobacteria, which might be an important contributor of organic matter in shallow-marine phosphorites (Papineau et al., 2016).

Another interesting feature noted in the younger succession of the Aravalli Supergroup is the occurrence of drop-stone-type features marked by deposition of well-eroded boulders embedded in the phyllitic matrix in the Udaipur region (Fig. 9.13). A suggestion is muted that these could represent diamictite-type rocks representing 'glacial erratics'.

The Aravalli Supergroup in the Bhilwara Sector shows bimodal distribution of lithologies. The eastern belt (also described as the Hindoli belt by some author, Gupta et al., 1997) shows dominance of

FIG. 9.13

Occurrence of well-rounded very large boulders and other larger clasts in phyllite matrix in the younger succession of the Aravalli Supergroup in a section at Udaipur.

Reproduced from Roy and Paliwal. (1981) with permission.

metavolcanics with minor metasediments that have undergone a low-grade metamorphic reconstitution. The western belt of the Bhilwara Sector on the other hand is characterized by shale-sand-carbonate-type lithoassociations showing an upper-amphibolite-facies metamorphism. The typical lithology includes quartzite, dolomitic marble, and sillimanite-almandine garnet rock along with graphite. The metamorphic and mineralogical character of these banded rocks appears quite similar to the khondalites of the Eastern Ghats Granulite Belt. There are several zones of lead-zinc and copper mineralization in this belt of the Aravalli Supergroup. The largest deposit of this type occurs at Rampura Agucha where massive lead-zinc-bearing sulphide ore body occur interlayered with the shale-sand-carbonate-bearing lithological association (Fig. 9.14a and b).

The Aravalli Supergroup characteristically shows a complex deformation pattern marked by two phases of major folding. The earlier fold phase produced tight to isoclinal folding having an extremely penetrative deformation pattern accompanied by finely foliated slaty cleavage or schistosity (Fig. 9.15a). The style and geometry of the superposed folds varied from place to place depending on the orientation of basins and the direction of shortening. In the southern belt between Nathdwara and Udaipur, the most common feature is the occurrence of curved lineation over the fold hinges (Fig. 9.15b). In other cases, folding of extremely stretched pebbles (Fig. 9.15c) provided evidence of multiphased deformation. There are also examples heterogeneous shortening producing detached fold limbs along zones of strong ductile shearing (Fig. 9.15d).

Regionally, Zawar mineralized belt south of Udaipur presents an example of deformation along slices bound between shear zones in a narrow belt constrained between two relatively rigid basement blocks, the Sarara block in the south and the Babarmal block in the north (Fig. 9.16).

An instructive example of basement-cover relationship between the Banded Gneissic Complex and the Aravalli Supergroup is observed at Kherwara, south of the Zawar Mineralised belt. The basement rocks (granite-gneiss along with the supracrustals like quartzite) occur as inlier that is bounded by faults (Fig. 9.17). The inlier block shows complex outcrop pattern compared to the Aravalli supracrustals.

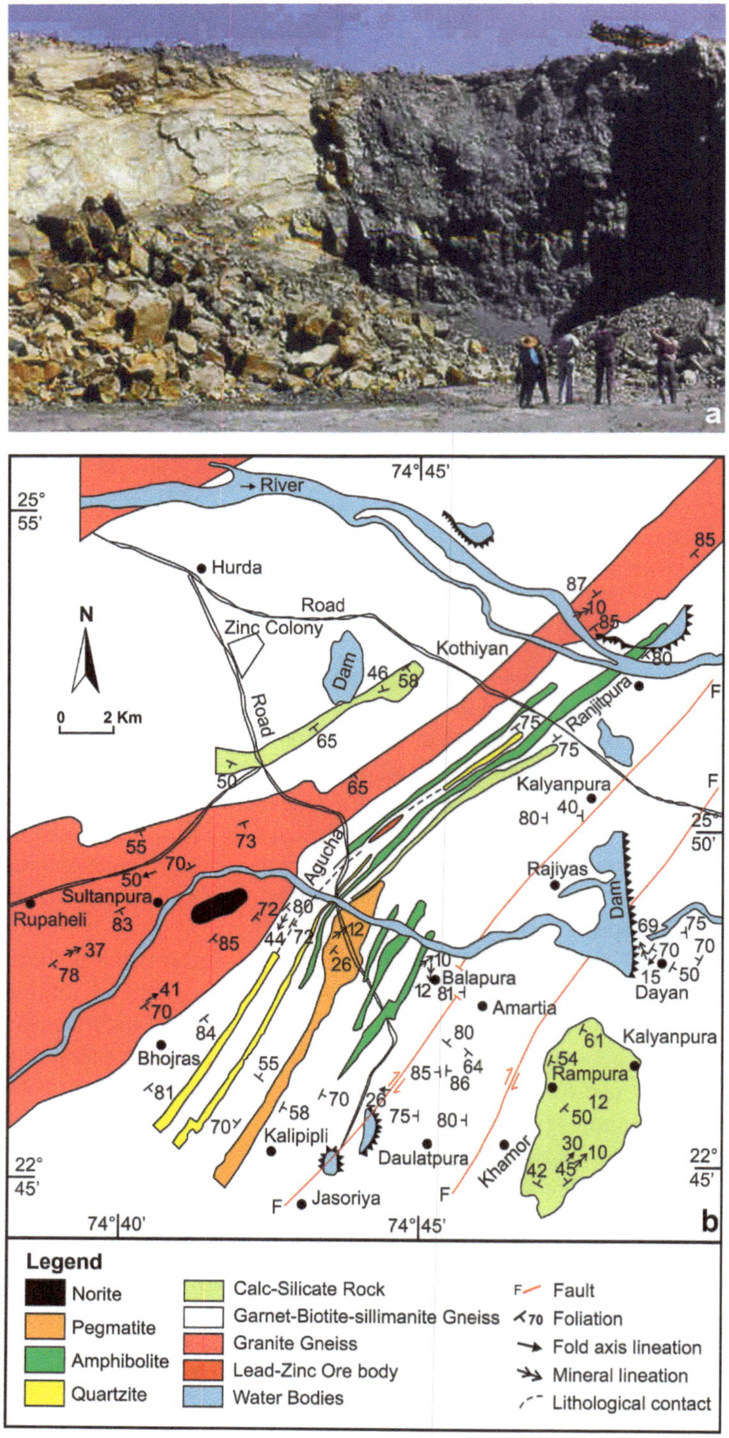

FIG. 9.14

(a) Massive zinc-lead-sulphide-bearing ore band between calc-silicate rocks (left) and mica schists (right, the shaded zone) from the Rampura Agucha Mine, southeast of Ajmer. (b) Geological map of the Rampura Agucha area showing lithological association of the massive lead-zinc-bearing sulphide ore body. Mapped by Sanjay Kumar, Vivek Laul, and N.K. Chauhan.

FIG. 9.15

(a) Development of penetrative slaty cleavage in lowly metamorphosed phyllitic rocks. (b) Open upright fold of second-generation deforming early lineation formed during early deformation. (c) Asymmetric folding deforming extremely stretched pebbles formed during the first-generation fold. (d) Heterogeneous strain pattern indicated by shearing along stretched folds.

In the southernmost part of the Aravalli Mountains, a complex structural geometry has resulted due to interference of several phases of folding deformation. The intricate fold interference pattern shown by the bands of quartzite occurring within mica-schist-phyllite-bearing country rocks is unknown in any other part of the Aravalli Mountains (Fig. 9.18).

Primarily based on the distribution pattern and the stratigraphic correlation of lithologies, a multistaged evolution of the Aravalli rift basins has been proposed (Roy and Jakhar, 2002). The basins opened as 'FRR' rifts having a triple junction with one or more failed arms (Fig. 9.19). The sequence of basin evolution is not clearly known. However, it appears possible that all the different basins did not open simultaneously.

Unequivocal evidence of Palaeoproterozoic age of the Aravalli sedimentary-volcanic succession is indicated by the presence of features, like (i) the dolomitic character of the major carbonate-bearing formations (Roy and Jakhar, 2002), (ii) prolific growth of cyanobacterial colonies depositing stromatolitic rock phosphate in the carbonate (Fig. 9.12a and b) (Banerjee, 1971a,b;

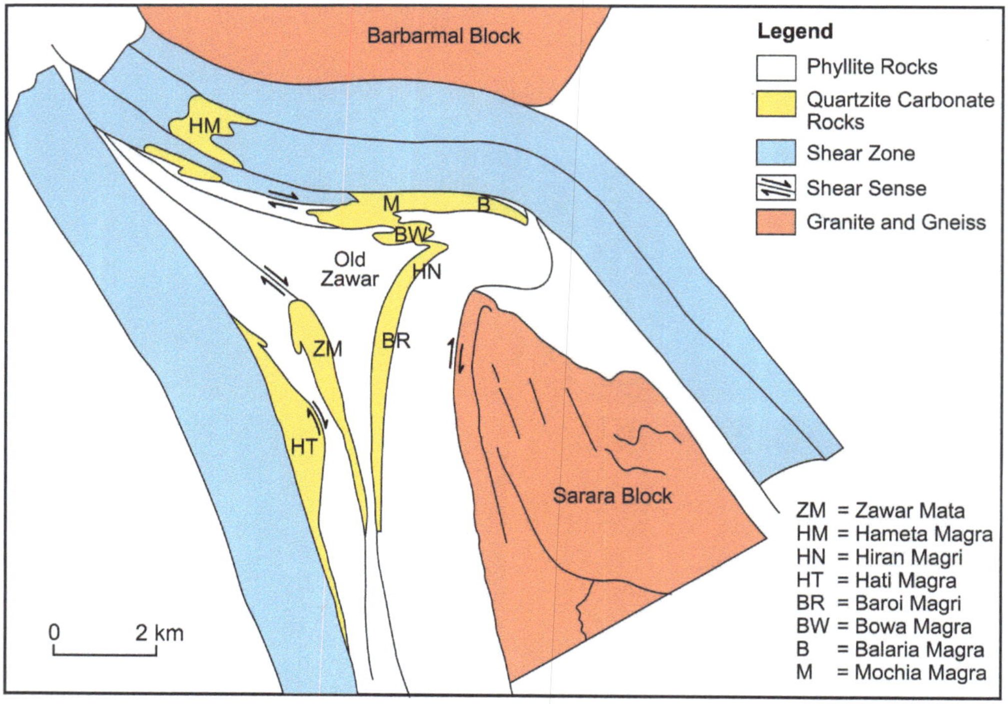

FIG. 9.16

Sketch map showing extremely complex pattern of large-scale regional deformation noted in the Zawar mineralized belt, south of Udaipur marked by shear-zone-bounded tectonic slices constrained between two relatively rigid basement blocks, the Sarara 'Inlier' block in the south and the Babarmal block in the north.

Chauhan, 1979; Roy et al., 1980; Roy and Paliwal, 1981), (iii) occurrence of *syn*-sedimentary lead-zinc sulphide mineral deposits (Fig. 9.14a and b) (Roy, 2000; Roy et al., 2004), (iv) extensive development of black-shale facies rocks indicating the presence of biogenic carbon, (v) isotopic character of carbon in the lead-zinc-bearing sulphide ore bodies (Deb, 2000), and (iv) the nature of $\delta^{13}C$ enrichment in the carbonate rocks (Purohit et al., 2010). The geochronological control over the basin opening and depositional age of the Aravalli Supergroup is poor except the Pb–Pb model age of ca. 2075 and 2150 Ma determined from galena associated with barite in the basal volcanics (Deb and Thorpe, 2004). Somewhat similar, ~2.06 Ga age is suggested from the study of the heavy ^{13}C enrichment in the carbonates correlating with the Paleoproterozoic Lomagundi-Jatuli event (Purohit et al., 2010). The closing of the Aravalli orogeny is correlated with 1900 ± 80 Ma Rb/Sr isochron age of the synkinematically emplaced Darwal Granite (Choudhary et al., 1984). Confirmation of the age data comes from 1921 + 67 Ma Pb–Pb isochron age from the tectonically reconstituted dolomitic carbonate of the Aravalli Supergroup of the Udaipur region (Sarangi et al., 2006).

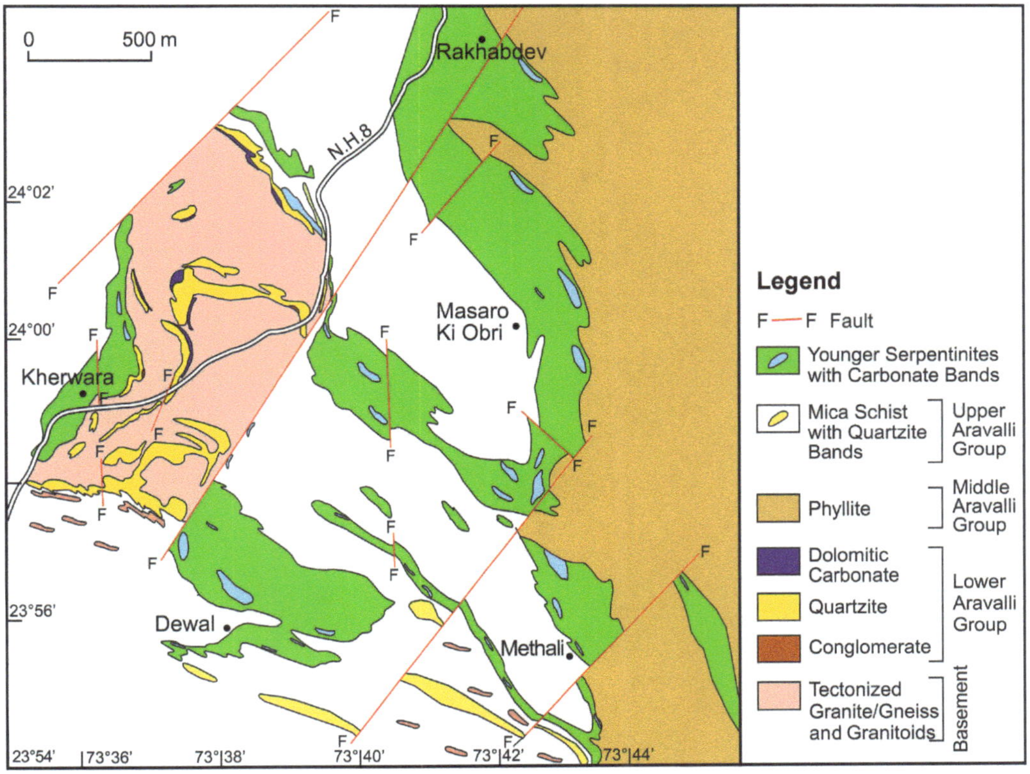

FIG. 9.17

Geological map of the Kherwara area showing fault-bounded inlier of the basement rocks surrounded by the rocks of the Aravalli Supergroup. Mapped by Ritesh Purohit Jogaram and Harsh Bhu.

A large number of serpentinite bodies occur with the Aravalli basin of the Udaipur sector both in the shelf facies and the deepwater regions. A group of such bodies define a linear trend following the separation zone of the two depositional subbasins. No such trend in the occurrence of ultramafic bodies is visible in the western belt of deepwater facies sediments. The ultramafics of both the belts show similar mineral assemblages comprising mainly antigorite and serpentine, both of which have undergone extensive chloritization. Serpentine often shows alteration to dolomite and talc. Dominance of olivine and lizardite is observed in the Rakhabdev ultramafics, whereas magnetite and actinolite-tremolite are predominantly present in the Jharol ultramafics. Coherence is observed in the 'O' and 'H' stable isotope of ultramafics in both the belts, which suggests uniformity in the process of serpentinization (Purohit et al., 2015).

From the nature of deformation and the thermal alternation of the carbonates occurring close to these bodies, the serpentinite intrusion appears to have postdated the Aravalli Supergroup (Roy and Jakhar, 2002; Purohit et al., 2015).

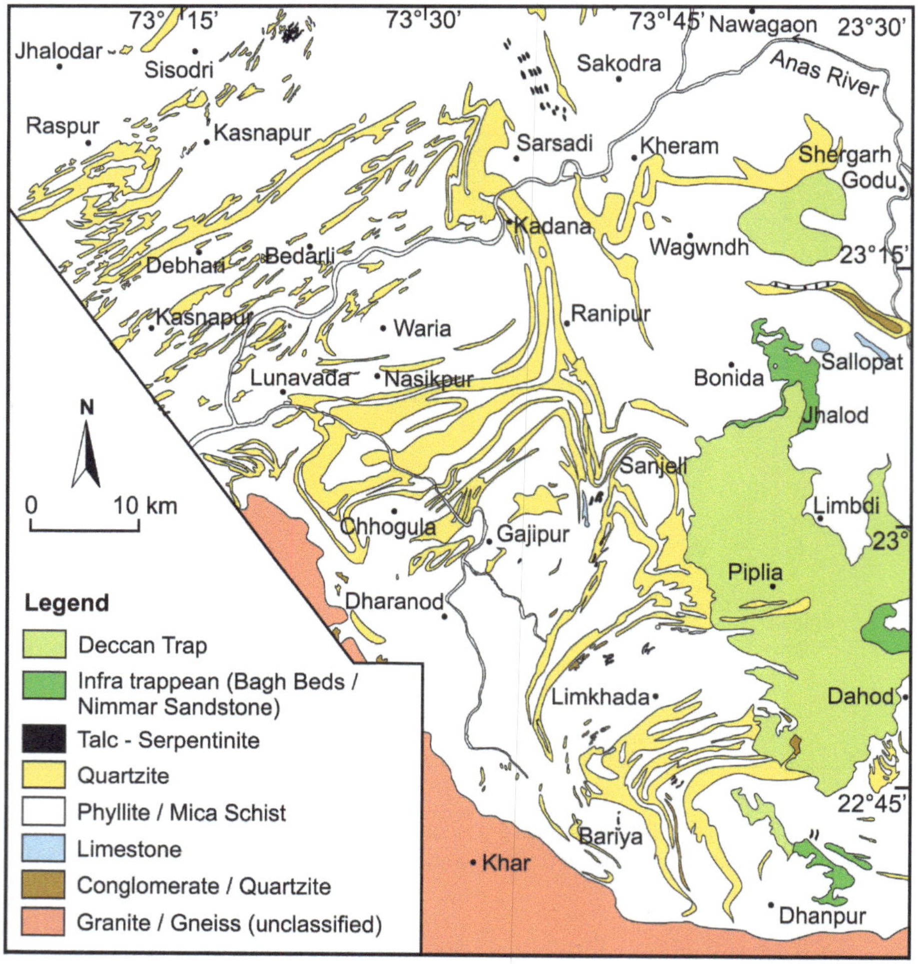

FIG. 9.18

Geological map of the Lonavada region, southern part of the Aravalli Mountains showing extremely complex pattern of outcrop because of interference of multiple phase of folding deformation.

Reproduced from Roy and Jakhar (2002) with permission.

9.4 DELHI SUPERGROUP: HISTORY OF MESOPROTEROZOIC RIFT BASIN EVOLUTION

Outcrops of the Delhi Basin replicate the shape of a huge fan having a long handle (Fig. 9.1). The lithological associations occurring in this basin, barring the large bodies of 'inlier rocks' and the younger anorogenic granites, diorites, and gabbro-norite-charnockite, constitute the Delhi Supergroup.

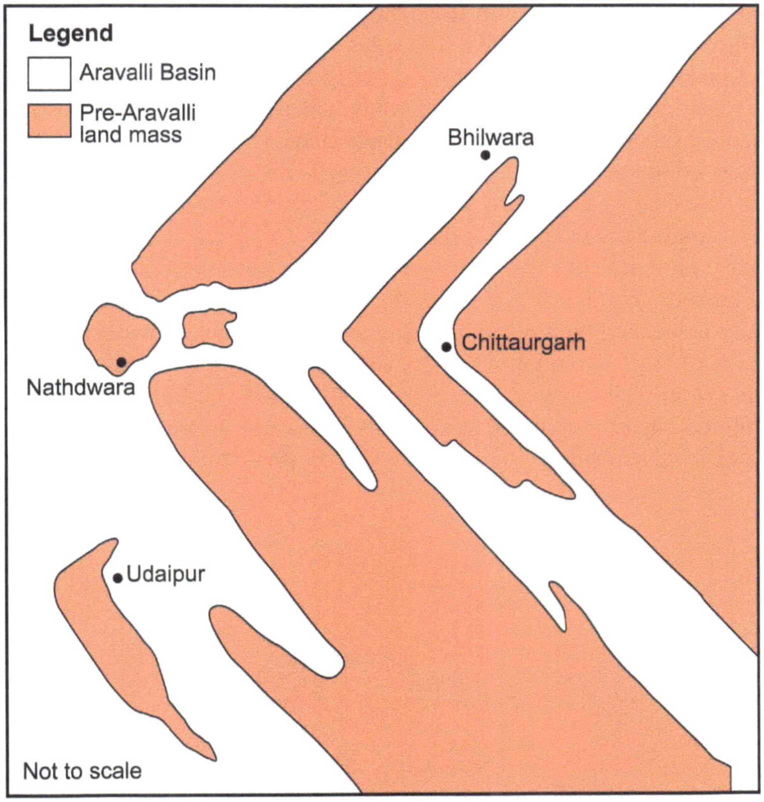

FIG. 9.19

Model of evolution of the Aravalli Basins along FRR rifts having failed limbs.

Reproduced from Roy and Jakhar (2002) with permission.

Unconformable relationship between the Delhi Supergroup and the pre-Delhi rocks (constituted of Aravalli Supergroup and the Banded Gneissic Complex) has been recorded in the northeastern part of the Delhi Basin (Roy and Jakhar, 2002). For the rest of the area, the field relationship does not provide much information on the stratigraphic relationship with the older or the younger successions. The Delhi Basin margins over a large part are demarcated by zones of shear (Roy and Jakhar, 2002).

A unique feature in the stratigraphic classification suggested by Heron (1953) is the recognition of a Raialo 'series' supposed to have been deposited during Archaean-Algonkian (*the earlier used term Algonkian signifies Proterozoic*) interval within and outside the Delhi Basin. More recent tectonostratigraphic studies (Naha and Halyburton, 1974; Roy et al., 1988), however, failed to corroborate any separate stratigraphic status for the so-called Raialo rocks reported from different parts of the Aravalli Mountains. In the present description, we preferred to retain the nomenclature Raialo Group (Group replacing series of Heron, 1953) but only for the sequence that occurs in the northeastern part of the Delhi Basin.

The Delhi Basin comprises several subbasins that include Alwar-Bayana Basin in the northeast and the Main Delhi Basin in the west and south. The latter extends from north of Khetri in the north to the southwest of Mt. Abu in the south. Heron's (1917a) stratigraphic scheme has been adopted by later workers (Singh, 1988) with some modification in conformity with the 'Standard Code of Stratigraphic nomenclature'. The following lithostratigraphic groups have been recognized in the Alwar-Bayana Subbasin:

Ajabgarh Group—dominantly pelitic and psammopelitic rocks and volcanics.

Alwar Group—dominantly arenitic rocks and volcanics.

Raialo Group—dominantly volcanics and carbonates with pelites.

The sedimentary attributes of different lithostratigraphic units indicate that their deposition in a number of rifted grabens (Fig. 9.20) (Singh, 1988). The presence of volcanic assemblages, mostly of mafic flows locally associated with felsic tuffs and ash beds, provides evidence of active rifting during different phases of basin evolution. Overall character of the lithologies suggests their deposition under the stable platform-type basins. A host of primary sedimentary features including cyanobacteria-induced stromatolitic structures recorded in rocks indicate shallow-water environment of deposition, which varied from fluvial to shallow-marine palaeoenvironment (Singh, 1988).

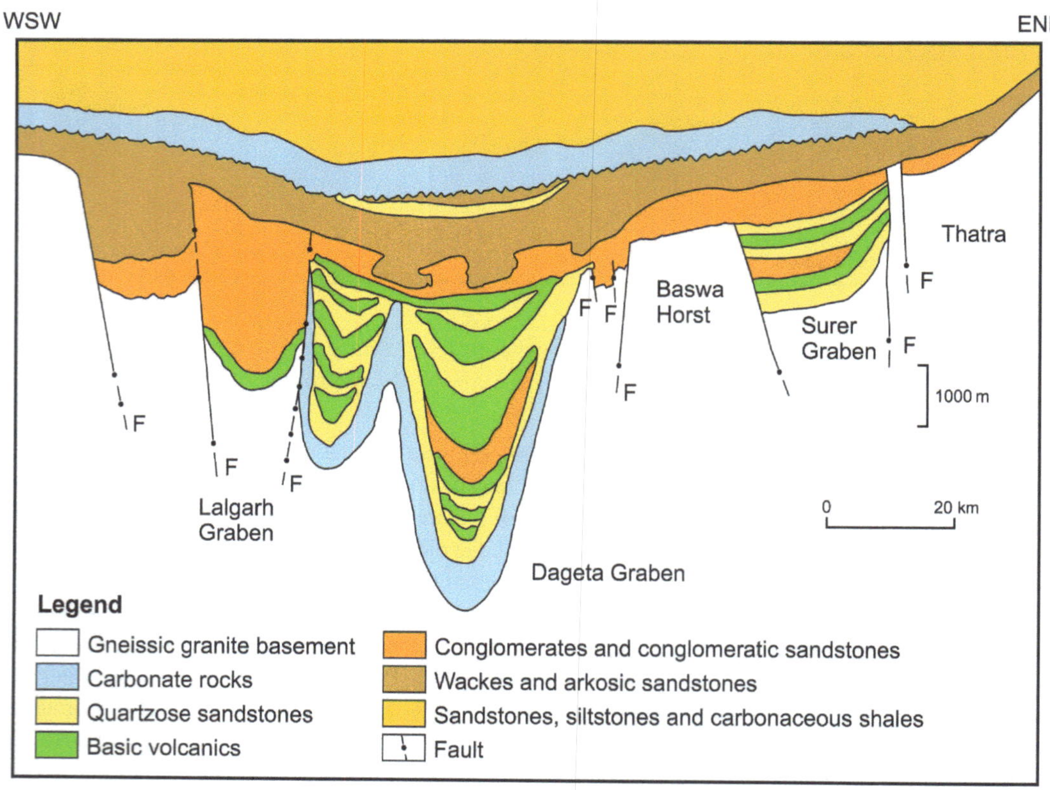

FIG. 9.20

The model of rift basin evolution of the Delhi Basin in the Bayana region.

Reproduced from Singh (1988) with permission.

The lithostratigraphic correlation of the different formations in the Main Delhi Basin remains enigmatic. The earlier proposed stratigraphic scheme of the Main Delhi Basin by Heron (1953) was based on the model that was established in the Alwar-Bayana Basin. In most of the recent studies in the Khetri-Saladipura belt, the Heron's stratigraphic scheme erected in the Alwar-Bayana Subbasin has been followed (Fig. 9.21).

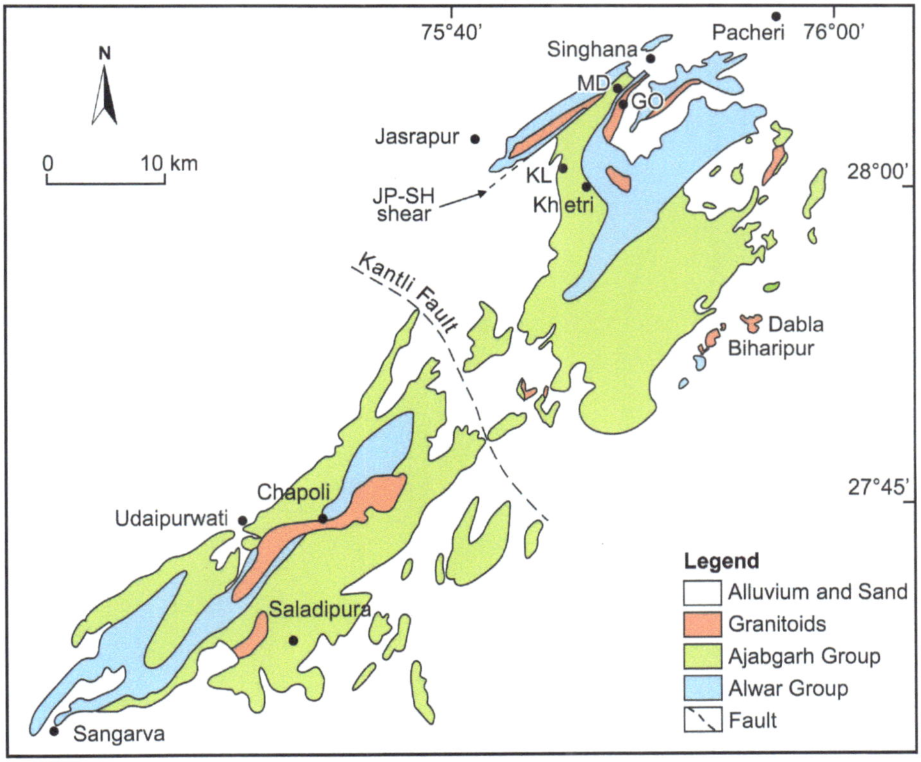

FIG. 9.21

Geological map of the Khetri Complex showing major rock units. JP-SH shear, Jasrapur-Singhana shear; GO, Gothara; MD, Madan-Kudhan; KL, Kolihan.

Reproduced from Gupta et al. (1998) with permission.

Between the Khetri-Saladipura belt in the north and the Pisangan-Bithur Line south of Ajmer, little study has been made to understand the lithostratigraphy of the region. However, some of the more recent workers in the Main Delhi Basin did not find Alwar-Bayana model suitable for to erect stratigraphic succession in other parts. A suggestion has been made to divide the Main Delhi Basin into two different fold belts: an older North Delhi Fold Belt and a younger South Delhi Fold Belt (Sinha-Roy et al., 1998). The division, apparently relying on unconstrained granites ages, does not specify the line (trace of the surface) of separation of the two blocks. However, Roy and Jakhar (2002) have hinted about the possibility that the separation between the southern and the northern blocks could represent a folded unconformity (Fig. 9.22).

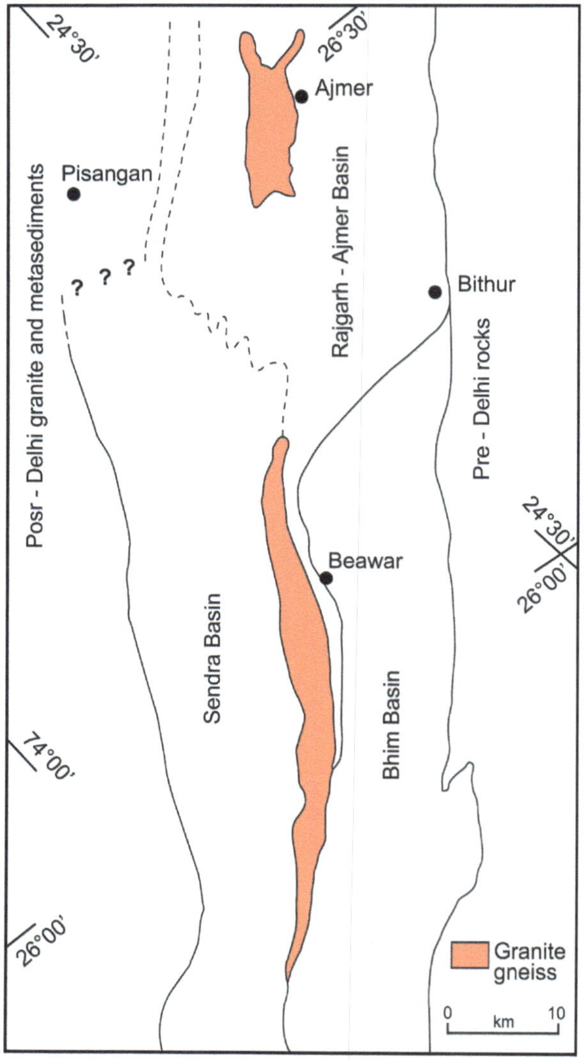

FIG. 9.22

Geological sketch map of the area between Ajmer and south of Beawar, showing relationship between northern Rajgarh-Ajmer Basin and the two southern basins: the eastern Bhim-Shyamgarh Basin and the western Sendra-Barotia Basin separated by a median inlier of granite-gneiss.

Reproduced from Roy and Jakhar (2002) with permission.

The Main Delhi Basin over its major part is a shear-zone-bounded body. Evidence of strong shearing is noted at several places both in the eastern and northern part. A classic example is noted in east of Shyamgarh a little south of Beawar (Fig. 9.23).

The narrow belt of inlier of basement gneissic rocks separates two different lithological associations in the southern belt, south of the Pisangan-Bithur Line (Fig. 9.22). Gupta et al. (1995) considered these

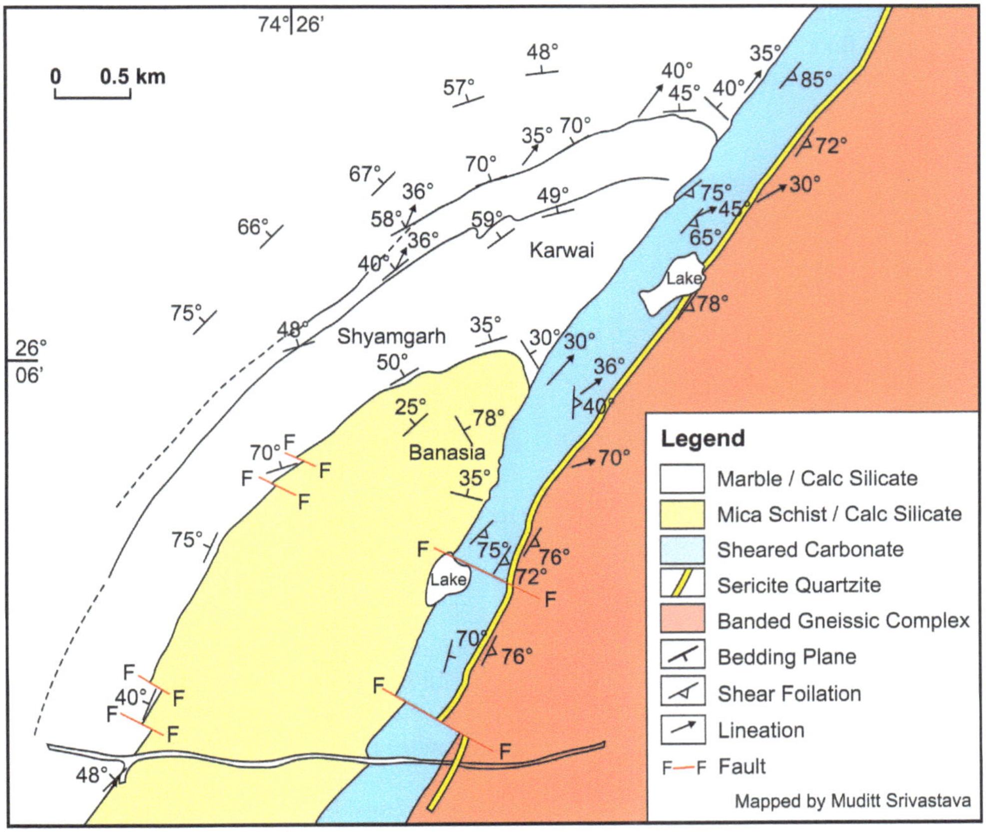

FIG. 9.23

Geological map of the Shyamgarh Antiform in the east-central part of the Main Delhi Basin showing evidence of a prominent shear zone (marked by the disposition of the rocks of carbonate association).

Reproduced from Roy and Jakhar (2002) with permission.

two belts as representing two separate subbasins: the eastern Bhim-Shyamgarh Basin and the western Sendra-Barotia Basin. The Bhim-Shyamgarh Basin is devoid of any volcanics. These rocks truly represent platformal association of metamorphosed shale-shale-carbonate assemblage deposited in a passive basin (Roy and Jakhar, 2002).

The western Sendra-Barotia (Barotia is also spelt as Biranthia or Birantia in literature) shows dominance of volcanic rocks (Gupta et al., 1995). Heron (1953), however, divided the Sendra-Barotia as two separate basins: the western Barotia sequence and the eastern Sendra Complex. Some interesting effusive and plutonic igneous associations have been reported along a narrow zone of about 100 km passing through Basantgarh (24°44′;76°00′) in the southern part of the Barotia belt. Based on the occurrence of low-K pillowed tholeiite along with dacite, andesite, banded gabbro, and massive pyroxenite (Fig. 9.24a and b) along a narrow strip in the southern part of the Delhi Basin, suggestions have been made about the occurrence of an ophiolite sequence (Sinha-Roy and Mohanty, 1988; Khan et al., 2005) representing a slice of obducted slice of palaeo-ocean Crust (Sugden et al., 1990). The occurrence of such an exotic ensemble in an ensialic basin, according to Roy and Jakhar (2002), could be due to the limited 'opening of the ocean floor' within the Aulacogen that constituted the Main Delhi Basin, which in a way is similar to the ocean opening observed in the case of the present-day Red Sea between Africa and the Arab Peninsula.

FIG. 9.24

(a) Massive pyroxenite occurring in a milieu of exotic rocks in the southern part of the Main Delhi Basin.
(b) Pillow structure in low-K basalt in little deformed low-K basalt in the southern part of the Main Delhi Basin.

Reproduced from Roy and Jakhar (2002) with permission.

In spite of the controversies on stratigraphic correlation of rocks occurring in different subbasins, the structural and metamorphic history of the entire Delhi Basin suggests a single-stage orogenic evolution (Roy and Das, 1985). According to these authors, the closing of the Delhi Basin took place at c.1450 Ma as suggested for the major synkinematic granites in the northern part of the Delhi basins (Gopalan and Chaudhary, 1984).

The pattern of deformation and metamorphism varies considerably over the entire Delhi Basin. However, for the major part of the Delhi Basin, a polyphase deformation pattern has been recorded (Roy and Jakhar, 2002). Characteristically, the early folds have isoclinal fold geometry having NNE-SSW fold trends (Fig. 9.25a and b). These early folds are superimposed coaxially by the second-generation NNE-SSW trending upright folds (Fig. 9.25c and d). Cross-cutting early folds at high angles occur in the third phase of folds in different scales of study. On an average, middle-to-upper-amphibolite-facies metamorphism is recorded over the major part of the terrane. However, in the extreme northeastern part of the Delhi Fold Belt, the rocks generally show a very low-grade greenschist facies metamorphism. In addition, the evidence of high-temperature plutonic magmatism has been recorded in the southwestern end, where extensive melting and thermal perturbation is noted which is correlated with the phase of Malani magmatism (Roy et al., 2005b) (Fig. 9.26).

It is possible that all the different components of the Delhi lithostratigraphic formations did not evolve simultaneously in different parts of the Delhi Basin. In spite of all the prevailing uncertainty about the internal stratigraphic relationship of the rocks occurring in different areas, the structural and metamorphic history of rocks clearly indicates a single-stage orogenic evolution of the Delhi rocks (Roy and Das, 1985). However, isotope data indicated that the Delhi Fold Belt witnessed repeated (tectono-) thermal reconstitution at later periods (Roy and Jakhar, 2002). The single-zircon age of the thermal perturbation noted in the Balaram area yielded 757.8 ± 0.9 Ma (Roy et al., 2005b) making it correlatable with the Malani magmatism.

FIG. 9.25

(a) Profile of a recumbent, first-generation fold in Ajabgarh 'calc-schist' from the Main Delhi basin.
(b) Refolding of early isoclinal folds by NNE-SSW trending folds second-generation upright folding.
(c) second-generation upright folds from near Kishangarh. (d) Upright, vertical fold from the Barotiya belt.

(d) Image courtesy: Sadhana M. Chatterjee.

Like the Aravalli Supergroup, there is also no direct geochronological control over the age when the earliest Delhi Basin opened. Fareeduddin and Kröner (1998) reported several detrital zircon ages from the western margin of the Delhi Basin. The youngest of these dates is about 1700 Ma. Similar ages (1.78–1.72 Ga) have also been reported by Kaur et al. (2011) from northern part of the Delhi Basin. Roy and Jakhar (2002) considered the youngest of these (~1700 Ma) as the maximum age of the Delhi Basin opening. These ages correlate well with the single-zircon ages ranging between 1725 and 1622 Ma ages of the granulites of the Bhinai region in the Sandmata Complex reported by Roy et al. (2005a, 2012). Assuming that the granulite facies metamorphism and the process of exhumation took place under an extensional regime, Roy et al. (2005b, 2012) linked these tectonic processes to the rift opening stage of the Delhi Orogenic cycle west of the outcrops of the Sandmata Complex.

There are a number of Rb–Sr isochron ages between 1340 and 1480 Ma of apparently synkinematic granites from the northern part of the Delhi Basin (Choudhary et al., 1984; Gopalan et al., 1979). Relying on these ages, Roy (1990) and Roy and Das (1985) suggested c.1450 Ma as the closing age of the Delhi Orogeny.

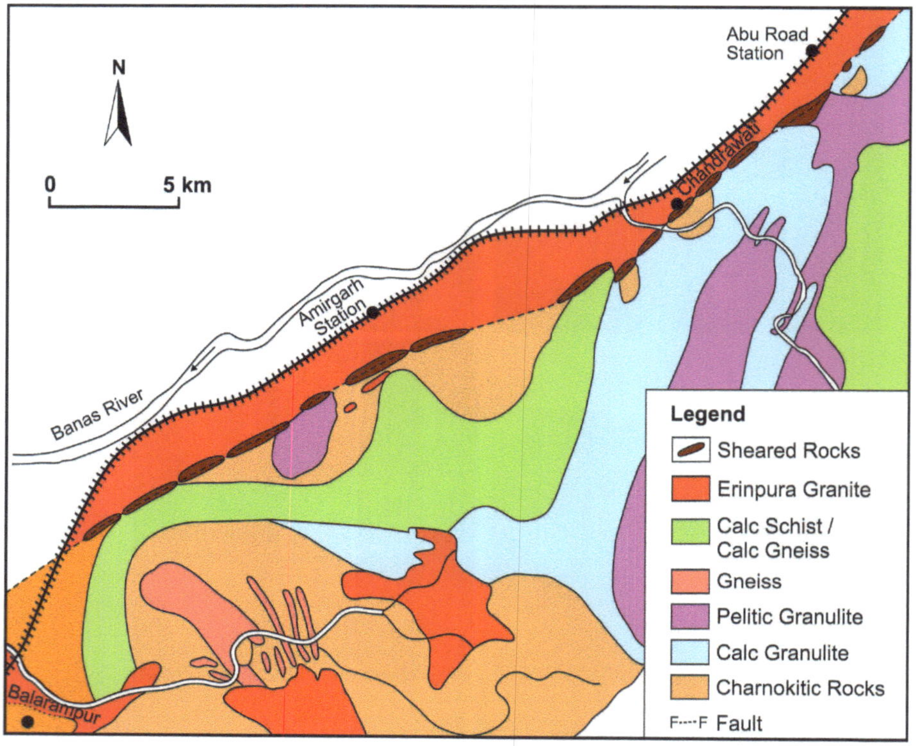

FIG. 9.26

Geological map of the area between Kui (southeast of Abu Road) and Balarampur.

Reproduced from Desai et al. (1978) with permission.

Apart from these, the Delhi Fold Belt rocks were affected by a number of anorogenic thermal events, and because of these, the rocks have reconstituted at places repeatedly that are noted primarily in the southern part of the Delhi Fold Belt. The major events recorded in this part of the Delhi Basin are the following (Roy and Jakhar, 2002):

~750 Ma—granulite emplacement in the southernmost part of the Delhi Basin.

~850 Ma—emplacement of granites post-Sirohi in age (Erinpura Granite, sensu stricto).

~1000 Ma—emplacement of diorite and granodiorite mainly in the southern part of the Delhi Basin.

The geological history of the Delhi Basin is marked with diversity in opinions, especially about the stratigraphic correlation of rocks of the different parts. However, the following broad generalization can be made in regard to the tectonostratigraphic relationship of the rocks of the Main Delhi Basin:

1. A broadly uniform pattern of structural and metamorphic evolution of the supracrustal ensembles is observed in the major part of the Delhi Basin including the Main Delhi basin.
2. Besides the occurrences of inliers of the gneiss-granite basement at Beawar and at Ajmer, some recent geochronological data indicate the presence of pre-Delhi metasedimentary bodies in the northern part of the Main Delhi Basin (Gupta et al., 1995; Gupta et al., 1998; Kaur et al., 2011b, 2013).

3. No direct stratigraphic correlation of appears possible between the rocks of the Main Delhi Basin and the Alwar-Bayana Basin or even between the northern and the southern part of the Main Delhi Basin.

In spite of all these, some nagging uncertainty remains about the occurrence of an 'ophiolite belt' in the southern part of the Delhi Basin. These 'exotic' rocks have not undergone any deformation and coeval metamorphism with any of the Delhi Supergroup rocks, nor there is any precise isotope data to correlate these with the Delhi Basin rocks or with the Delhi Orogeny. We cannot rule out the possibility that these undeformed unmetamorphosed bodies formed at a later date much after the closing of the Delhi Basin.

9.5 NEPHELINE SYENITE OF KISHENGARH

An interesting suite of alkaline rocks with nepheline syenite as the most dominant component occurs in the area around Kishangarh, northeast of Ajmer (Fig. 9.27). The main body of alkaline syenite occurs east of the Main Delhi Basin and is intimately associated with the metasedimentary-metavolcanic assemblages that occur between the Delhi Basin rocks in the west and the banded gneisses of the Sandmata Complex in the east. Apart from the main body at Kishangarh, there are also some thin, extremely sheared bodies of nepheline syenite at the contact of the Delhi Basin rocks. According to Roy et al. (2016), the nepheline syenite forms an elongate flattened dome, refolded into a hook-shaped along the boundary of the Delhi Basin rocks. There are no reliable isotope age data for the emplacement of the nepheline syenite body. The lithostratigraphic correlation of the associated metasedimentary and metavolcanic rocks (mapped as the Kishangarh Group) with those of the Aravalli Supergroup suggests a post-Aravalli age of the Kishangarh Nepheline Syenite. On the other hand, evidence of strong shearing along the margins of the Delhi Basin constrains the emplacement age of the alkaline body as the post-Aravalli but pre-Delhi in age.

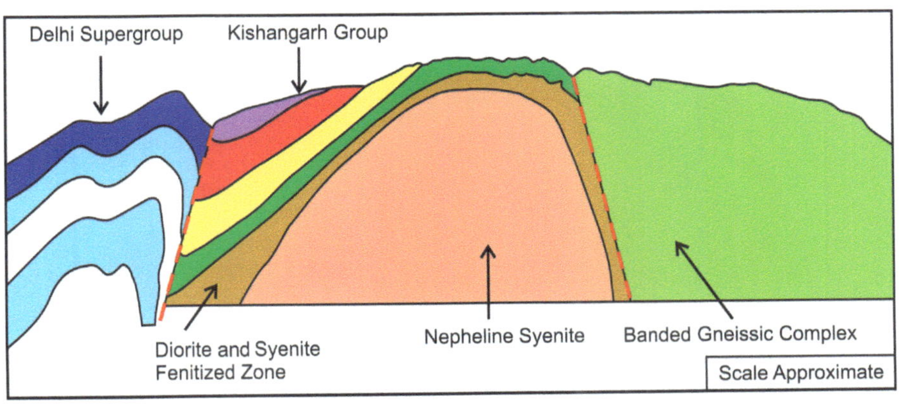

FIG. 9.27

Schematic cross section across the diapirically uplifted nepheline syenite exhibiting antiformal dome-like structure within the rocks of the Kishangarh Group.

Reproduced from Roy et al. (2016) with permission.

9.6 SIROHI GROUP

A group of deformed metasedimentary rocks occur west of the Main Delhi Basin with an intervening belt of gneiss-granite ensemble of Mount Abu and in the surrounding region in the north (Fig. 9.28). The narrow belt of metasediments that passes through Sirohi and earlier correlated with the 'Aravalli system' of Heron (1953) has been remapped as the Sirohi Group by Roy and Sharma (1999). Isolated outcrops of this metasediments are also traceable further north of Sirohi, continuing along the line of occurrence of marble outcrops of Ras and Makrana. The marbles of the latter outcrops were correlated earlier as the 'Raialo series' by Coulson (1933) following the usage of Heron (1953).

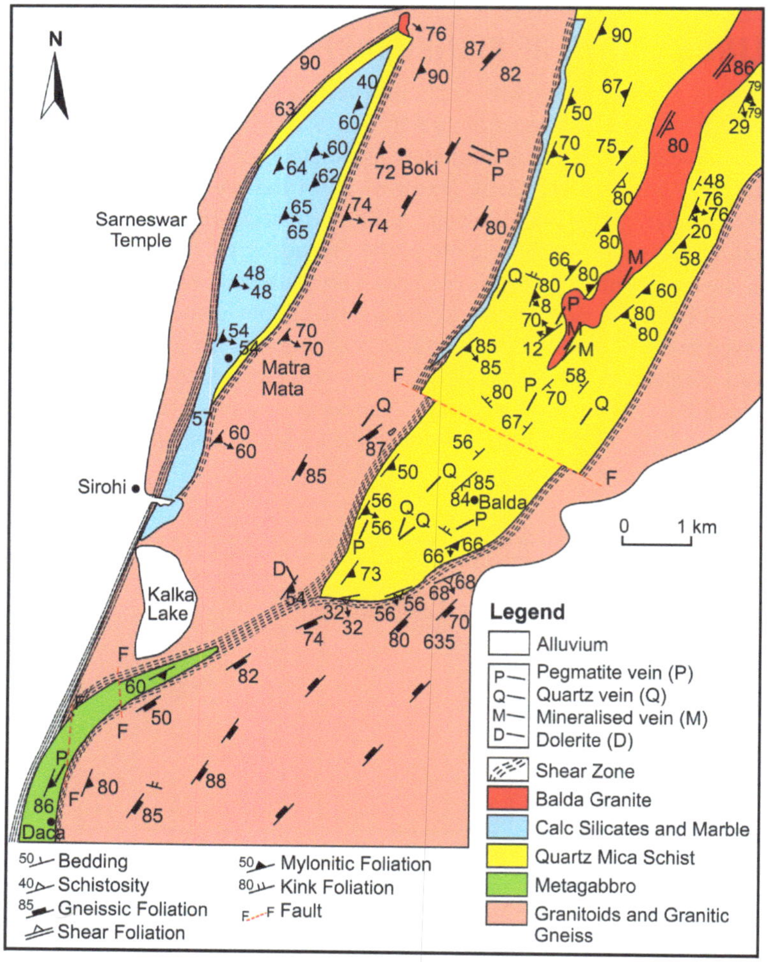

FIG. 9.28

Geological map of the Sirohi region showing two subparallel bands of metasediments of the Sirohi Group.

Image courtesy: K.K. Sharma.

The Sirohi Group is now considered the youngest of the three Proterozoic orogenic supracrustal ensembles that constituted the edifice of the Aravalli Mountains in northwestern India (Fig. 9.1). Comprising dominantly a 'shale-carbonate' association with minor bodies of fine-grained arenites (Roy and Jakhar, 2002), these low-grade metamorphic rocks underwent a single-phase shear-related deformation showing strong development of penetrative foliation and lineation, especially along their margins with the basement granite-gneisses. Locally, the Sirohi rocks show development of upright folds deforming the pre-existing lineations (Fig. 9.29a). Imprint of later thermal metamorphism is indicated in the haphazard development of large 'porphyroblasts' of andalusite in the pelitic components of the Sirohi sediments (Fig. 9.29b). Sirohi Group is devoid of any volcanic component, which suggests that the basin opened during passive rifting in an ensialic environment. But for the some local occurrence, the very local development of arenites and virtual absence of coarse 'clastics' (especially conglomerates) in the Sirohi rocks are manifestations of highly peneplained hinterland (Provenance) that supplied the sediments (Roy and Sharma, 1999).

FIG. 9.29

(a) Northerly plunging upright folds in quartzite deforming the pre-existing lineations, Didwana, north of Sirohi. Photo, B.S. Paliwal. (b) Haphazard development of large 'porphyroblasts' of andalusite in the pelitic components of the Sirohi sediments.

(b) Reproduced from Roy and Jakhar (2002) with permission.

According to Roy and Jakhar (2002), the single-phase deformation pattern observed in the Sirohi rocks is quite distinctive and cannot be correlated with the Delhi Basin rocks that show evidence of multiphase folding. Available geochronological data and C isotope studies confirm the Neoproterozoic age of the Sirohi Group (Purohit et al., 2012b). Reports are now abounding about the ages of around 1000 Ma from different parts of the Delhi Basin (Deb et al., 1989, 2001; Fareeduddin and Kröner (1998); Volpe and Macdaugall, 1990). According to Roy and Jakhar (2002), this thermal event in the Delhi Basin could be correlated with the rift opening of the Sirohi Basins west of it. The basin inversion of the Sirohi Group has been envisioned as the final phase of cratonization of the Precambrian Aravalli Crust during 835–820 Ma. The event is marked by coeval intrusion of granites not only in the Sirohi Basin but also in the southern part of the Delhi Basin.

Sirohi Group is the youngest of the three Proterozoic orogenic supracrustal ensembles that constituted the edifice of the Aravalli Mountains in northwestern India (Fig. 9.1). Occurring along a linear belt west of the Main Delhi Basin with an intervening belt of granites of diverse types and ages, the Neoproterozoic age of the Sirohi Group is based on the reports of younger tectonothermal events between 1000 and 950 Ma from the region surrounding the Sirohi basins (Deb et al., 1989, 2001; Fareeduddin and Kröner, 1998; Just et al., 2011; Volpe and Macdaugall, 1990). Maximum age of the Sirohi Basin is indicated by the age of single zircon collected from granite-gneiss outcrop from near the village of Veerwara, 10 km east of Sirohi town, which unconformably underlie the Sirohi Group. $^{207}Pb/^{206}Pb$ age indicated 920.4 ± 0.8 Ma with one 'xenocryst' grain indicating 992.6 ± 1.3 Ma age (Purohit et al., 2012b). The 992.6 Ma 'age' obtained from the xenocrystic zircon, considering the expected error, appears quite similar to the age of the thermal event at c.1000 Ma reported from different parts of the Aravalli Mountains (Buick et al., 2006; Deb et al., 2001; Just et al., 2011; Pandit et al., 2001; Sarkar et al., 1989; Volpe and Macdaugall, 1990). We interpret that the Sirohi Basin opening must have taken place later than c.993 Ma and more possibly later than 920 Ma. The single-zircon age of sample collected from Jawai Bandh region, 40 km north of Sirohi, is 822.8 ± 0.8 Ma (Purohit et al., 2012b). The younger age is interpreted as the minimum age of the Sirohi Basin on the ground that the granite body from which the sample was collected show features indicating its synkinematic character in relation to the deformation of the Sirohi rocks. The ~820 Ma age obtained from the study appears a little younger than the earlier reported ~835 to ~850 Ma age from different granite bodies occurring within or close to the outcrops of the Sirohi Group (Choudhary et al., 1984). Based on the single-zircon ages, a time span between c.920 and 820 Ma is suggested for the opening and closing of the Sirohi Basin (Roy and Jakhar, 2002).

9.7 MALANI GROUP

The closing of the Sirohi Basin by mid-Neoproterozoic marks the complete cratonization of Precambrian Crust of the Aravalli Mountains. During the succeeding period till the opening of the large stable platform-type basins in the western part of the Aravalli Mountains, the terrain witnessed effusive magmatism with development of volcanic lava flows (dominantly acidic and minor mafic bodies) and pyroclastics like the ash beds and 'welded' tuffs (ignimbrites) and plutonic intrusions of granites of diverse types and composition (Fig. 9.30a and b).

FIG. 9.30

(a) Columnar jointed Malani Rhyolite at the base of the Jodhpur Sandstone (top capping of hills), at Meherangarh Fort, Jodhpur. (b) Packs of pillow structures at Bombolai, a component of Malani Group.

Image courtesy: Surja Ram Jakhar.

Out of all different constituents, acid lava flows and pyroclastics (including the ash beds, welded tuffs or ignimbrites, and volcanic breccias) occur most extensively over the entire region, which is also referred to as the Malani Igneous Province (Coulson, 1933). The dominant lithology of the acid flows is rhyolite. Rhyodacite, dacite, trachyte, and pitchstone are present in some of the acid volcanic succession.

A number of granites of different composition occur in the Malani Group, some of which show close association with rhyolite and tuffs. Broadly speaking, two different types of granite occur in the milieu of Malani ensembles, the prototypes of which are the Jalore Granite and the Siwana Granite. The Jalore Granite that forms a huge boss-like outcrop is predominantly peraluminous, subsolvus type presumably formed from melting of lower Crust (Dhar et al., 1996). The Siwana Granite, on the other hand, is peralkaline and hypersolvus, which has resulted from a higher temperature (between 660 and 740 °C) thermal event having a melt fraction (Venkataraman et al., 1968). Based on Sr, Pb, and Nd isotope studies of the Siwana Granite, Dhar et al. (1996) suggested mantle derivation of the magma, which show significant modification in composition due to crustal contamination. Apart from these two types, a third type of granite is recorded that has resulted from the melting of upper Crust at low temperatures. Some of these granite bodies show evidence of tungsten mineralization and are impregnated with lithophile elements with extensive metasomatism of the pre-existing crustal components.

The field relationship, in spite of the isolate occurrences of the rock bodies, provides evidence that the diverse rock types are stratigraphic or more precisely timewise interconnected rock assemblages (La Touche, 1902; Bhushan, 2000; Kochhar, 1984; Pareek, 1981). The entire ensemble of rocks have been classified as the Malani Group by Roy (1998), in preference to the different nomenclatures like Malani Rhyolite, Malani Volcanics, Malani Igneous Suite, and Malani Igneous Complex.

Almost simultaneously with the development of igneous rocks masses, some linear sedimentary basins evolved in which conglomerate, grits, and arkose were deposited mainly as interlayered bodies within the effusive bodies. This unmetamorphosed sedimentary sequence has been described as the Sindreth Group considered equivalent to the Malani Group (Fig. 9.31a and b) (after Roy and Jakhar, 2002).

FIG. 9.31

(a) Basal conglomerate in the Sindredth Basin showing assorted, angular larger clast in arkosing matrix; (b) folded uniformly bedded arkose overlying the rhyolite in the Sindhreth region. The precipitous hill in the background represents the rhyolite-bearing massif near Sirohi.

Reproduced from Roy and Jakhar (2002) with permission.

Anorogenic character of the Malani magmatism is well established. The chemistry of the different volcano-plutonic components of the Malani Group is a clear reflection of magma generation from different levels of the Crust and Mantle, which according to Kochhar (1984) resulted from typically anorogenic hotspot activities developing over a Mantle Plume (Roy and Jakhar, 2002).

Considerable number of well-constrained isotope ages is now available for the essentially tectono-thermal event related to the development of the volcano-sedimentary-plutonic ensemble, called the Malani Group. Rathore et al. (1999) reported Rb/Sr isochron age of 779 ± 10 Ma of the felsic volcanics from the eastern margin of the occurrence of the Malani Group. Apart from the oldest volcanics and 698 ± 10 Ma Rb/Sr isochron age of the of the Siwana Granite (Rathore et al., 1999), a great majority of ages (mostly granite ages) reported from the Malani rocks fall within the age bracket between 750 and 720 Ma (Aswathanarayana, 1964; Crawford and Compston, 1970; Gregory et al., 2009; van Lente et al., 2009; Torsvik et al., 2001a,b).

9.8 MARWAR SUPERGROUP

Marwar Supergroup is constituted of about 2 km thick sedimentary sequence (without any volcanics) overlying the eroded surfaces of the Malani Group (Fig. 9.32). Representing a typical association of sand-shale-carbonate with evaporates, the ensemble shows faunal affiliation similar to the Himalayan Basin of Salt Range (in Pakistan) (Awasthi and Prakash, 1981; Barman, 1980). The Himalayan affiliation makes the Marwar Supergroup a distinctive sedimentary lithostratigraphic formation that was earlier described as the 'Trans-Aravalli Vindhyans' (Heron, 1953).

The Marwar Supergroup developed over two different basins: the Main Marwar Basin in the north and the Birmania Basin in the southwest. The Main Marwar Basin characteristically shows development of a typical platformal shelf facies rock association of sand-shale-carbonate with evaporites. The ensemble consists of three successive lithostratigraphic units: Jodhpur Group, Bilara Group, and Nagaur Group.

The opening of the Main Marwar Basin is linked with deposition of a thin boulder bed at the base of the succession. The main lithology includes boulders, pebbles, and cobbles mainly of felsic volcanics, granitoids, and quartzite enclosed in reddish and brownish matrix of clay and silt-sized sand mass. Known as the Pokhran Boulder Bed, this assorted basal succession has been recognized as the glacial and fluvioglacial erratics (Kumar, 1999) (Fig. 9.33a). The overlying siliciclastic sediments constituting the Jodhpur Group show characteristics of cratonic sheet sediments deposited predominantly under deltaic conditions, which during the terminal phases changed to beach environment and fluvial settings (Fig. 9.33b). The Bilara Group that overlies the dominantly arenitic sequence of Jodhpur Group is a carbonate sequence having an estimated thickness of about 300 m. The outcrops of the Bilara Group are, however, quite scanty. Recent studies, based on the detailed geological, geophysical, and subsurface borehole data, helped to recognize a typical evaporite succession, named Hanseran Evaporite Group (Dasgupta et al., 1984). The Hanseran Group is considered homotaxial (time equivalent) of the Bilara Group recognized on the surface (Kumar, 1999). Borehole data revealed the presence of seven cycles of halite-bearing evaporites with rare potash salt deposit.

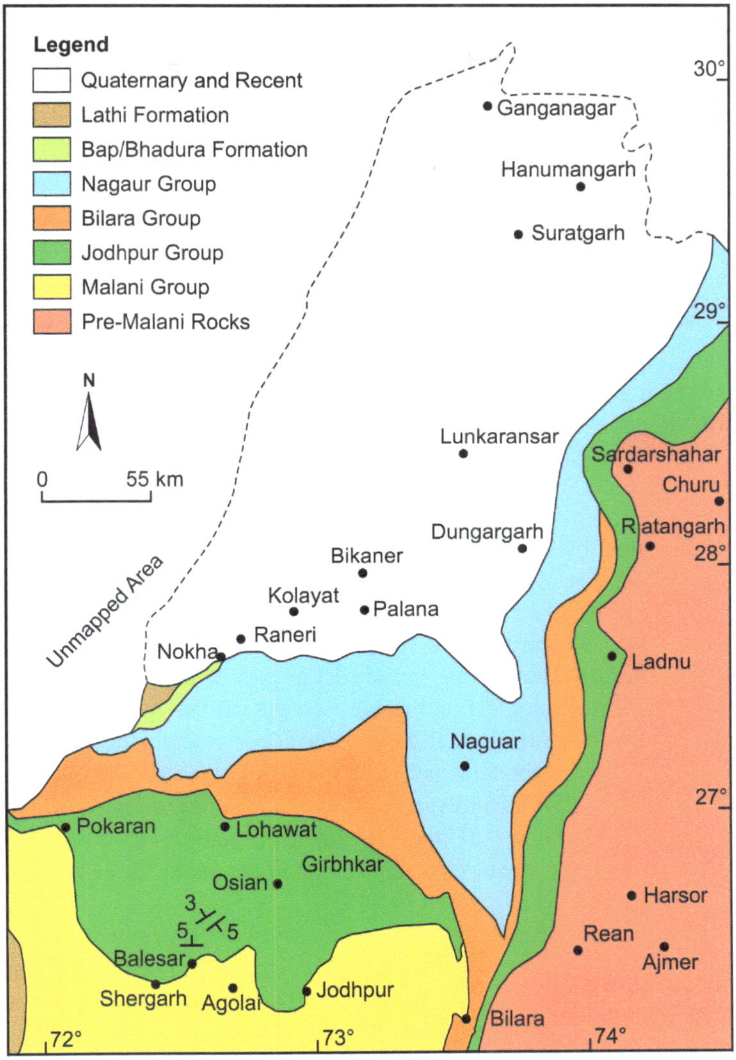

FIG. 9.32

Geological map of the Marwar Supergroup in western Rajasthan.

Reproduced from Roy and Jakhar (2002) with permission.

Pandit et al. (2001), based on study of carbon isotopic evolution curve, suggested an end-Neoproterozoic (Vendran-Tommatian) age of the Bilara Group. Extremely low $\delta^{13}C$ values in the Bilara carbonates has been interpreted as indicator of glacial related cold climatic condition. On the other hand, the marked oscillations in the $\delta^{13}C$ values showing cyclic negative and positive shifts are considered to indicate sequential changes from cold-to-warmer climatic conditions (Pandit et al., 2001).

FIG. 9.33

(a) Scattered boulders from the Pokhran Boulder bed, forming the base of the Marwar Supergroup.
(b) Subhorizontal layered sandstone-shale succession of the Jodhpur Sandstone from Khatu, east of Jodhpur.

Image courtesy: Surja Ram Jakhar.

The Nagaur Group that overlies the Bilara carbonates is dominantly an arenite succession. Culmination of the Main Marwar Basin with the inversion of the Nagaur Subbasin marks the end of the Precambrian depositional history in the region.

Carbonate rocks of late Neoproterozoic-early Cambrian age occur in two extremities of the Nagaur-Ganganagar Basin in western India: in the east as part of Bilara hills and in the west as part of Baghewala oil field. Sedimentological and stable isotopic studies of these rocks show prevalence of arid and evaporitic conditions during their deposition. Close resemblance of C-isotopic profiles of Hanseran evaporites and Bilara carbonates helps to suggest that these two are coeval facies variants (Mazumdar and Bhattacharya, 2004). Highly depleted carbon isotopic composition of the organic matter in subsurface organic-rich carbonates (Upper Carbonate Formation) has been ascribed to incorporation of biomass synthesized by chemoautotrophic bacteria under anoxic or eutrophic basinal condition caused by salinity stratification.

The Birmania Basin occurring southwest of the Main Marwar Basin covers relatively much smaller area compared with the Main Marwar Basin (Fig. 9.34). The rocks occurring in the basin belong to two lithostratigraphic formations. The basal Randha Formation is dominantly an arenitic succession correlatable with the Jodhpur Group of the Main Marwar Basin. Overlying this occurs the carbonate dominant Birmania Formation that characteristically shows development of stromatolitic and bedded phosphorite deposits. Unlike the northern Main Marwar Basin, the Birmania Basin does not show any arenite sequence overlying the carbonates. The absence of arenite beds overlying the Birmania Formation is thought to be the result of extensive postdepositional erosion.

The Marwar Supergroup marks the last event in the Precambrian crustal growth in Rajasthan and adjoining regions. However, assuming the correlation of the Pokhran Boulder Bed of possible glaciogenic sequence (Kumar, 1999) with the Blaini Formation of the Krol Group in the Himalayas, an age corresponding to Marinoan glaciation phase at ca. 635 Ma (Kaufman et al., 2006) has been suggested for the initiation of the Marwar depositional basins. The inversion of Marwar Basins is possibly linked with the major stratigraphic break at 543 Ma marking the Proterozoic-Cambrian boundary in this terrain (Roy and Jakhar, 2002).

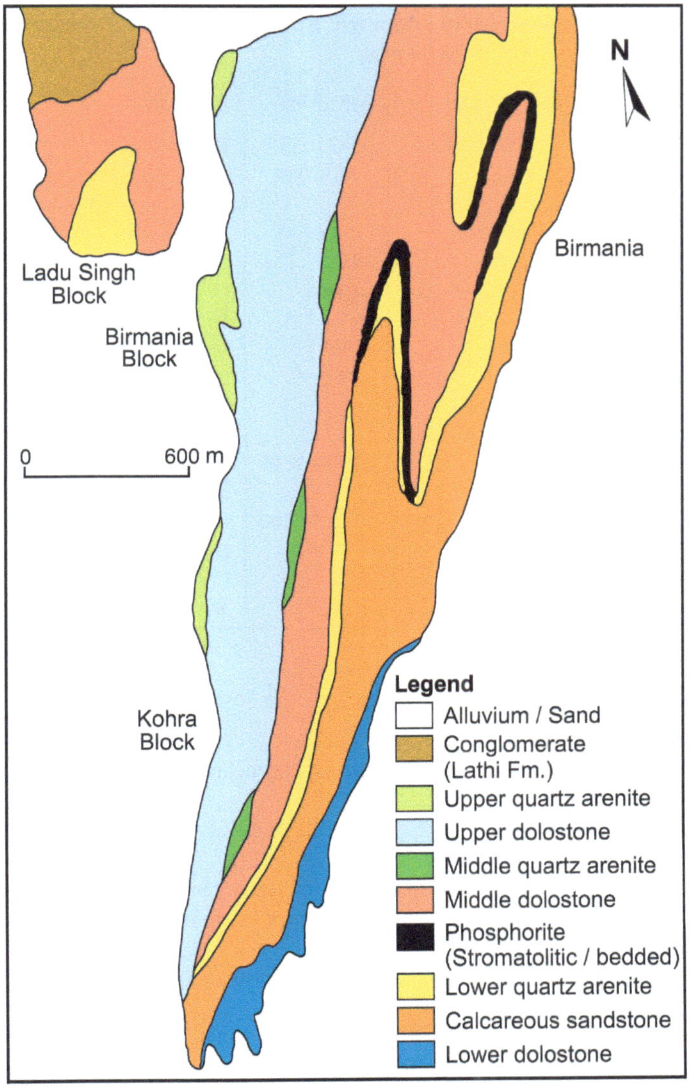

FIG. 9.34

Geological map of the Birmania Basin

Image courtesy: S.C. Mathur.

9.9 GEOCHRONOLOGIC FRAMEWORK OF THE CONSTITUENTS OF THE ARAVALLI PROTOCONTINENT

Lithostratigraphic and geochronologic framework of the Precambrian formations of the Aravalli Mountains and adjoining areas is shown in Table 9.3.

Table 9.3 Lithostratigraphic and Geochronologic Framework of the Precambrian Formations of the Aravalli Mountains and Adjoining Areas

Lithostratigraphic Units	Tectonic Setting	Age (Ma)	References
Marwar Supergroup	Stable cratonic basins	~543–635	Kaufman et al. (2006)
Malani Group = Sindredth Group	Plume related magmatism; opening of shallow basins	~720–780	Aswathnarayan (1964), Crawford and Compston (1970), Gregory et al. (2009), Rathore et al. (1999)
Sirohi Group	Basin opening	~852	Bhowmik et al. (2010), Choudhary et al. (1984), Deb et al. (1989), Fariduddin and Kröner (1998), Volpe and Macdaugall (1990)
Sirohi Orogenic Cycle	Basin opening	~1000	
Delhi Supergroup	Basin closing	~1450	Choudhary et al. (1984), Roy and Das (1985)
Delhi Orogenic Cycle	Basin opening (Emplacement of Sandmata Granulites)	~1725	Sarkar et al. (1989), Buick et al. (2006), Fareeduddin and Kröner (1998), Roy et al. (2005a, 2012)
Aravalli Supergroup	Basin closing	~1900–1850	Choudhary et al. (1984), Roy et al. (2012), Sarangi et al. (2006)
Aravalli orogenic Cycle	Basin opening	~2100–2200	Deb and Thorpe (2004), Schleicher et al. (1997)
Banded Gneissic Complex and Granitoids (Archaean Basement)	~2500 to ~3300		Gopalan et al. (1990), Roy and Kröner (1996), Wiedenbeck and Goswami (1994), Wiedenbeck et al. (1996)

REFERENCES

Aswathanarayana, U., 1964. In: Age determination of rocks and geochronology of India, Brochure. 22nd Int. Geol. Cong., New Delhi, p. 23.

Awasthi, A.K., Prakash, B., 1981. Depositional environments of unfossiliferous sediments from Jodhpur Group, Western India. Sediment. Geol. 30, 15–42.

Banerjee, D.M., 1971a. Aravallian stromatolites from Udaipur, Rajasthan. J. Geol. Soc. India 12, 349–355.

Banerjee, D.M., 1971b. Precambrian stromatolitic phosphorites of Udaipur, Rajasthan, India. J. Geol. Soc. America 82, 2319–2380.

Barman, G., 1980. An analysis of the Marwar Besin, western Rajasthan. In: The Light of Stromatolitic Study. 44. Misc. Publ. Geol. Surv. India, Kolkata, pp. 292–297.

Bhowmik, S.K., Bernhardt, H.-J., Dasgupta, S., 2010. Grenvillian age high-pressure upper amphibolite-granulite metamorphism in the Aravalli-Delhi Mobile Belt, Northwestern India: new evidence from monazite chemical age and its implication. Precambrian Res. 178, 168–184.

Bhushan, S.K., 2000. Malani Rhyolites—a review. Gondwana Res. 3, 65–77.

Buick, I.S., Allen, C., Pandit, M., Rubatto, D., Herman, J., 2006. The Proterozoic magmatic and metamorphic history of the Banded Gneissic Complex, central Rajasthan, India: La-ICP-MS U-Pb zircon constraints. Precambrian Res. 151, 119–142.

Chauhan, D.S., 1979. Phosphate-bearing stromatolites of the Precambrian phosphorite deposits of Udaipur region, their environmental significance and genesis of phosphorite. Precambrian Res. 8, 95–126.

Choudhary, A.K., Gopalan, K., Sastry, C.A., 1984. Present status of the geochronology of the Precambrian rocks of Rajasthan. Tectonophysics 105, 131–140.

Coulson, A.L., 1933. The Geology of Sirohi State, Rajputana. 63 Mem. Geol. Surv. India, Kolkata. 166p.

Crawford, A.R., 1970. The Precambrian geochronology of Rajasthan and Bundelkhand, Northern India. Can. J. Earth Sci. 7, 91–110.

Crawford, A.R., Compston, W., 1970. The age of Vindhyan system of Peninsular India. Quart. J. Geol. Soc. Lond. 125, 351–372.

Dasgupta, S.P., Chandra, R., Jairam, M.S., Kumar, V., 1984. Potash occurrences in Nagaur-Ganganagar Evaporite in Northwestern Rajasthan. Indian Minerals 42 (2), 1–14.

de Wall, H., Pandit, M.K., Chauhan, N.K., 2012. Paleosol occurrences along the Archean–Proterozoic contact in the Aravalli craton, NW India. Precambrian Res. https://doi.org/10.1016/j.precamres.2012.06.017.

Deb, M., 2000. VMS deposits: Geological characteristics, genetic models, and a review of their metallogenesis in the Aravalli Range. In: Deb, M. (Ed.), Crustal evolution and metallogeny in northwestern Indian shield. Narosa, New Delhi, pp. 328–363.

Deb, M., Thorpe, R.I., 2004. Geochronological constraints in the Precambrian Geology of Rajasthan and their metallogenic implications. In: Deb, M., Goodfellow, W.D. (Eds.), Sediment-hosted Lead- Zinc Sulphide Deposits. Narosa Publishing House, New Delhi, pp. 246–263.

Deb, M., Thorpe, R.I., Cumming, G.L., Wagner, P.A., 1989. Age, source and stratigraphic implications of Pb-Isotope data for conformable, sediment hosted, base metal deposits in the Proterozoic Aravalli-Delhi orogenic belt, northwestern India. Precambrian Res. 43, 1–22.

Deb, M., Thorpe, R.I., Krstic, D., Davis, D., Corfu, F., 2001. Zircon U-Pb and galena Pb isotope evidence for an approximate 1.0GA terrane along western margin of the Aravalli-Delhi orogenic belt, northwestern India. Precambrian Res. 108, 195–213.

Desai, S.J., Patel, M.P., Merh, S.S., 1978. Polymetamorphites of Balarampur-Abu Road area, north Gujarat and southwest Rajasthan. J. Geol. Soc. India 19, 383–394.

Dhar, S., Frei, R., Kramers, J.D., Nagler, T.F., Kochar, N., et al., 1996. Sr, PB, and ND isotope studies and their bearing on the paragenesis of Jalore and Siwana Complexes, Rajasthan, India. J. Geol. Soc. India 48, 151–160.

Fareeduddin, A., Kröner, Paliwal, B.S., 1998. Single zircon age constraints on the evolution of Rajasthan granulite. In: The Indian Precambrian. Scientific Publishers (India), Jodhpur, pp. 547–556.

Gopalan, K., Chaudhary, A.K., 1984. The crustal records in Rajasthan. Proc. Indian Acad. Sci. (Earth Planet. Sci.) 93, 337–342.

Gopalan, K., Macdaugall, J.D., Roy, A.B., Murali, A.V., 1990. Sm-Nd evidence for 3.3 Ga old rock in Rajasthan, north-western India. Precambrian Res. 48, 287–297.

Gopalan, K., Trivedi, J.R., Balasubramanyaum, M.N., Ray, S.K., Sastri, C.A., 1979. Rb-Sr geochronology of the Khetri belt, Rajasthan. J. Geol. Soc. India 20, 450–456.

Gregory, L.C., Meert, J.G., Bingen, B.A., Pandit, M.K., Torsvik, T.H., 2009. Palaeomagentism and geochronology of Malani igneous suite, NW India: implications for the configuration of Rodinia and assembly of Gondwana. Precambrian Res. 170, 13–26.

Gupta, B.C., 1934. The geology of central Mewar. Mem. Geol. Surv. India 65, 107–168.

Gupta, S.N., Arora, Y.K., Mathur, R.K., Iqbaluddin, Prasad, B., Sahai, T.N., Sharma, S.B., 1980. Lithostratigraphic Map of the Aravalli Region, Southern Rajasthan and Northeastern Gujarat. Geological Survey of India, Calcutta. Scale 1:1,000,000.

Gupta, S.N., Arora, Y.K., Mathur, R.K., Iqbaluddin, Prasad, B., Sahai, T.N., Sharma, S.B., 1997. The Precambrian geology of the Aravalli Region, southern Rajasthan and northeastern Gujarat. Mem. Geol. Surv. India 123, 262 p.

Gupta, P., Guha, D.B., Chattopadhyay, B., 1998. Basement-cover relationship in the Khetri Copper Belt and the emplacement mechanism of the granite massifs, Rajasthan. J. Geol. Soc. India 52, 417–432.

Gupta, B.C. and Mukherjee, P.N. 1938. Geology of Gujarat and southern Rajputana. Rec. Geol. Surv. India, 73 (2), 103–208.

Gupta, P., Mukhopadhyay, K., Fareeduddin, M.S., Reddy, 1995. Stratigraphy and structure of Delhi Supergroup of rocks in central part of Aravalli Range. Rec. Geol. Surv. India 120, 12–26.

Heron, A.M., 1917a. Geology of northeastern Rajputana and adjacent districts. Mem. Geol. Surv. India 45 (1). 128p.

Heron, A.M., 1917b. The Byana-Lalsot hills eastern Rajputana. Rec. Geol. Surv. India 48, 181–203.

Heron, A.M., 1936. Geology of Pre-Vindhyan Geology of Rajputana. Trans. Natl. Inst. Sci., India 1 (2), 17–33.

Heron, A.M., 1953. Geology of central Rajputana. Mem. Geol. Surv. India 79, 339p.

Just, J., Schulz, B., de Wall, H., Jourdan, F., Pandit, M.K., 2011. Monazite CHIME/EPMA dating of Erinpura granitoid deformation: Implications for Neoproterozoic tectono-thermal evolution of NW India. Gondwana Res. 19, 402–412.

Kaufman, A.J., Jian, G., Christic-Blick, N., Banerjee, D.M., Rai, V., 2006. Stable isotope record of the thermal Neoproterozoic Krol platform in the Lesser Himalayas of northern India. Precambrian Res. 147, 156–185.

Kaur, P., Chaudhri, N., Raczek, I., Kröner, A., Hofmann, A.W., Okrusch, M., 2011. zircon ages of late Palaeoproterozoic (ca. 1.72–1.70 Ga) extension-related granitoids in NE Rajasthan, India: regional and tectonic significance. Gondwana Res. 19, 1040–1053.

Kaur, P., Zeh, A., Chaudhri, N., Gerdes, A., Okrusch, M., 2011. Archaean to Palaeo-proterozoic crustal evolution of the Aravalli orogen, NW India, and its hinterland: the U–Pb and Hf isotope record of detrital zircon. Precambrian Res. 187, 155–164.

Kaur, P., Zeh, A., Chaudhri, N., Gerdes, A., Okrusch, M., 2013. Nature of magmatism and sedimentation at a Columbia active margin: Insights from combined U–Pb and Lu–Hf isotope data of detrital zircons from NW India. Gondwana Res. 23, 1040–1052.

Khan, M.S., Smith, T.E., Raza, M., Huang, J., 2005. Geology, Geochemistry and Tectonic Significance of Mafic-ultramafic Rocks of Mesoproterozoic Phulad Ophiolite Suite of South Delhi Fold Belt, NW Indian Shield. Gondwana Res. (4), 553–566.

Kochhar, N., 1984. Malani Igneous Suite: Hot-spot magmatism and cratonization of the northern part of the Indian shield. J. Geol. Soc. India 25, 155–161.

Kumar, V., 1999. Evolution and geological set-up of the Nagaur-Ganganagar Basin, northwestern Rajasthan. In: Paliwal, B. (Ed.), Geological Evolution of Northwestern India. Scientific Publishers (India), Jodhpur, pp. 34–60.

La Touche, T.D.H., 1902. Geology of western Rajputana. Mem. Geol. Surv. India 35, 116p.

Mazumdar, A., Bhattacharya, S.K., 2004. Stable isotopic study of late Neoproterozoic-early Cambrian (?) sediments from Nagaur-Ganganagar basin, western India: Possible signatures of global and regional C-isotopic events. Geochem. J. 38, 163–175.

Naha, K., Halyburton, R.V., 1974. Early Precambrian stratigraphy of central and southern Rajasthan, India. Precambrian Res. 1, 55–73.

Naha, K., Roy, A.B., 1983. The problem of Precambrian basement in Rajasthan, Western India. Precambrian Res. 19, 217–223.

Pandit, M.K., Sial, A.N., Jamrani, S.S., Ferreira, V.P., 2001. Carbon isotopic profile across the Bilara Group rocks of trans-Aravalli Marwar Supergroup in western India: implications for Neoproterozoic-Cambrian transition. Gondwana Res. 4, 387–394.

Papineau, D., Gregorio, B., De Fearn, S., Kilcoyne, D., McMahon, G., Purohit, R., Fogel, M., 2016. Nanoscale petrographic and geochemical insights on the origin of the Palaeoproterozoic stromatolitic phosphorites from Aravalli Supergroup, India. Geobiology 14, 3–32.

Papineau, D., Purohit, R., Fogel, M.L., Shields-Zhou, G.A., 2013. High phosphate availability as a possible cause for massive cyanobacterial production of oxygen in the Paleoproterozoic atmosphere. Earth Planet. Sci. Lett. 362, 225–236.

Pareek, H.S., 1981. Petrochemistry and petrogenesis of the Malani Igneous Suite. India. Bull. Geol. Soc. America 92 (67–70), 206–273.

Purohit, R., Bhu, H., Mehta, P., Avadich, P., 2012a. Tectonostratigraphic Status of the Proterozoic Babarmal Pink Marble from the Aravalli Supergroup. J. Geol. Soc. India 79, 235–250.

Purohit, R., Bhu, H., Sarkar, A., Ram, J., 2015. Evolution of the ultramafic rocks of the Rakhabdev and Jharol Belts in southeastern Rajasthan, India: new evidences from imagery mapping, petro-minerological and O-H stable isotope studies. J. Geol. Soc. India 85, 331–338.

Purohit, R., Papineau, D., Kröner, A., Sharma, K.K., Roy, A.B., 2012b. Carbon isotope geochemistry and geochronological constraints of the Neoproterozoic Sirohi Group from northwest India. Precambrian Res. 220–221, 80–90.

Purohit, R., Sanyal, P., Roy, A.B., Bhattachrya, S.K., 2010. δ^{13} C enrichment in the Palaeoproterozoic carbonate rocks of the Aravalli Supergroup NW India: influence of depositional environment. Gondwana Res. 18, 538–546.

Rathore, S.S., Venkatesh, T.R., Srivastava, R.K., 1999. Rb-Sr isotope dating of Neoproteroozic (Malani Group) magmatism from southwest Rajasthan, India: evidence of younger Pan-African thermal event by ^{40}Ar-^{39}Ar studies. Gondwana Res. 2, 271–281.

Roy, A.B., 1988. Stratigraphic and tectonic framework of the Aravalli Mountain Range. In: Roy, A. B. (ed.), Precambrian of the Aravalli Mountain, Rajasthan, India. Mem. Geol. Soc. India, Bangalore, 7, 3–31.

Roy, A.B., 1990. Evolution of early Precambrian crust of the Aravalli Mountain range. In: Naqvi, S.M. (Ed.), Precambrian Continental Crust and Its Economic Resources. Development of Precambrian Geology, Elsevier, pp. 327–348.

Roy, A.B., 1998. Malani Igneous Suite of rocks (Correspondence). J. Geol. Soc. India 51, 556–557.

Roy, A.B., 2000. Geology of Palaeoproterozoic Aravalli Supergroup of Rajasthan and northern Gujarat. In: Deb, M. (Ed.), Crustal Evolution and Metallogeny in Northeastern Indian Shield. Narosa Publishing House, New Delhi, pp. 87–114.

Roy, A.B., Das, A.R., 1985. A study of time relation between movements, metamorphism and granite emplacement in the middle-Proterozoic Delhi Supergroup of Rajasthan. J. Geol. Soc. India 26, 726–733.

Roy, A.B., Dutt, K., Rathore, S., 2016. Development of ductile shear zones during diapiric magmatism of nepheline syenite and exhumation of granulites—examples from central Rajasthan, India. 2016. Curr. Sci. 110, 1094–1101.

Roy, A.B., Jakhar, S.R., 2002. Geology of Rajasthan: Precambrian to Recent. Scientific Publishers (India), Jodhpur. 421p.

Roy, A.B., Kröner, A., 1996. Single zircon evaporation ages constraining the growth of the Archaean Aravalli craton, northwestern Indian shield. Geol. Mag. 133, 333–342.

Roy, A.B., Kröner, A., Bhattacharya, P.K., Rathore, S., 2005a. Metamorphic evolution and zircon geochronology of early Proterozoic granulites in the Aravalli Mountains of northwestern India. Geol. Mag. 142, 287–302.

Roy, A.B., Kröner, A., Laul, V., 2001. Detrital zircons constraining basement age in a late Archaean greenstone belt of south-eastern Rajasthan, India. Curr. Sci. 81, 407–410.

Roy, A.B., Kröner, A., Laul, V., Purohit, R., 2005b. Single Zircon dating of hypersthene bearing granitoid from Balarampur-Abu Road area, southern part of the Aravalli Mountains, NW India-Implications for Malani magmatism. In: Tomas, H. (Ed.), Metamorphism and Crustal Evolution. Atlantic, Publishers & Distributors, New Delhi, pp. 339–346.

Roy, A.B., Kröner, A., Rathore, S., Laul, V., Purohit, R., 2012. Tectono-metamorphic and geochronologic studies from Sandmata Complex, northwest Indian Shield: implications on exhumation of late-Palaeoproterozoic granulites in an Archaean-early Palaeoproterozoic granite-gneiss terrane. J. Geol. Soc. India 79, 323–334.

Roy, A.B., Kumar, S., Laul, V., Chauhan, N.K., 2004. Tectono- stratigraphy of the lead- zinc bearing metasedimentary rocks of the Rampura-Agucha mine and its neighbourhood, district Bhilwara, Rajasthan: implications on metallogeny. In: Deb, M., Goodfellow, W.D. (Eds.), Sediment-hosted Lead-Zinc Sulphide Deposits: Attributes and Models of some Major Deposits in India, Australia and Canada. Narosa Publishing House, New Delhi, pp. 273–289.

Roy, A.B., Nagori, D., Golani, P.R., Dhakar, S.P., Choudhuri, R., 1980. Structural geometry of the phosphorite bearing Aravalli rocks around Jhamarkotra Mines area, Udaipur dist., Rajasthan. Indian J. Earth Sci. 7, 191–202.

Roy, A.B., Paliwal, B.S., 1981. Evolution of lower Proterozoic epicontinental deposits: Stromatolite bearing Aravalli rocks of Udaipur, Rajasthan, India. Precambrian Res. 14, 49–74.

Roy, A.B., Paliwal, B.S., Shekhawat, S.S., Nagori, D.K., Golani, P.R., Bejarniya, B.R., 1988. Stratigraphy of the Aravalli Supergroup in the type area. In: Roy, A.B. (Ed.), Precambrians of Aravalli Mountain, Rajasthan, India. 7. Mem. Geol. Soc. India, Bangalore, pp. 121–131.

Roy, A.B., Sharma, K.K., 1999. Geology of the region around Sirohi Town, western Rajasthan—a story of Neoproterozoic evolution of the Aravalli crust. In: Paliwal, B.S. (Ed.), Geological Evolution of North-Western India. Scientific Publishers, Jodhpur, pp. 19–33.

Sarangi, S., Gopalan, K., Roy, A.B., Sreenivas, B., Das Sharma, S., 2006. Pb-Pb age of the carbonates of Jhamarkotra Formation constraining the age of the Aravalli Supergroup, Rajasthan. J. Geol. Soc. India 67 (4), 442–446.

Sarkar, G., Barman, T.R., Corfu, F., 1989. Timing of continental arc magmatism in northwest India: evidence from U–Pb geochronology. J. Geol. 97, 607–612.

Schleicher, H., Todt, W., Viladkar, S.G., Schimdt, F., 1997. Pb-Pb age determinations of the Newania and Savattur carbonatites of India: evidence for multiple stage histories. Chem. Geol. 140, 261–273.

Singh, S.P., 1988. Stratigraphic and sedimentation patterns of the Proterozoic Delhi Supergroup, northeastern Rajasthan, India. In: Roy, A.B. (Ed.), Precambrians of the Aravalli Mountain, Rajasthan. 7. Mem. Geol. Soc. India, Bangalore, pp. 193–206.

Sinha-Roy, S., Malhotra, G., Mohanty, M., 1998. Geology of Rajasthan. Geol. Soc. India, Bangalore. 278p.

Sinha-Roy, S., Mohanty, M., 1988. Blue schist facies metamorphism in the ophiolite melange of the late Proterozoic Delhi Fold Belt, Rajasthan, India. Precambrian Res. 42, 97–105.

Sreenivas, B., Roy, A.B., Srinivasan, R., 2001. Geochemistry of sericite deposits at the base of the Proterozoic Aravalli Supergroup, Rajasthan, India: Evidence for metamorphosed Precambrian palaeosols. Proc. Indian Acad. Sci. (Earth Planet. Sci.) 110, 39–61.

Sugden, T.J., Deb, M., Windley, B.F., 1990. The tectonic setting of mineralization in the Proterozoic Aravalli-Delhi orogenic belts, NW India. In: Naqvi, S.M. (Ed.), Precambrian continental Crustand its econiomic resources. Elsevier, pp. 367–396.

Torsvik, T.H., Ashwal, L.D., Tucker, R.D., Eide, E.A., 2001a. Neoproterozoic geochronology and paleogeography of the Seychelles microcontinent: the India link. Precambrian Res. 110, 47–59.

Torsvik, T.H., Carter, L.M., Ashwal, L.D., Bhushan, S.K., Pandit, M.K., Jamtveit, B., 2001b. Rodinia refined or obscured: paleomagnetism of the Malani Igneous Suite (NW India). Precambrian Res. 108, 319–333.

Upadhyaya, R., Sharma Jr., B.L., Sharma, B.L., Roy, A.B., 1992. Remnants of greenstone sequence from the Archaean rocks of Rajasthan. Curr. Sci. 63, 87–92.

van Lente, B., Ashwal, L.D., Pandit, M.K., Bowring, S.A., Torsvik, T.H., 2009. Neoproterozoic hydrothermally-altered basaltic rocks from Rajasthan, northwest India; implications for late Precambrian tectonic evolution of the Aravalli Craton. Precambrian Res. 170, 202–222.

Venkataraman, P.L., Ray, S.M., Vishwanathan, T.V., Murthy, M.V.N., 1968. Comagmatic relationship among rhyolites and high level granites of the Malani Suite, Rajasthan. India. Misc. Publ. Geol. Surv. India 8, 93–105.

Vinogradov, A.P., Tugarinov, A.I., Zhukov, C.I., Stanikova, N.I., Bibikova, E.V., Khorre, K., 1964. In: Geochronology of Indian Precambrians. Report, 22nd Int. Geological Congress, New Delhi. 10. pp. 553–567.

Volpe, A.M., Macdaugall, J.D., 1990. Geochemistry and isotope characteristics of mafic (Phulad Ophiolite) and related rocks in the Delhi Supergroup, Rajasthan, India: implications for rifting in the Proterozoic. Precambrian Res. 48, 167–191.

Wiedenbeck, M., Goswami, J.N., 1994. An ion-probe single zircon $^{207}Pb/^{206}Pb$ age from Mewar Gneiss at Jhamarkotra, Rajasthan. Geochim. Cosmochim. Acta 58, 2135–2141.

Wiedenbeck, M., Goswami, J.N., Roy, A.B., 1996. Stabilization of the Aravalli craton of the northwestern India at 2.5 Ga.: an ion-microprobe zircon study. Chem. Geol. 129, 325–340.

CHAPTER

SOUTHERN GRANULITE BELT

10

10.1 INTRODUCTION

In the southern part of the (Peninsular) Indian Shield, an irregular, broadly east–west-lying Charnockite Line was drawn by Fermor (1936) to demarcate the contact between the Dharwar Protocontinent in the north and the dominantly charnockite-bearing granulite rocks in the south, known as the Southern Granulite Belt (Fig. 10.1). Considering that the granulitic rocks occurring south of the Charnockite Line are tectonostratigraphically different from the granite-greenstone belt sequence in the north, the 'Line' was considered a terrane boundary between the Dharwar in the north and the dominantly charnockite-rich granulites in the Southern Granulite Terrane. Subsequent studies in the belt, however, indicated that the Fermor's Charnockite Line does not truly define a terrane boundary but merely marks a metamorphic transition from the low-grade greenstone-amphibolite facies rocks in the north to the dominantly charnockite-bearing granulite facies rocks in the south (Janardhan, 1983). The new thought that 'Charnockite Line' is just a transition line in the regional metamorphism of the granite-greenstone belt rocks implied continuity of the Dharwar rocks further south. Field confirmation about the continuity of the Dharwar rocks further south of Fermor's Charnockite Line is evidenced from the occurrence of the trails of the coevally metamorphosed Archaean granite-greenstone belt rocks (most importantly, the metamorphosed banded iron formation) showing a metamorphic state compatible with the enclosing granulite bodies along with the 'tonalite-trondjhemite'-bearing Peninsular Gneiss in the south (Rama Rao, 1940; Janardhan, 1983). The reports of end-Archaean ages ranging between 2500 and 2900 Ma both from the charnockitic granulites and also from the Peninsular Gneiss from the region provide further proof of extension of the Dharwar belt in the south. Additional proof of the continuity of the Archaean Dharwar belt south of Fermor's Charnockite Line comes from the study of Palaeomagnetic poles of the mafic dykes south of the 'Line' that are similar to the coeval Dharwar dykes north of the Palghat-Cauvery Shear Zone (Dash et al., 2013).

The actual boundary between the Archaean Dharwar Protocontinent and the Southern Granulite Belt has now been demarcated along the prominent Lineament defining the southern boundary of the wide shear zone, the Palghat-Cauvery Shear Zone (Fig. 10.1). Confirmation for this comes from the evidence of Pan-African tectonothermal reconstitution ages centring around 500 ± 50 Ma from the granulites and associated rocks south of the Palghat-Cauvery Shear Zone (Chaudhary et al., 1992; Bartlett et al., 1995; Unnikrishnan-Warrier et al., 1995; Santosh et al., 1992, 2003; Shabeer et al., 2004; Collins et al., 2007). The Palghat-Cauvery Shear Zone has also been described as a Suture Zone marking accretion (or welding) of two distinctive crustal blocks in the north and the south, respectively (Radhakrishna and Naqvi, 1986).

Indian Shield. https://doi.org/10.1016/B978-0-12-809839-4.00010-2

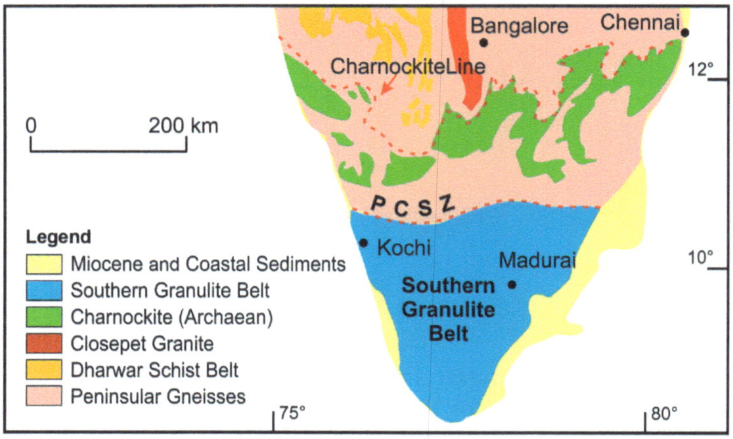

FIG. 10.1

Geological map of the southern tip of Indian Subcontinent showing 'Charnockite Line' of Fermor (1936) and the revised northern boundary of Southern Granulite Belt. *PCSZ*, Palghat-Cauvery Shear Zone.

Reproduced from Roy (2014) with permission.

Apart from the geochronologic evidence, Ramakrishna and Vaidyanadhan (2008) enumerated several lithologic and tectonic features marking differences between the two charnockitic terranes. These are the following:

(1) The dominant trend in the Dharwar Procontinent is north–south, whereas a conspicuous east–west trend is observed in the southern belt, south of the Palghat-Cauvery Shear Zone.

(2) The Dharwar charnockites include several linear inclusions of greenstone belt rocks comprising the mafic-ultramafic ensembles that also comprise metamorphosed banded iron formation. By contrast, the granulites include strips and patches of supracrustals like quartzite and metapelites which include garnet-sillimanite-cordierite-spinel-bearing gneiss (popularly described as khondalite), calc-silicate rocks, and calcitic marble. There are, however, only a few minor bodies of mafic-ultramafic rocks, and virtually no banded iron formation of greenstone belt association.

(3) The charnockitic granulite and associated basement gneisses of the Dharwar exhibit a dominant thermal event at 2500–2600 Ma. On the other hand, Southern Granulite Belt shows strong evidence of 500 ± 50 to 750 ± 50 Ma Pan-African thermal overprints.

10.2 PALGHAT-CAUVERY SHEAR ZONE: A ZONE OF CRUSTAL INTERACTION

Drury and Holt (1980) identified a wide zone of complex shear system along the southern margin of the Archaean Dharwar Protocontinent and named that as the Noyil-Cauvery Shear Zone. Recognizing the southernmost boundary of the shear zone as the terrane boundary, Drury et al. (1984) described it as the Palghat-Cauvery Shear Zone (Fig. 10.2a). Structurally, the Palghat-Cauvery Shear Zone is a highly

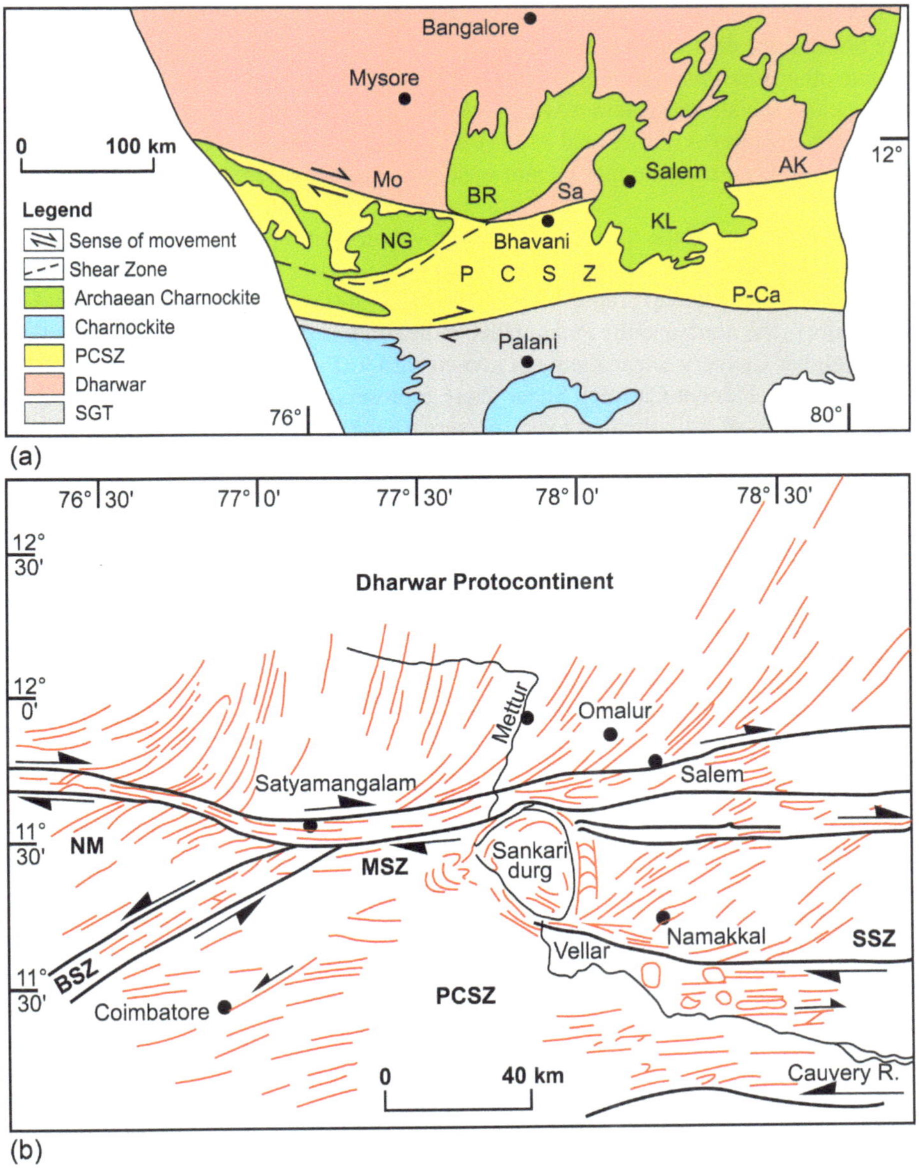

FIG. 10.2

(a) Geological map showing Palghat-Cauvery Shear Zone, which demarcate the southern marginal zone of the Dharwar Protocontinent. The Southern Granulite Belt (SGT) occurs south of Palghat-Cauvery Lineament (P-Ca). *NG*, Nilgiri Hill; *BR*, Biligiriranga Hill; *KL*, Kollimalai Hill. (b) A schematic map of the Palghat-Cauvery Shear Zone showing distribution of the shear zones and foliation trend lines. *NM*, Nilgiri mountains; *MSZ*, Moyar Shear Zone; *BSZ*, Bhavani Shear Zone; *PCSZ*, Palghat-Cauvery Shear Zone; *SSZ*, Sewta Nadi Shear Zone.

Reproduced from Jain et al. (2003) with permission.

tectonized belt, criss-crossed by narrow shear zones, and marked by the development of strong planar fabrics. The Shear Zone encloses less deformed belts of massifs, like the Nilgiri and Kollimalai Hills. There are some other small masses of granulites and granite-gneisses along with swathes of high-grade metamorphic pelitic schists and gneisses which also include bands and strips of 'layered' anorthosite–gabbro–ultramafic complexes (Drury et al., 1984). The Shear Zone is broadly an east–west-striking belt having intersecting planar tectonic fabrics into which the north–south fabric of the Dharwar curves in (Fig. 10.2b).

The southern boundary of the Shear Zone is largely concealed beneath the monotonous soil cover. Sparse outcrops close to the boundary indicate rotation of earlier planar fabrics indicating a dextral sense of movement. Detailed map prepared by Chetty et al. (2003) along the linear transect between the Biligiriranga Hills in the north and the Palani Hills in the south is very revealing as it provides structural details of the Palghat-Cauvery Shear Zone not known earlier (Fig. 10.3).

Geometrically, the Palghat-Cauvery Shear Zone represents a collage of narrow shear zones that enclose lenticular bands of little sheared rocks, preserving the earlier history of deformation and evolution. Some of the sheared bands occur south of the terrane boundary, which indicate the continuity of the shear deformation beyond the southernmost Palghat-Cauvery dislocation line.

BOX 10.1

In spite of the proposed change in the domain boundary, the dominantly granulite terrane south of the Palghat-Cauvery Shear Zone continued to be described as the 'Southern Granulite Terrain' in the literature. Considering this as anomalous, Ramakrishnan (1988) introduced a new term 'Pandyan Mobile Belt' for the southernmost granulite terrane. However, this new nomenclature, though may appear rational, did not find general acceptance in literatures (Sharma, 2009). In the present description, therefore, the traditionally accepted nomenclature is retained but with a rider constraining the northern limit of the southernmost crustal block of the Indian Shield only to the south of the Palghat-Cauvery Shear Zone, and changing the spelling from 'Terrain' to 'Terrane'.

In the northernmost part, immediately south of narrow zone marked by the development of planar fabrics occurs a zone that differs structurally from the zones of planar fabric that delineates this. The belt, known as the Moyar-Salem-Attur Shear Zone which occurs between Bhavani (11°27′: 77°41′) in the north and Chennimalai (11°10′: 77°36′) in the south (Fig. 10.3), shows several oval to irregular (swirling) closed structures, which simulate structures of plutonic diapirism. Chetty et al. (2003), however, preferred to describe these structures as large-scale sheath folds. The domain of planar fabrics returns in the successive shear-zone-bounded lensoid belts south of the Chennimalai.

Another interesting feature manifested in the structural map of Chetty et al. (2003) is that the zone of intense flattening deformation seems to continue southward beyond the dislocation zone marked as the Palghat-Cauvery Lineament.

A host of small-scale ductile deformation structures have developed in these shear zones as illustrated in Figs. 10.4a–d, 10.5a–d, and 10.6a and b. The images provide evidence of complex geometry and pattern of deformation along discrete belts of the Palghat-Cauvery Shear Zone.

Drury et al. (1984) based on Satellite Imagery studies reported a prominent westerly bending of the geomorphotectonic features north of the Palghat-Cauvery Shear Zone (Fig. 10.2b). The authors interpreted the features as dextrally rotated trend lines implying right-lateral strike-slip movement along the shear zone. Similar dextral bending of trend lines has also been recorded in the granulite massifs of Nilgiri and Kollimalai Hills that have been separated from main bodies in the north.

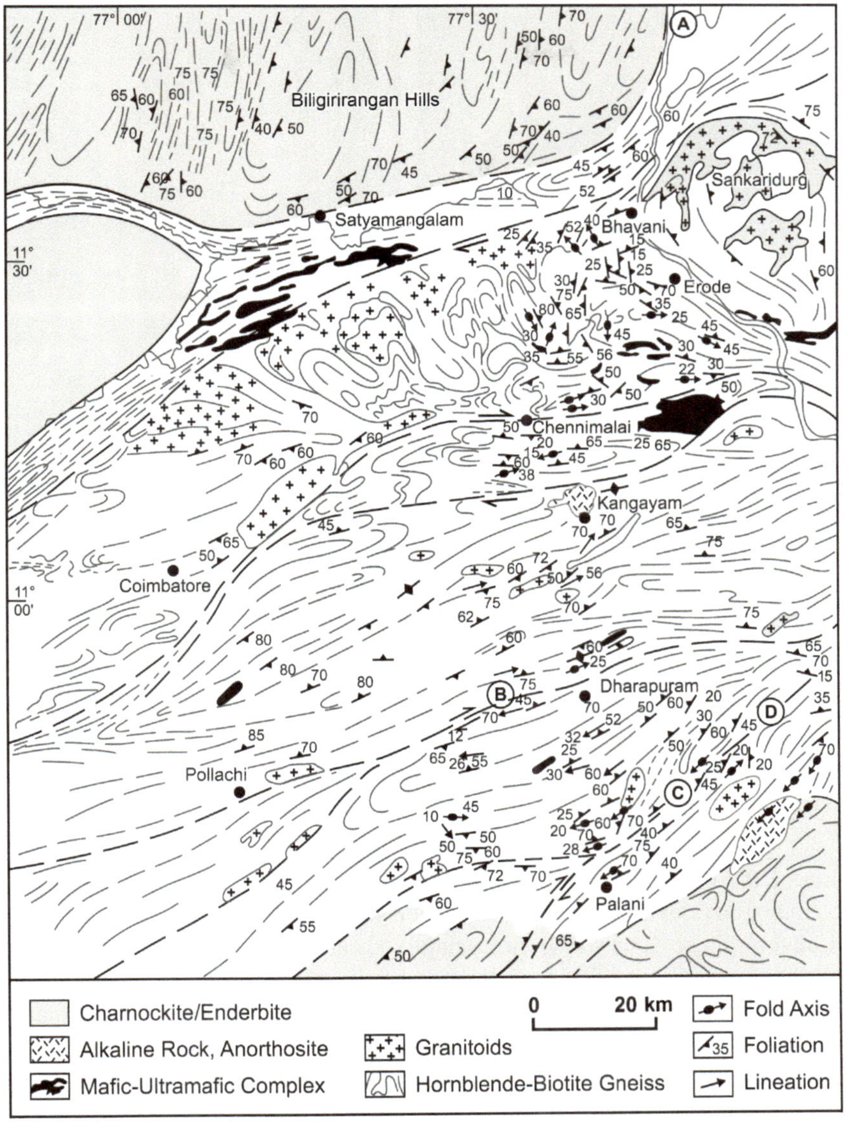

FIG. 10.3

A synoptic structural map across the Palghat-Cauvery Shear Zone based on 'Landsat TM' data, published geological maps and field data showing complex interlacing of shear zone bounded blocks.

Reproduced from Chetty et al. (2003) with permission.

FIG. 10.4

Field photographs from the Palghat-Cauvery Shear Zone south of Namakkal (11°13′: 78°10′). (a) Z-shaped folding in stretched and boudinaged mafic body in a felsic granulite. (b) Development of a thin shear zone in massive granulite. (c) Small-scale sheath fold showing complex outcrop pattern. (d) Extremely flattened subhorizontal isoclinal fold showing sinistral geometry in mafic granulite.

Reproduced from Mukhopadhyay et al. (2003) with permission.

The separation of the two ranges is estimated to be of around 70 km measured along the narrow shear zones. Analogous dextral rotations have been replicated in small scales in different part of the Palghat-Cauvery Shear Zone shown in miniature shear zones illustrated in different small-scale structures. All these provide proof of overall dextral strike-slip movement resulting from horizontal shear stresses between the sheared blocks. The modern analogue of similar deformation pattern is also observed along several zones where transform fault system is seen to cross over the continental Crust. The best example is that observed in the present-day San Andreas Fault in the southwestern part of North America (Fig. 10.7).

Before concluding on the issue of the terrane boundary between the northern Dharwar Protocontinent and the Southern Granulite Terrain, it may be worth mentioning about the alternate views proposed by some authors suggesting changes in the orientation 'terrane boundary line' along different shear zones other than Palghat-Cauvery Shear Zone, based mainly on the geochronological consideration (Ghosh et al., 2004; Plavsa et al., 2012; Brandt et al., 2014). However, the conventionally accepted terrane boundary along southern margin of the Palghat-Cauvery Shear Zone is so vividly documented in the field data focusing on lithologic character, geological history, and change in the tectonic style that any other suggestions based on somewhat dubious geochronological data may appear a futile exercise.

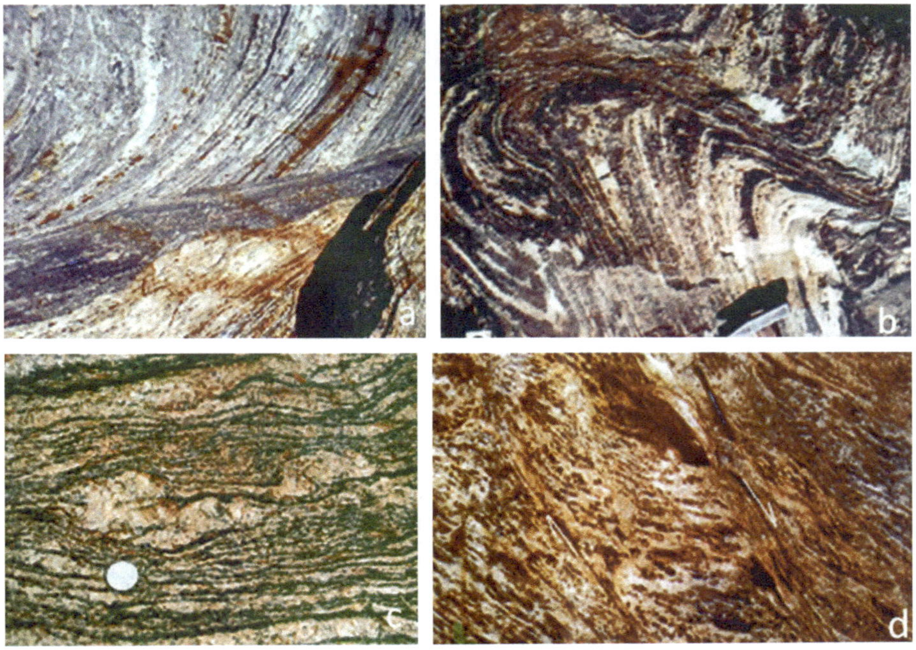

FIG. 10.5

Field evidence of small-scale deformation pattern along the Palghat-Cauvery Shear Zone. (a) Evidence of dextral drag of earlier planar fabric along the dislocation surface. (b) Refolding of an isoclinal fold, illustrating formation of secondary shear planes along the attenuated limb of the later fold. (c) Isoclinal folding of stretched and boudinaged felsic layer in mylonitized biotite gneiss. (d) A dextrally deformed shear band within mylonitized biotite gneiss.

(a,b) Reproduced from Chetty et al. (2003) with permission. (c,d) Reproduced from Srikantappa et al. (2003) with permission.

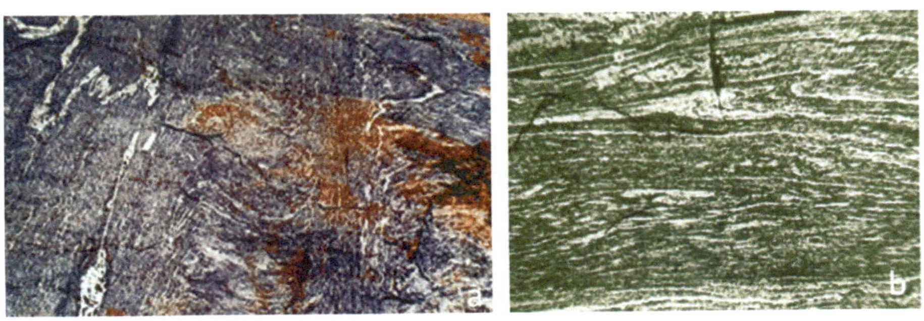

FIG. 10.6

(a) Wide zone of shear formed along limbs dextrally folded block. (b) Penetrative heterogeneous deformation in the Palghat-Cauvery Shear Zone showing development of strongly asymmetric dextral folds.

(a) Reproduced from Mukhopadhyay et al. (2003) with permission. (b) Reproduced from Jain et al. (2003) with permission.

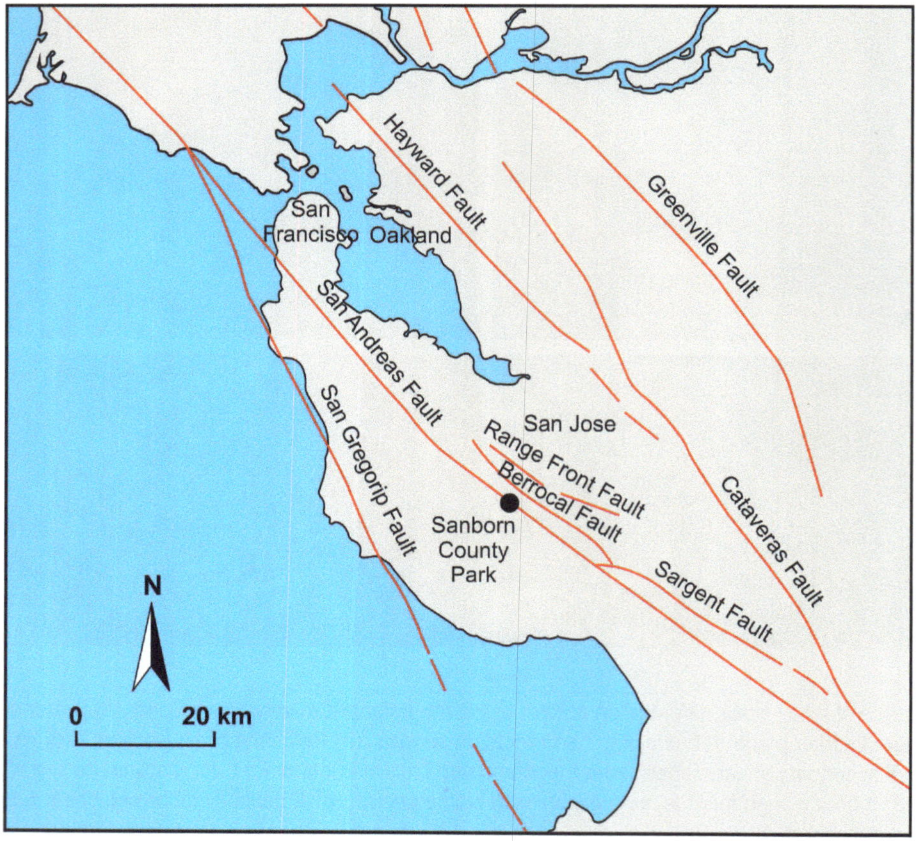

FIG. 10.7

Anastomosing pattern formed cluster of strike-slip faults along the San Andreas Fault in southwest of North America.

10.3 LITHOLOGY AND LITHOSTRATIGRAPHY OF THE SOUTHERN GRANULITE BELT

Southern Granulite Terrane is traditionally divided into two blocks, the Madurai and Trivandrum. The northern Madurai Block includes the region south of the Palghat-Cauvery Shear Zone in the north and the Achankovil Shear Zone in the south (Fig. 10.8). The southern Trivandrum Block also known as the Kerala Khondalite Belt is separated from the Madurai Block by a major Shear Zone called Achankovil Lineament or Shear Zone.

The Madurai Belt includes a variety of rocks apart from the different types of granulites including charnockites and some high-grade metasedimentary rocks. Lithostratigraphically, an important component but least discussed lithoassociation is the granitic gneiss that varies in composition from biotite gneiss, biotite hornblende gneiss, diorite gneiss, etc. along with granitoids of different composition and character. Most of these granite-gneisses and granitoids in the Southern Granulite Belt include

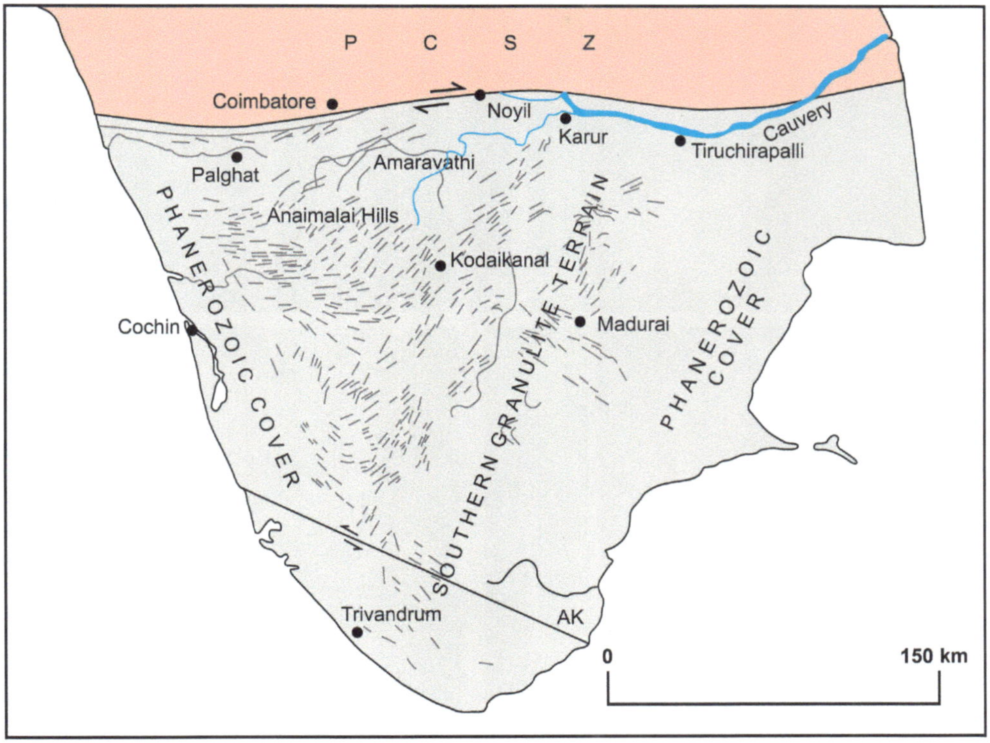

FIG. 10.8

Geological map of the Southern Granulite Belt. Irregular discontinuous lines indicate varying trends of foliation surfaces. *PCSZ*, Palghat-Cauvery Shear Zone. *AK*, Achankovil Shear Zone.

Compiled from different sources.

well-crystalline garnet that indicates postdeformational metamorphic character of basement bodies. Structurally, the gneissic rocks show evidence of multiple deformation (Fig. 10.9a–d). The basement character of the gneissic rocks appears to be a foregone conclusion especially looking into the physical resemblance of these granite-gneisses with the Peninsular gneisses and other gneissic rocks from different other terrains. Geochemical composition of some of the biotite gneisses indicates tonalite-trondjhemite-granodiorite type.

Granitoids rocks in the Southern Granulite Terrain can be distinguished according to their chemical composition and their mode of formation (Chandra Sekaran et al., 2016). The porphyritic granites occurring as coarse-grained augen gneisses developed at amphibolite-facies conditions (Braun, 2006). Garnet- and biotite-bearing leucogranites are abundant throughout the Kerala Khondalite Belt and also at several places in the Madurai Block. The fine-grained granites, having distinct alkaline affinities, are abundant in the area, especially along the Achankovil Shear Zone.

Apart from the granitoids and gneissic rocks, there are also hosts of metasediments belonging to the shale-sand-carbonate association. The field geologists describe these rocks as ‘QPC’ suite that is an abbreviation of quartzite-phyllite-carbonate marble (Ramakrishna and Vaidyanadhan, 2008).

FIG. 10.9

(a–d) Biotite gneisses from the Karur–Namakkal sector in northeastern part of the Madurai Block of Southern Granulite Belt. (a,b) Biotite gneiss showing strong ductile deformation marked by repeated folding and ductile shearing. Growth of garnet indicates metamorphic character postdating the deformation. (c) Folding of strongly developed mylonitic spaced cleavage showing evidence of syntectonic melting. (d) Complexly deformed granite-gneiss showing repeated syn- to postkinematic melting.

Images courtesy, Rajnees Bhutani.

Several enclaves of supracrustal rocks occur as swathes and linear patches within the biotite gneiss and 'massive-to-gneissic' charnockitic rocks. These supracrustal rocks include quartzite, Mg–Al-rich pelites, garnet-sillimanite-bearing gneisses, and calc-silicate rocks along with calcitic marble. All these metasedimentary rocks occur within the vast expanse of biotite gneiss and charnockites of 'massive-to-gneissic' varieties (Fig. 10.10). In the Southern Trivandrum belt (also described as Kerala Khondalite Belt), the high-grade varieties of the supracrustals comprising 'sand-shale-carbonate' occur in close association of massive charnockites.

In the northern Madurai Block, the metasedimentary bands show sharp facies change from north to south. Between Dindigul and Madurai, there is a large exposure of folded graphitic garnet-sillimanite gneiss (described as khondalite) and quartzite. Numerous small linear bands of quartzite occur south of Madurai in the Tirunelveli area along with bands of crystalline marble associated with calc-silicate rocks (Fig. 10.10).

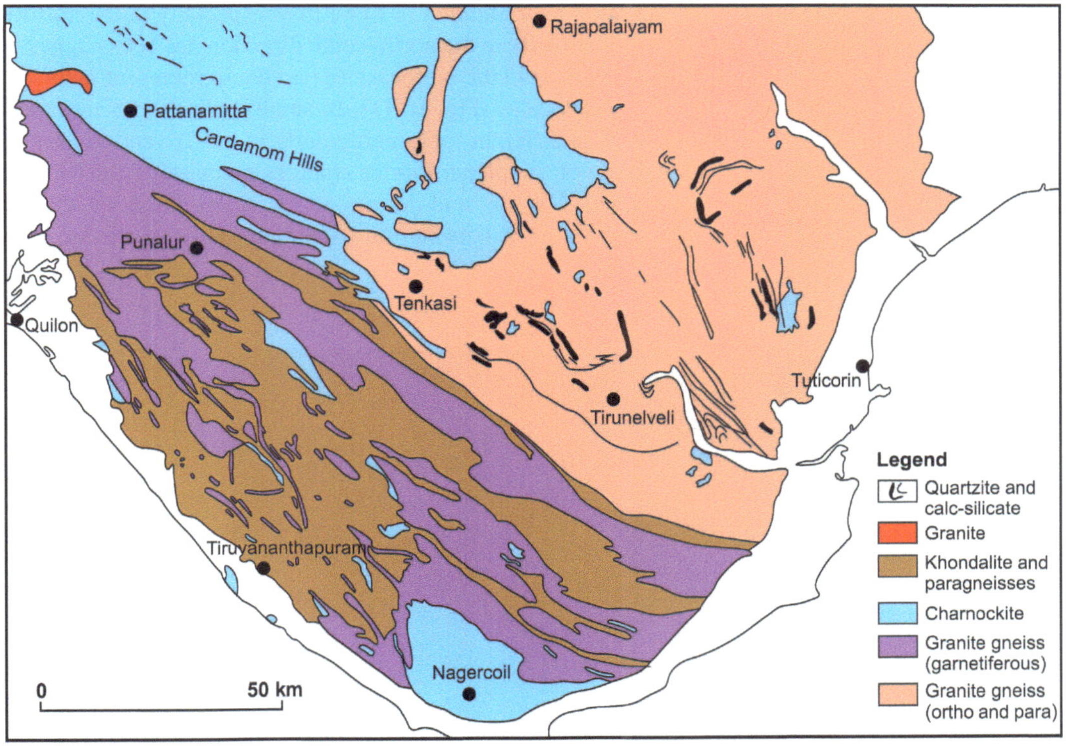

FIG. 10.10

Geological map of the southernmost part of the Southern Granulite Belt showing distribution of different rock types and structural pattern.

Reproduced from Ghosh et al. (2004) with permission.

The Trivandrum Block of the Kerala Khondalite Belt shows large exposures of tightly folded pelitic khondalite, interbanded with garnet-bearing biotite gneiss (leptynite) and charnockite. Regionally, a broad lithological difference exists between dominantly khondalite rocks in the northern part and large outcrops of charnockite in the south. These are intruded by granitic rocks that show evidence of partial melting and migmatization and transformed into augen gneisses and fine-grained biotite gneisses.

The charnockitic rocks are dominant in the western part forming high topographic relief. Eastwards, the metasedimentary bands become more prominent and occur in intricately folded forms, generally interbanded with biotite gneisses and charnockitic rocks (Mahadevan, 2003).

Both the Madurai and Trivandrum Blocks are distinctive in that they host extensive alkaline granite magmatism of around 500 ± 50 Ma which are correlated with the anorogenic magmatism accompanying the Pan-African thermal event in the region.

No detailed structural map is available to illustrate the large-scale geometry and deformation pattern of the Southern Granulite Terrane as a whole. In this regard, the reconnaissance structural map based on limited field data and mapping on digitally enhanced Landsat images by Drury et al. (1984)

appears quite handy and useful. Looking into the orientation of trend lines especially in the northern part, the presence of regional-scale open fold broadly in the north–south direction has been suggested. Some oval outlines observed in the Sirumalai Hills in the northeastern part could represent map-scale outcrop of granite/anorthosite bodies in some locales (Fig. 10.11). In small scales, there are reports of development of recumbent isoclinal folds in different rocks particularly in the Trivandrum region. In the Palani Hills area, the regional attitude of the planar fabrics changes eastwards from virtually flat-lying to a NNE-SSW strike with steep WNW dip and defines an extremely large wavelength, southward-closing and northward-plunging synform between the Anaimalai Hills in the west and Kodaikanal in the east.

FIG. 10.11

Google Satellite image showing complex outcrop pattern developed due to repeated folding. The oval outcrops in the northeast of Dindigul are because of intrusion of anorthosite body (Kadavur anorthosite).

To the east of the Kodaikanal massif, a major lithological and structural change occurs where supracrustal rocks become dominant, although they are interlayered with acid to intermediate granulite sheets. The supracrustals are represented by a layered sequence of metapelites, quartzites, calc- silicates, and thin mafic-ultramafic horizons. A key structural feature of this supracrustal-rich zone is an array of isoclinal folds, up to 15 km across, west of Madurai (Fig. 10.12a). Series of oval patterns observed in the map pattern the Sirumalai Hills region that could be due to diapiric intrusion of some magmatic bodies, including anorthosite in the region. More detailed structural trend map has been prepared by Chenki and Kriegsman (2005) (Fig. 10.12b) that differs considerably from that of the Drury et al. (1984). The new structural map shows a contrasting pattern of deformation in the Madurai and Trivandrum belts (Fig. 10.12b).

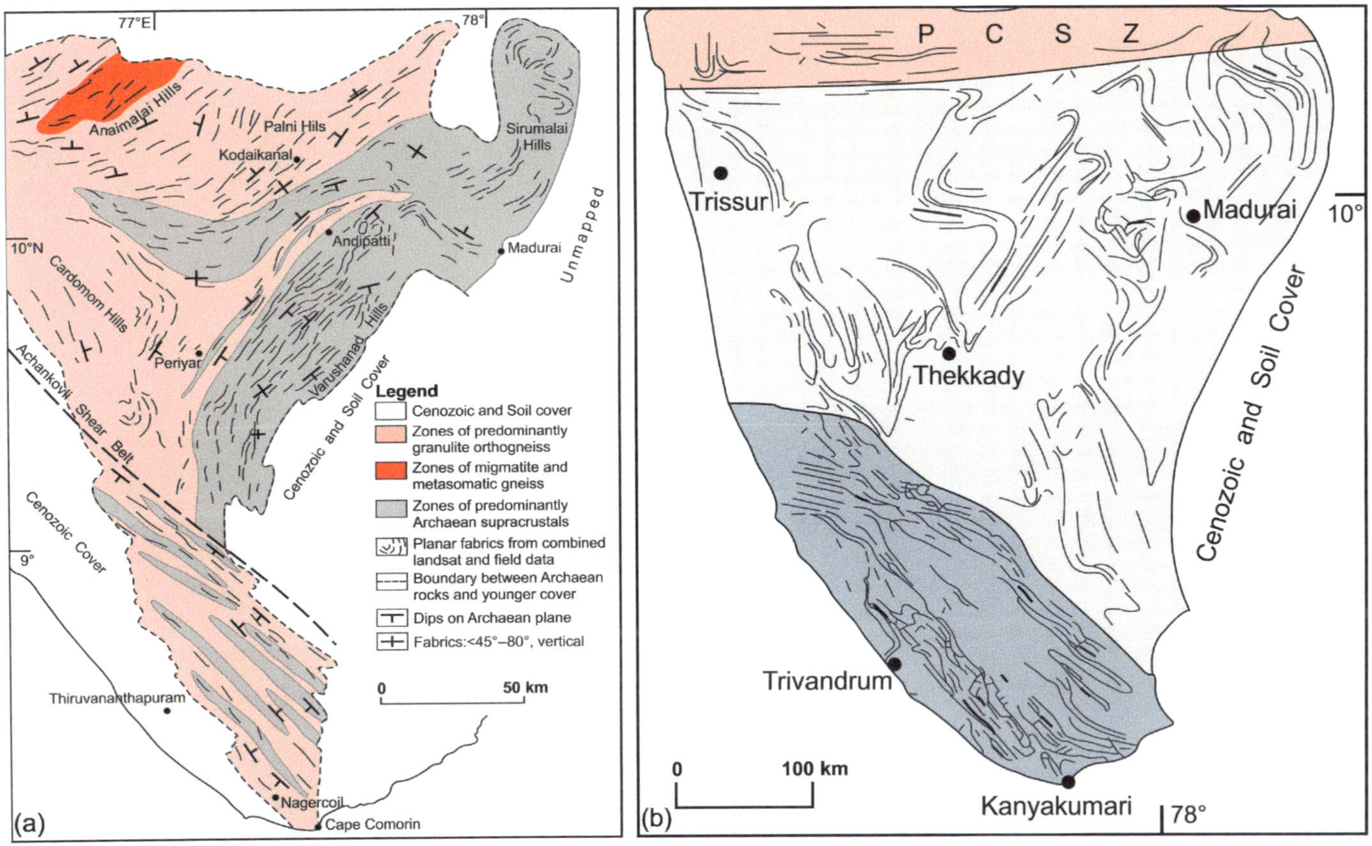

FIG. 10.12

(a) Reconnaissance structural map of the Southern Granulite Belt based on field data and interpretation of digitally enhanced Landsat image. (b) Simplified structural trends *(thin black lines)* showing contrasting pattern of deformation in two different structural domain of Madurai and Trivandrum Belt. *Light grey* represents Madurai Block, while *dark grey* represent the Trivandrum Block.

(a) Reproduced from Drury et al. (1984) with permission. (b) Modified after Chenki and Kriegsman (2005) with permission.

10.4 GEOCHRONOLOGY AND TECTONO-MAGMATIC/METAMORPHIC EVOLUTIONARY HISTORY

A number of geochronological data are now available on the rocks of the Southern Granulite Belt, and some of these may appear geologically meaningful. Careful analysis of the geochronological data helps to decipher the evolutionary history of the rocks constituting the southernmost crustal block of the Peninsular India. We critically analysed the set of more than 100 compiled isotope data showing apparently disparate ages listed in Chandra Sekaran et al. (2016) and Ghosh et al. (2004). These ages, as our analysis suggests, are partitioned into four age-wise subgroups. This division of the isotope age data appears strikingly similar to that of the 'probability density plot' of Chandra Sekaran et al. (2016), barring the description about the 'origin' of the rocks. The earliest ones are those that show Archaean ages. Except for the couple of older ages between 3000 and 2875 Ma (Crawford, 1969; Brandt et al., 2014), most of the older dates represent end-Archaean ages centring around 2500 Ma (taking into consideration the error factor and interlaboratory discriminations) which in all likelihood mark the closure of the Archaean Crust-building history.

There is absolutely no information, field or isotopic, on the post-Archaean geological history of the rocks of the region, except the fact that the 'sand-shale-carbonate' association, now represented by the quartzite-phyllite/mica schist/gneiss (often described as sillimanite-bearing paragneiss or khondalite) and marble, represents shallow marine shelf deposits. We correlate this metasedimentary succession as Palaeoproterozoic predating 1760 Ma, as the oldest post-Archaean formations in the Madurai Block. The deformation and metamorphism of these Proterozoic sediments resulted in the development of planar fabric (foliation) that defines the large- and small-scale folds of the region. Because of intense deformation, marked by the development of tight/isoclinal folding, the bedding orientation has been transposed subparallel to the axial planar foliation. This deformation seems to have affected both the cover and basement rocks showing development of outcrop-scale recumbent folds showing isoclinal or near isoclinal geometry (Fig. 10.13a and b). In view of the common pattern of deformation both in the basement gneisses and in the pelitic metasediments of Proterozoic age, it is possible to link the tectonothermal changes that took place between 1660 and 1760 Ma period.

FIG. 10.13

(a) Development of isoclinal recumbent folds with strong thinning of the limbs, Alamparai, Nagercoil area.
(b) Recumbent folding on the foliation planes in sillimanite-cordierite-bearing metapelite, near Trivandrum.

(a) Reproduced from Chenki et al. (2005) with permission.

Reports of somewhat older ages come from the single zircon evaporation and EPMA monazite data which indicate a thermal event in the early Proterozoic (ca. 1.8 Ga), possibly under upper-amphibolite-facies conditions, especially in the Kerala Khondalite Belt (Bartlett et al., 1995; Braun, 2006). The first evidence for a Palaeoproterozoic (ca. 1.85 Ga) Crust-forming event has been documented in the Kerala Khondalite Belt by Rimša (2007). Broadly similar 1893 ± 13 Ma age is reported by Kröner et al. (2012) from the zircon evaporation study. Whitehouse et al. (2014) reported 1853 ± 21 Ma based on SHRIMP and SIMS single zircon in situ analysis, which mark the protracted felsic magmatism-related Crust-forming event during the Palaeoproterozoic. Incidentally, it may be interesting to note that these, ~1853–1893 Ma ages have not been reported from the Madurai Block of the Southern Granulite Belt. No penetrative deformation features seem to have formed subsequent to the end-Palaeoproterozoic tectonothermal event.

There is virtually no geological record of any event in the Southern Granulite Belt during Mesoproterozoic, and only a very few data are available that suggest ushering in of the Neoproterozoic at ~1000 Ma. Out of three available age data suggesting ‘emplacement ages’ (Chandra Sekaran et al., 2016), two are from the granodiorite indicating 989 ± 21 Ma and quartz monzonite indicating 1007 ± 23 Ma (Plavsa et al., 2012). The third one suggesting 1024 ± 23 Ma (Brandt et al., 2014) needs some elucidation. Firstly, the age is described as metamorphic/thermal event, and the dated zircon has a record of prehistory of end-Archaean age. The rock has been described as ‘migmatitic charnockites’ (Chandra Sekaran et al., 2016). A critical examination of the field photograph (Brandt et al., 2014) (Fig. 10.14a), however, does not indicate any possible migmatitic character of the rock in the absence of any evidence of melting feature in the rock. On the other hand, the leucocratic bands show well-preserved tectonic deformation history of stretching and shearing (showing sinistral geometry) that could have developed during the Archaean or Palaeoproterozoic orogenic event. We infer that 1024 ± 69 Ma age of Brandt et al. (2014) is indicative of thermal event linked with granodiorite/quartz monzonite magmatism in the region.

FIG. 10.14

(a) Field photo of charnockite from Kerala Khondalite Belt described as ‘migmatitic charnockites’. (b) A slab of Kadavur anorthosite, indicating undeformed magmatic fabric.

(a) Reproduced from Brandt et al. (2014) with permission.

There are quite a few ages indicating Neoproterozoic dates between 780 and 850 Ma. The ages mark an important thermal event related to the emplacement/intrusion of anorthosite-gabbro (Kadavur gabbro-anorthosite, e.g. Fig. 10.14b) coeval with the ultrahigh-temperature metamorphism responsible for dry melting and formation of charnockites.

The youngest ages are in the 500 ± 50 Ma age range. The age is traditionally marked as the closing event of the Gondwana assembly, marked by emplacement of granitoids originating from crustal melting. The related low thermal metamorphism noted in the rocks has preserved some retrogression-related structures (Chandra Sekaran et al., 2016). There are a few data showing older ages (i.e. $>500 \pm 50$ Ma). It is difficult to comment on those data in the absence of any other information.

10.5 THE CHANOCKITE RIDDLE

One of the fundamental problems associated with the occurrence of charnockite in the Southern Granulite Belt is the lack of information about the precise time of formation of the rock and the lithotectonic/lithostratigraphic relationship of the associated rock types that make up the granulite terrane. In spite of its wide occurrence in different parts of the region, coupled with ample reference in different literatures, very little is known on the field relationship of this high-grade granulite facies rocks, especially the charnockite along with the other associated rocks. The most disquieting feature is the absence of any reliable geological map indicating lithological and structural details of the rocks.

Small patches and pods of charnockites having irregular, diffused boundaries occur more commonly within biotite gneisses and in some other schistose rocks of pelitic and psammopelitic composition in several parts of the Southern Granulite Belt. Several reports mention the occurrence of veinlike intrusion of 'arrested/incipient' charnockitic into the gneissic and associated schistose rocks cross-cutting the regionally developed foliation in different rocks including biotite gneisses and pelitic/psammopelitic schists (Fig. 10.15a and b).

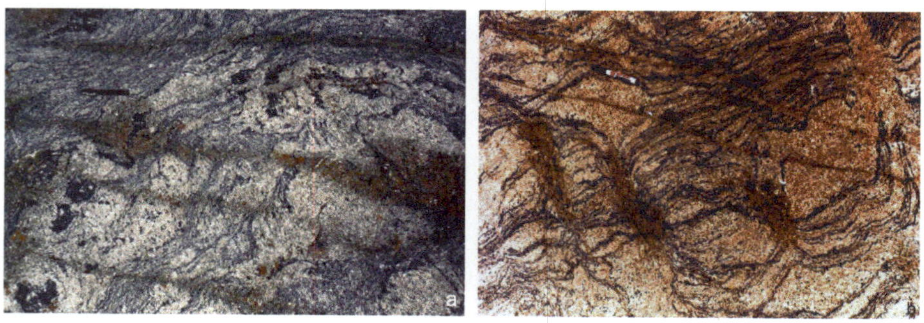

FIG. 10.15

(a) Field photograph showing occurrence of arrested charnockite forming narrow linear band cross-cutting the fabric in the host gneiss. (b) Formation of arrested charnockites along neck zones and fracture surfaces of the pink coloured biotite gneiss. There are also some narrow, linear veinlike bodies of arrested charnockites.

Reproduced from Ravindra Kumar (2004) with permission.

Several large bodies of charnockites occur in different parts of the Southern Granulite Belt. All such bodies show evidence of patchy or incipient charnockite formation along their margins. Most of the massive charnockite bodies occur in the western and southern part of the Southern Granulite Belt, for example, in the Cardamom and Nagercoil Hills (Fig. 10.12a). The commonly occurring charnockites belong to the charnockite-enderbite series rocks, which are interpreted to have magmatic origin based on the geochemical evidence (Chacko et al., 1992, 1996). Ghosh et al. (2004), however, contradicted the concept of magmatic origin of these granulite bodies, suggesting the presence of gradational contact inferred from the field observation indicating continuity of foliation traces from the country rocks (usually biotite gneiss and, also in some instances, the pelitic and psammopelitic rocks) into the massive charnockites. Describing the foliation traces in the charnockites as the 'ghost foliation', Ghosh et al. (2004) suggested that the foliation traces (appearing only on surface markers) merely indicate the presence of preexisting gneissic foliation in the protolith rocks, which are now completely replaced by coarse-grained recrystallized minerals. Field studies indicated that the disappearance of the preexisting foliation is linked with the progressive transformation of the foliated rocks into homogenous, massive charnockites.

In literature, there are several reports suggesting presence of two different types of charnockites: the massive charnockite and the gneissic charnockite. The field description of the 'gneissic charnockites' however uncovers the fallacy that the described foliation appearing as gneissic banding is nothing but traces of relic fabric in the protolithic mass that transformed into massive charnockites (Fig. 10.16a and b).

FIG. 10.16

(a,b) Massive, coarse-grained charnockites showing traces of relict gneissic foliation. Except the relic traces, the rock does not show the presence of any megascopic or microscopic planar fabric marked by recrystallized minerals.

In the absence of any large- and intermediate-scale map showing lithologic boundary between the charnockites of different composition and the associated country rock, it is hard to suggest that the massive charnockites are intrusive bodies emplaced into the host rocks and have not formed through the process of 'patchy' charnockite formation involving metasomatic transformation.

Based on geochronological data, several authors reported occurrence of charnockites of different ages from Archaean to latest Neoproterozoic. Most of the data represent U–Pb zircon ages using LA-ICP-MS systematics (Chandra Sekaran et al., 2016), and reported these dates either as the age of crystallization or as a metamorphic/thermal event. Most of the zircon dates are interpretative as it preserves records of multiple tectonothermal events. As far as the age of charnockites formation is concerned, we preferred to rely on the field data that unequivocally indicate the development of charnockites from the patchy, arrested types to large massive bodies invariably postdate the fabric forming deformation in the rocks, especially the penetrative gneissic foliation. Using this field criterion, we can shortlist the time of the charnockites formation to different thermal events in the Neoproterozoic. Our analysis indicated that all the tectonic deformation must have been completed by the end of Palaeoproterozoic period, and there was complete absence of record of any tectonothermal event during the entire Mesoproterozoic period. Combining this information with the field evidence that the charnockites formation postdated most important tectonically induced planar fabric (foliation) in different rocks, we suggest that the granulite facies metamorphism (resulting in the development of charnockites) was a Neoproterozoic event. We can, however, omit the ~1000 Ma age for the charnockite formation as only a couple data of this age are known that mark the emplacement ages of 981 ± 21 Ma age of granodiorite and 1007 ± 23 Ma of the quartz monzodiorite of the Madurai Block of the Southern Granulite Belt (Plavsa et al., 2012). We can also omit the youngest Neoproterozoic ages as the date of charnockite formation for two reasons: (a) this was the period when the youngest massive granitoids were emplaced at several places generally yielding 500 ± 50 Ma age, and (b) there are reports of retrogressive changes in the preexisting charnockites resulting from the youngest Neoproterozoic tectonothermal event indicating preexistence of charnockite (Chandra Sekaran et al., 2016). It may be interesting to note that there are several reports of crystallization/emplacement age of charnockites from different localities by different authors within a short range between 768 and 894 Ma (Chandra Sekaran et al., 2016). Within this age group, the report of 829 ± 14 Ma emplacement age of gabbro-anorthosite (Teale et al., 2011) appears quite significant. Several emplacement ages of rocks were recognized as charnockites close to this age. The charnockites of the Southern Granulite Belt are reported to be of ultrahigh-temperature (UHT) type (Santosh et al., 2006, 2008). This information matches with the occurrence of anorthosite in the Kadavur and Oddanchatram area in the Madurai Block of the Southern Granulite Belt.

In summary, we conclude that the charnockites of the Southern Granulite Belt have preserved some records about their formation, tectonic relationship with the associated rocks, and age of their formation that are likely to resolve the geological riddle associated with this controversial rock type, the charnockite.

REFERENCES

Bartlett, J.M., Harris, N.B.W., Hawkesworth, C.J., Santosh, M., 1995. New isotope constraints on crustal evolution of south India and Pan-African Granulite metamorphism. In: Yoshida, M., Santosh, M. (Eds.), Indian and Antarctica During the Precambrian. 34. Mem. Geol. Soc. India, Bangalore, pp. 391–397.

Brandt, S., Raith, M.M., Schenk, V., Sengupta, P., Srikantappa, C., Gerde, A., 2014. Crustal evolution of the Southern Granulite Terrane, south India: new geochronological and geochemical data for felsic orthogneisses and granites. Precambrian Res. 246, 91–122.

Braun, I., 2006. Pan-African granitic magmatism in the Kerala Khondalite Belt, southern India. J. Asian Earth Sci. 28, 38–45.

Chaudhary, A.K., Harris, N.B.W., Van Clasteren, P.C., Hawkesworth, C.J., 1992. Pan-African charnockite formation in Kerala, south India. Geol. Mag. 129, 257–264.

Collins, A.S., Santosh, M., Braun, I., Clark, C., 2007. Age and sedimentary provenance of Southern Granulites of Southern India: U-Th-Pb SHRIMP secondary ion mass spectrometry. Precambrian Res. 155, 125–138.

Chenki, B., Kriegsman, L.M., 2005. Tectonics of the Neoproterozoic Southern Granulite Terrain, south India. Precambrian Res. 138, 37–56.

Chacko, T., Ravindra Kumar, G.R., Meen, J.K., Rogers, J.J.W., 1992. Geochemistry of highgrade supracrustal rocks from the Kerala Khondalite Belt and adjacent massif charnockites, south India. Precambrian Res. 55, 469–489.

Chacko, T., Lamb, M., Farquhar, J., 1996. Ultra-high temperature metamorphism in the Kerala Khondalite Belt. In: Santosh, M., Yoshida, M. (Eds.), The Archaean and Proterozoic Terrains in Southern India Within East Gondwana. Gondwana Res. Group, Nagpur, Mem. 3. pp. 57–165.

Chandra Sekaran, M., Bhutani, R., Balakrishnan, S., 2016. Rb–Sr and Sm–Nd study of granite–charnockite association in the Pudukkottai region and the link between metamorphism and magmatism in the Madurai Block. J. Earth Syst. Sci. 125, 605–622.

Chetty, T.R.K., Bhaskar Rao, Y.J., Narayana, B.L., 2003. A structural cross section along Krishnagiri-Palani Corridor, Southern Granulite Terrain of India. In: Ramakrishnan, M. (Ed.), Tectonics of Southern Granulite Terrain: Kuppam-Palani Geotransect. Mem. 50. Geol. Soc. India, Bangalore, pp. 255–277.

Crawford, A.R., 1969. Reconnaissance Rb-Sr dating of the Precambrian rocks of southern Peninsular India. J. Geol. Soc. India 10, 117–166.

Dash, J.K., Pradhan, S.K., Bhutani, R., Balakrishnan, S., Chandrasekaran, G., Basavaiah, N., 2013. Paleomagnetism of ca. 2.3 Ga mafic dyke swarms in the northeastern Southern Granulite Terrain, India: constraints on the position and extent of Dharwar craton in the Paleoproterozoic. Precambrian Res. 228, 164–176.

Drury, S.A., Harris, N.B.W., Holt, R.W., Reeves-Smith, G.J., Wightman, T.T., 1984. Precambrian tectonics and crustal evolution in south India. J. Geol. 92, 3–20.

Drury, S.A., Holt, R.W., 1980. The tectonic framework of the south Indian craton: A reconnaissance involving Landsat imagery. Tectonophysics 65, T1–T15.

Fermor, L.L., 1936. An attempt at correlation of ancient schistose formations of peninsular India. Mem. Geol. Surv. India 70, 219–323.

Ghosh, J.G., Wit, M.J.d., Zartman, R.E., 2004. Age and tectonic evolution of Neoproterozoic ductile shear zonesin the Southern Granulite Terrain of India, with implications of Gondwana studies. Tectonics 23, 1–38.

Jain, A.K., Singh, S., Makickavasagam, R.M., 2003. Intracontinental shear zones in the Southern Granulite terrain: their kinematics and evolution. In: Ramakrishnan, M. (Ed.), Tectonics of Southern Granulite Terrain: Kuppam-Palani Geotransect. Mem.50. Geol. Soc. India, Bangalore, pp. 225–253.

Janardhan, A.S., 1983. Transformation of peninsular gneiss to charnockite in Southern Karnataka. In: Naqvi, S.M., Rogers, J.J.W. (Eds.), Precambrian of South India. 4. Mem. Geol. Soc. India, Bangalore, pp. 417–435.

Kröner, A., Santosh, M., Wong, J., 2012. Zircon ages and Hf isotopic systematics reveal vestiges of Mesoproterozoic to Archaean crust within the late Neoproterozoic–Cambrian high-grade terrain of southernmost India. Gondwana Res. 21, 876–886.

Mahadevan, T.M., 2003. Geological Evolution of south Indian shield-constraints on modeling. In: Ramakrishnan, M. (Ed.), Tectonics of Southern Granulite Terrain: Kuppam-Palani Geotransect. Mem. 50. Geol Soc. India, Bangalore, pp. 25–46.

Mukhopadhyay, D., Kumar, P.S., Srinivasan, R., Bhattacharya, T., 2003. Nature of Palghat-Cauvery Lineament in the region south of Namakkal, Tamil Nadu: implication for terrane assembly in South Indian Granulite Province. In: Ramakrishnan, M. (Ed.), Tectonics of Southern Granulite Terrain: Kuppam-Palani Geotransect Mem. 50. Geol. Soc. India, Bangalore, pp. 279–296.

Plavsa, D., Collins, A.S., Foden, J.F., Kropinski, L., Santosh, M., Chetty, T.R.K., Clark, C., 2012. Delineating crustal domains in Peninsular India: age and chemistry of orthopyroxene-bearing felsic gneisses in the Madurai Block. Precambrian Res. 198–199, 77–93.

Radhakrishna, B.P., Naqvi, S.M., 1986. Precambrian continental crust of India and its evolution. J. Geol. 94, 145–166.

Ramakrishnan, M., 1988. Tectonic evolution of the Archaean high-grade terrain of south India. J. Geol. Soc. India 31, 118–119.

Ramakrishna, M., Vaidyanadhan, R., 2008. Geology of India. Geoloical Society of India, Bangalore. 555 p.

Rama Rao, B., 1940. The Archaean complex of Mysore. Mysore Geol. Dept. Bull. 17, 1–101.

Ravindra Kumar, G.R., 2004. Mechanism of arrested charnockite formation at Nemmara, Palghat region, southern India. Lithos 75, 331–358.

Rimša, A., 2007. Understanding zircon geochronology—constraints from imaging and trace elements. (Unpublished Ph.D. Thesis, LITHOLUND Thesis No. 12). Lund University, Sweden. 87 p.

Roy, A.B., 2014. Indian Subcontinent, Reference Module in Earth Systems and Environmental Sciences. Elsevier. https://doi.org/10.1016/B978-0-12-409548-9.09372-6.

Santosh, M., Kagami, H., Yoshida, M., Nada-Kumar, V., 1992. Pan-African charnockite formation in East Gondwana: geochronologic (Sm-Nd and Rb-Sr) and petrologic constraints. Bull. Indian Geol. Assoc. 25, 1–10.

Santosh, M., Yokoyama, K., Biju-Sekhar, S., Rogers, J.J.W., 2003. Multiple tectonothermal events in the granulite blocks of southern India revealed from EPMA dating; implication for theory of supercontinent. Gondwana Res. 6, 29–63.

Santosh, M., Collins, A.S., Tamashiro, I., Koshimoto, S., Tsutsumi, Y., Yokoyama, M., 2006. The timing of ultrahigh-temperature metamorphism in southern India: U–Th–Pb electron microprobe ages from zircon and monazite in sapphirine bearing granulites. Gondwana Res. 10, 128–155.

Santosh, M., Yokoyama, M., Tsutsumi, Y., Shinichi, Y., 2008. Electron microprobe dating of monazites from an ultrahigh-temperature granulite in southern India: implications for the timing of Gondwana assembly. J. Mineral. Petrol. Sci. 103, 77–87.

Shabeer, K.P., Satish Kumar, M., Armstrong, M., Buick, I.S., 2004. Constraints on timingof Pan-African granulite facies metamorphism in the Kerala Khondalite Belt of south India: SHRIMP mineral ages and Nd isotope systematics. J. Geol. 113, 95–106.

Sharma, R.S., 2009. Cratons and Mobile belts of India. Springer 304.

Teale, W., Collins, A.S., Foden, J., Payne, J.L., Plavsa, D., Chetty, T.R.K., Santosh, M., Fanning, M., 2011. Cryogenian (830 Ma) mafic magmatism and metamorphism in the northern Madurai Block, southern India: a magmatic link between Sri Lanka and Madagascar? J. Asian Earth Sci. 42, 223–233.

Srikantappa, C., Srinivas, G., Basavarajappa, H.T., Prakash Narasimha, K.N., Basavalingu, B., 2003. Metamorphic evolution and fluid regime in the deep continental crust along the N-S Geotransact from Vellar to Dharapuram southern India. In: Ramakrishnan, M. (Ed.), Tectonics of Southern Granulite Terrain: Kuppam-Palani Geotransect. Mem. 50. Geol. Soc. India, Bangalore, pp. 319–373.

Unnikrishnan-Warrier, C., Santosh, M., Yoshida, M., 1995. First report of Pan-African Sm-Nd and Rb-Sr mineral isochron ages from regional charnockites of southern India. Geol. Mag. 132, 253–260.

Whitehouse, M.J., Ravindra Kumar, G.R., Rimša, A., 2014. Behaviour of radiogenic Pb in zircon during ultrahigh-temperature metamorphism: an ion imaging and ion tomography case study from Kerala Khondalite Belt, southern India. Contrib. Mineral. Petrol. 168 (1042), 1–18.

CHAPTER

EASTERN GHATS GRANULITE BELT

11

11.1 INTRODUCTION

The Eastern Ghats Granulite Belt represents a broadly linear terrain of discontinuous hill ranges facing the Bay of Bengal in the east. The belt is about 1000 km long and having a width of about 300 km in the northern part, tapering down to about 50 km in the south, south of Ongle (15°29′; 80°03′). Lying astride the structural grains of the three Precambrian Protocontinents, Dharwar, Bastar, and Singhbhum, this belt of high-grade metamorphic rocks comprises dominantly of granulites and charnockites (Fig. 11.1).

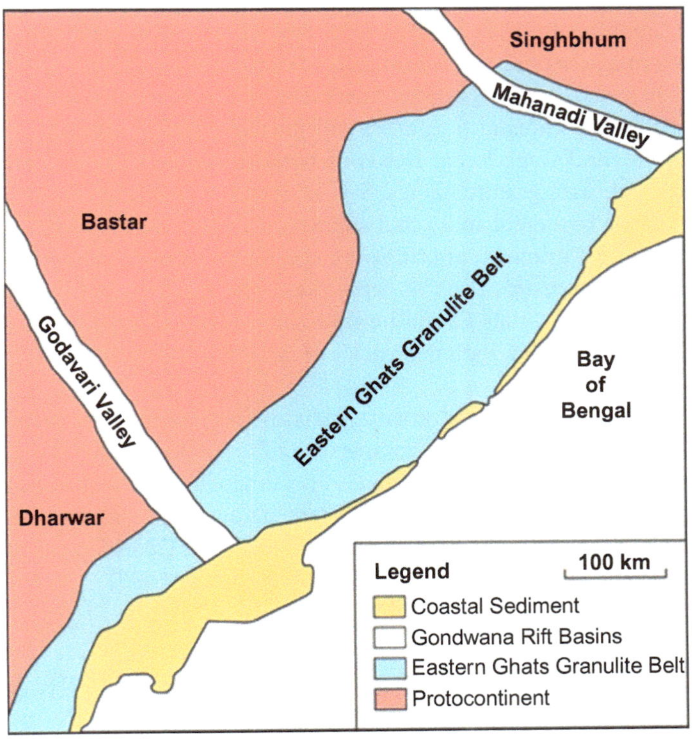

FIG. 11.1

Geological map of the Eastern Ghats Granulite Belt.

Reproduced from Roy (2014) with permission.

Indian Shield. https://doi.org/10.1016/B978-0-12-809839-4.00011-4

Tectonically, the Eastern Ghats Granulite Belt is known as an 'exotic block' or 'suspect tectonostratigraphic terrane' (Radhakrishna, 1989), which had accreted to the Indian Shield during the continental collision in response to the compression from the Enderby Land in the Antarctica. The intercontinental correlation of two high-grade terranes is based on the broad similarity in the history of evolution of both the Eastern Ghats Granulite Belt and the Rayner Province of East Antarctica (Rao et al. 1995). Understanding the process of accretion of this high-grade belt with the Proto-India holds key in apprehending the history of the Precambrian evolution and the process of reconstitution of the pristine Indian Shield.

BOX 11.1

Some authors have used the term Eastern Ghats Mobile Belt for this high-grade dominantly granulite facies terrane. We, however, would prefer to use more objective nomenclature, the Eastern Ghats Granulite Belt in the present discussion.

11.2 LITHO-TECTONIC CHARACTERIZATION OF THE EASTERN GHATS GRANULITE BELT

The Eastern Ghats Granulite Belt is conventionally known as a high-grade granulite terrane comprising rocks varying in age between Archaean and Neoproterozoic (Sarkar and Paul, 1998; Rickers et al., 2001). The belt consists of a gneissic basement of the Archaean age, overlain by hosts of granulitic supracrustals like khondalite (garnet–sillimanite–graphite gneiss), charnockitic gneiss, calc granulite, quartzite, and some linear bands of banded iron formation. The supracrustal rocks occur in close association with granite gneiss, S-type granites, and migmatites of different compositions and ages. Both the basement rocks and supracrustals are intruded by massive charnockites, enderbites, basic granulites, granites, anorthosites and alkali granites, and rocks derived from the partial melting of the supracrustals such as leptynites. The inclusion of the Nellore Schist Belts into the domain of the Eastern Ghats Granulite Belt necessitates change in the description that the Eastern Ghats Granulite Belt also includes rocks of amphibolite facies (Fig. 11.2) (Ramam and Murthy, 1997).

Apart from the supracrustal rocks and granite gneisses (representing the possible Archaean basement), several bodies of layered anorthosite complexes of the Archaean age (?) occur within this zone. Patches of the Proterozoic massifs of anorthosite also occur along the western and northern margins of the belt (Ramakrishnan and Vaidyanadhan, 2008). There are also some occurrences of deformed alkaline rocks and carbonatites along the western margin of the Eastern Ghats Granulite Belt, considered to be the marker of an ancient Suture Zone (Fig. 11.3) (Leelanandam, 1997).

11.3 DEFINING THE WESTERN CONTACT OF THE EASTERN GHATS GRANULITE BELT

Ramakrishnan et al. (1998) described the Eastern Ghats Granulite Belt as a zone of westerly directed thrust slices abutting against the three Protocontinents, Dharwar in the south, Bastar in the middle, and Singhbhum in the north. The western contact of the Eastern Ghats Granulite Belt follows the

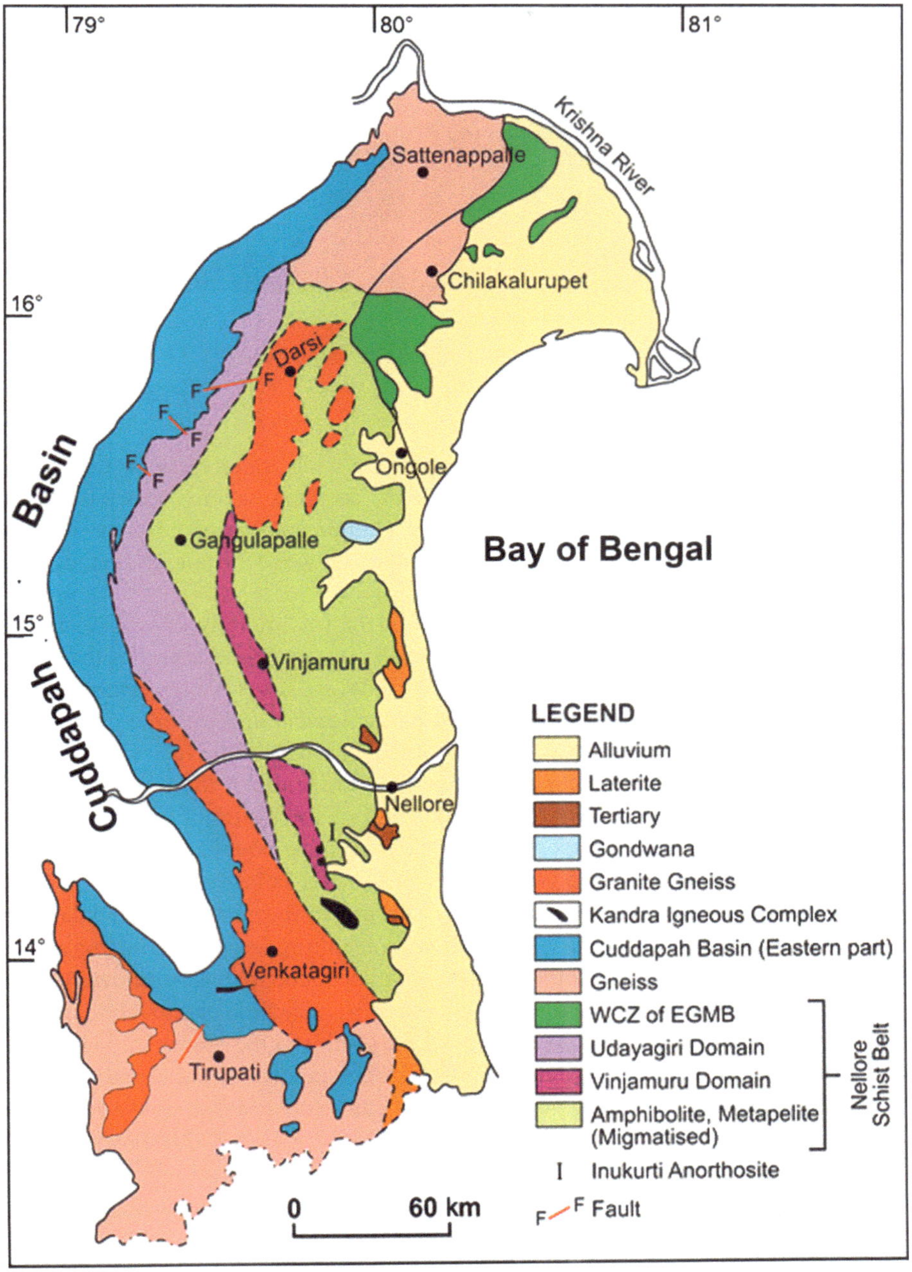

FIG. 11.2

Geology of the Nellore Schist Belt constituting the southwestern limit of the Eastern Ghats Granulite Belt close to the eastern margin of the Cuddapah Basin.

Reproduced from Ramam and Murthy (1997) with permission.

FIG. 11.3

Banded anorthosite in the Eastern Ghats Granulite Belt (a) at Chimalpad and (b) at Kondapalli.

Image courtesy, C. Leelanandam.

boundaries of the three Protocontinents and is marked by wide zones of shear traceable from the southernmost part southeast of the Cuddapah Basin to the northeast where the belt of the high-grade terrane dips into the Bay of Bengal near Bhubaneswar. Certain ambiguity, however, remains regarding the nature and precise location of the tectonic boundary between the autochthonous blocks of the Protocontinents and the thrusted exotic block.

Changes have also been proposed in the location of the traditionally known western boundary the Eastern Ghats Granulite Belt in some parts, especially in the southern part of the Eastern Ghats Granulite Belt. In the region south of the Pranhita-Godavari River Valley, Dobmeier and Reith (2003) based on the critical evaluation of the existing geological and isotopic data, suggested the inclusion of the Nellore-Khammam Schist Belt, a domain of amphibolite facies rocks and traditionally considered a part of the Dharwar Protocontinent into the realm of the Eastern Ghats Granulite Belt. The proposed correlation of these rocks with those of the Eastern Ghats implies shifting of the western boundary of the thrust-block close to the easternmost boundary of the Cuddapah Basin. In this context, the occurrence and evolution of the Nallamalai Fold Belt in the eastern part of the Cuddapah Basin assumes significance. Roy (2014) interpreted the eastern Nallamalai Fold Belt of the Cuddapah Basin as an 'indentation' tectonic feature resulting from the collision of the Eastern Ghats Granulite Belt with Dharwar Protocontinent of which Cuddapah Basin forms a part.

Further confirmation of the collisional tectonics comes from the field and Satellite Imagery studies in the region between the Pranhita-Godavari Valley in the south and the Mahanadi Valley in the north. A complex deformation pattern is recorded along the western contact of the Eastern Ghats Granulite Belt that according to Biswal et al. (1998) is compatible to those observed along the zones of continental collision. The geological map of the Deobhog region (Neogi and Das, 2000) showing the thrust contact between the Eastern Ghats Granulite Belt and the Bastar Protocontinent is marked by a prominent zone of mylonite and ultramylonite (Fig. 11.4). There are also patches of less sheared granitic rocks on the western side of the mylonite belt.

Biswal et al. (2004) described the major tectonized belt between the western Protocontinental block and the Eastern Ghats Granulite Belt as 'Terrane Boundary Shear Zone' that encompasses both the footwall and the hang-wall side of the dislocation surface (Fig. 11.5). The separation between the western Protocontinental block from the eastern exotic block is difficult to locate precisely for the occurrence of intensely mylonitized band of alkaline rocks in the east of the mylonitized belt (Fig. 11.6).

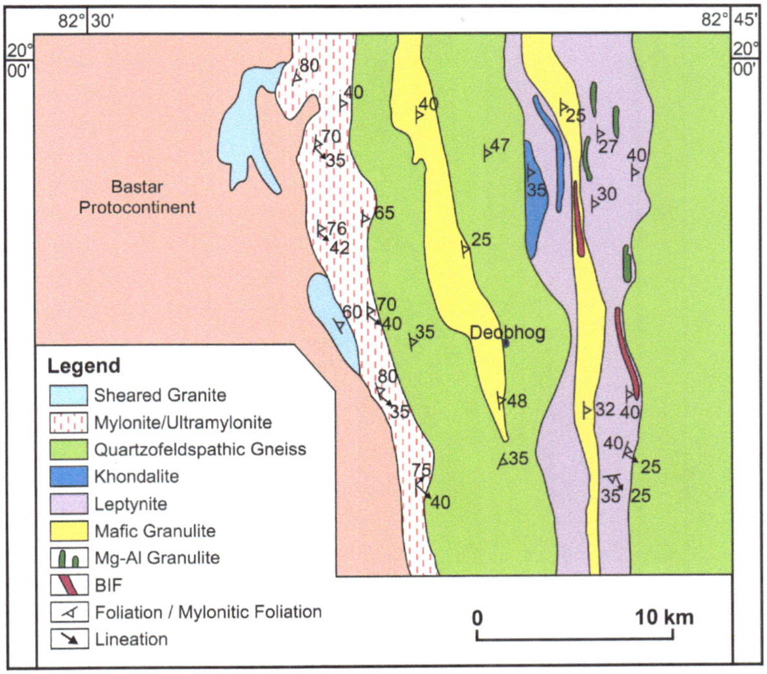

FIG. 11.4

Geological map of the Deobhog area showing thrust contact (marked by the zone of mylonite/ultramylonite) between the Bastar Protocontinent in the west and the Eastern Ghats Granulite Belt in the east.

Compiled from different sources.

The granite gneisses of the Protocontinents show evidence of repeated transposition of the mylonitic foliation providing evidence of repetitive folding and heterogeneous strain pattern illustrated in Fig. 11.7a–d.

Biswal and Sinha (2003) described the Eastern Ghats Granulite Belt as a fold-thrust belt, with the development of prominent 'Salient' on its northwestern margin north of the Deobhog area (19°59′; 80°59′). The Salient consists of Lathore Nappe and Turekela Klippe that have been thrust over the autochthonous belt in the northwest. The nature of the mylonitic structures developed along the Terrane Boundary Shear Zone suggests a low-temperature metamorphic condition causing retrogression of the granulite facies rocks into the greenschist facies rocks (Biswal et al. 2000). According to Biswal and Sinha (2003), the Salient structure of the fold-thrust belt is the combined result of lateral 'Ramps' on the decollement and differential displacement along the sole thrust due to lateral variation in detachment strength.

Like the southern part of the Eastern Ghats Granulite Belt, controversy persists about the northernmost part of belt occurring east of the Mahanadi River Basin as witnessed in the Rengali Domain (Crowe et al., 2003). According to the authors, the belt bound by major shear zones between the Singhbhum Protocontinent in the north, the Bastar Protocontinent in the northeast, and the Eastern

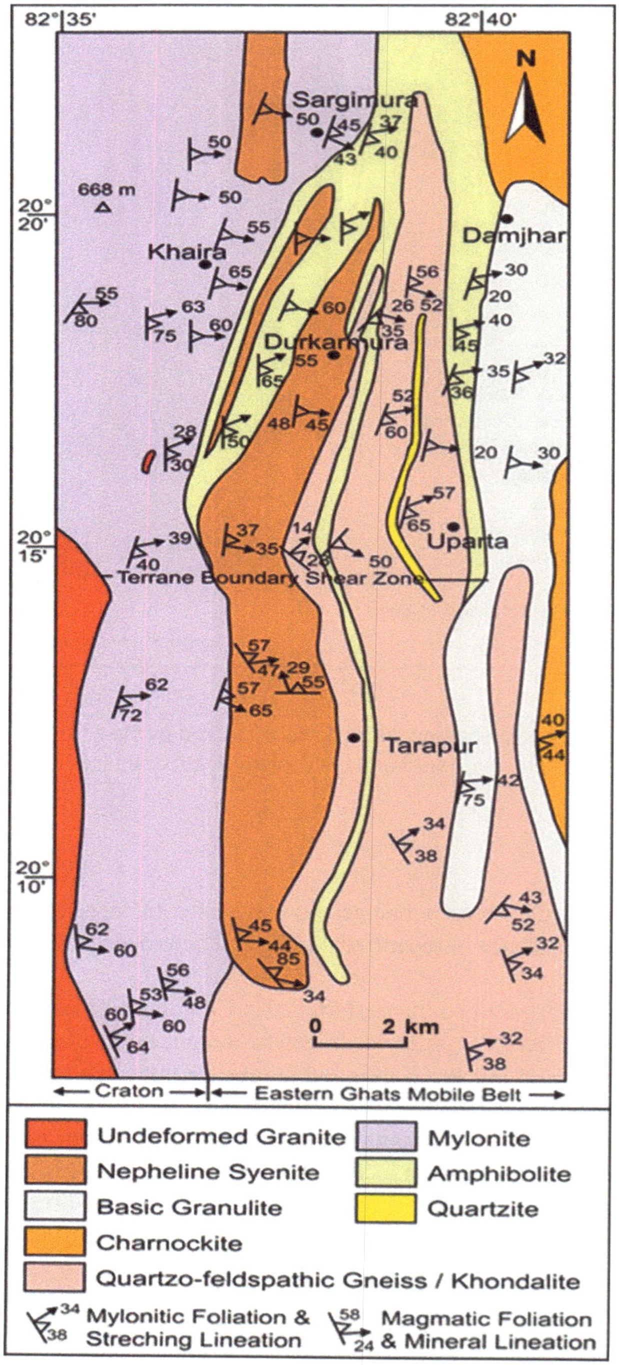

FIG. 11.5

Geology of the Terrane Boundary Shear Zone in the Khariar region showing occurrence of extremely sheared alkaline rocks in the eastern part of the mylonite belt.

Reproduced from Biswal et al. (2004) with permission.

FIG. 11.6

Mylonitized alkaline rock of the Khariar region occurring in the Terrane Boundary Shear Zone at Khariar.

Image courtesy, T. K. Biswal.

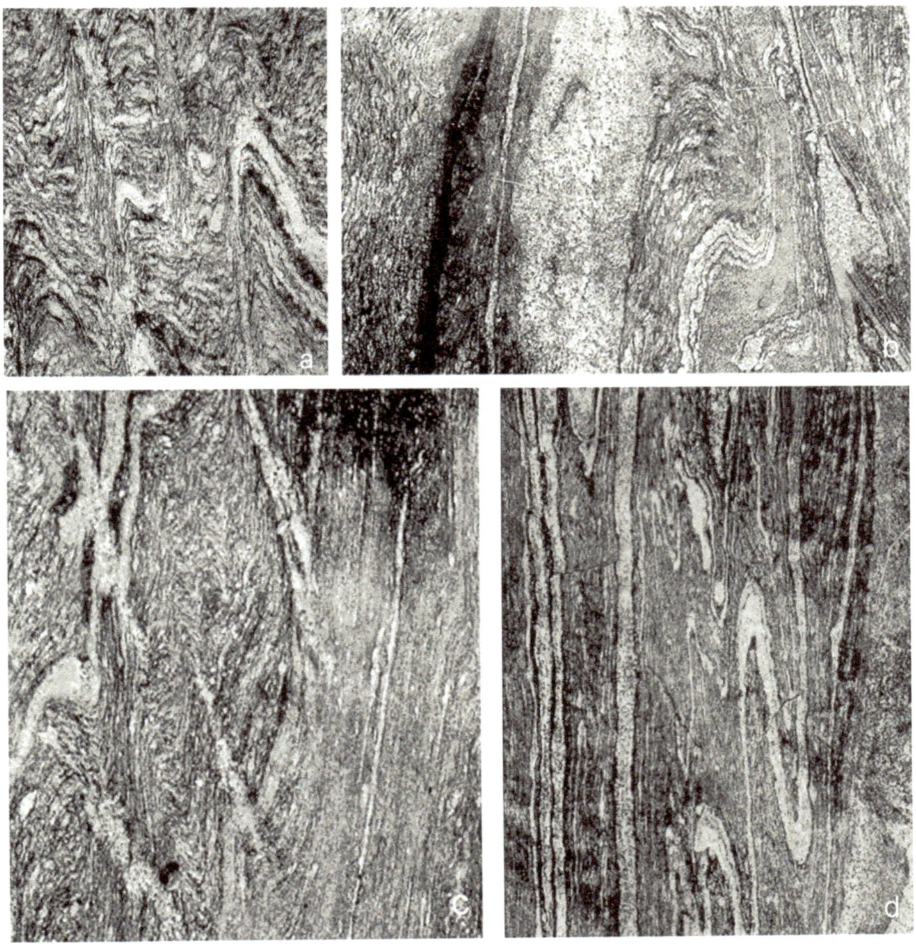

FIG. 11.7

(a–d) Outcrops of heterogeneously sheared, mylonitized granite gneisses at the contact of the Eastern Ghats Granulite belt, south of Deobhog, Chhattisgarh.

Reproduced from Bhadra et al. (2004) with permission.

Ghats Granulite Belt (*sensu stricto*) in the southeast shows a 'distinct lithological, structural, and metamorphic character' not witnessed in the southern part of the Eastern Ghats Granulite Belt. The Rengali Domain comprises a sequence of amphibolite facies rocks intercalated with basement gneiss and metavolcano-sedimentary lithologies. Deformation is characterized by a variation in strain from relatively low for much of the central and northeast part of the belt to a strongly deformed zone south of the Riamal Shear Zone (Fig. 11.8).

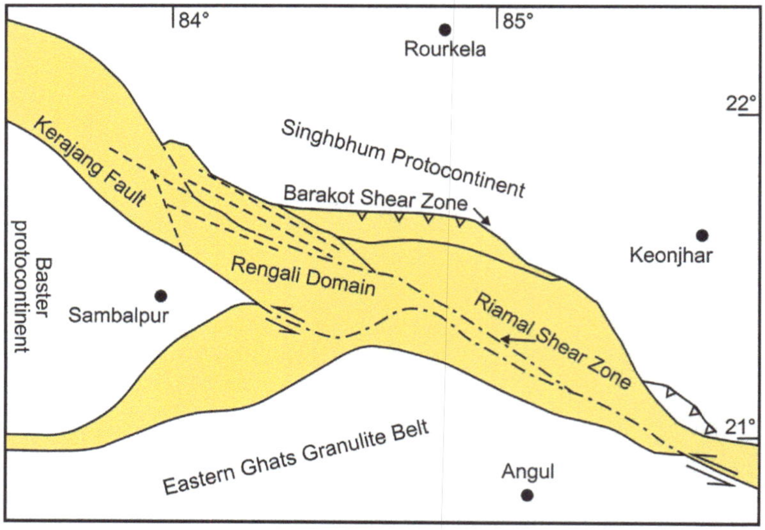

FIG. 11.8

Structural map of the Rengali Domain in the northern part of the eastern Ghats Granulite belt showing shear zones and interfingering blocks of both the Singhbhum Protocontinent and the Eastern Ghats Granulite Belt.

Modified from Crowe et al. (2003) with permission.

Geochronological data suggest that the dextral displacement along the bounding Kerajang and Barakot Shear Zones (Fig. 11.8) occurred prior to ~980 Ma within the Eastern Ghats Granulite Belt (Crowe et al. 2003). Deformation associated with the dextral displacement of the Singhbhum block relative to the Bastar shaped the Rengali Domain as a distinct and comparatively high-strain belt that accommodated large-scale dextral shearing at amphibolite facies condition. Progressive shearing and later reactivation associated with late retrogression were confined to the southern margin of the Rengali Province along the Kerajang Fault Zone.

11.4 LITHO-TECTONIC ZONATION OF THE EASTERN GHATS GRANULITE BELT

Based on the compiled geological map of the Eastern Ghats Granulite Belt, Ramakrishnan et al. (1998) reported occurrence of several linearly disposed lithologic zones parallel to the northeast-southwest elongation of the belt. From west to east, these zones are (i) Western Charnockite Zone, (ii) Western Khondalite Zone, (iii) Central Migmatite Zone, and (iv) Eastern Khondalite Zone (Fig. 11.9a).

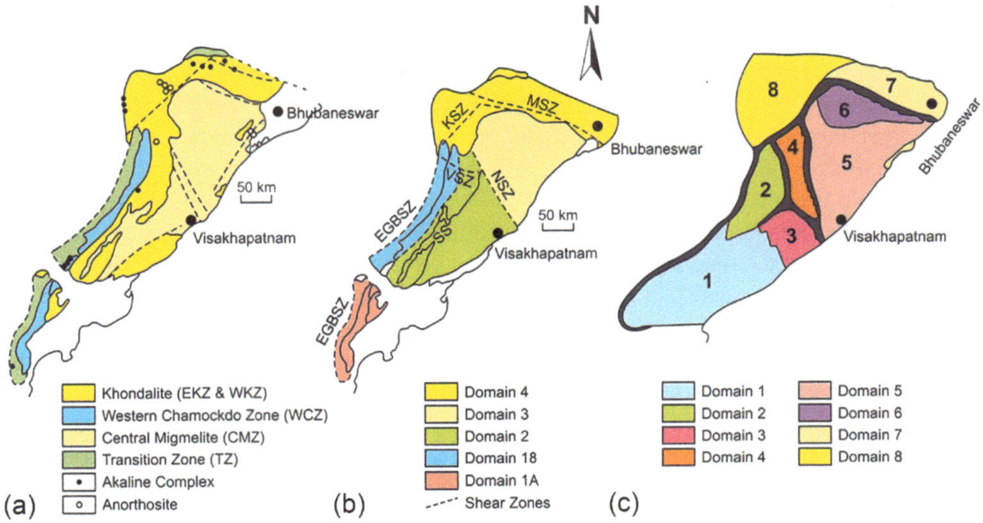

FIG. 11.9

Interpretative models of the Eastern Ghats Granulite belt: (a) Ramakrishnan et al. (1998), (b) Rickers et al. (2001), and (c) Chetty (2001).

Modified from Dasgupta et al. (2013) with permission.

Most of the Western Charnockite Zone includes massifs of charnockites, enderbites, and localized bands of mafic granulites. In addition to these four zones, Ramakrishnan et al. (1998) also recognized a Transition Zone marking the western boundary of the Eastern Ghats Granulite Belt with the autochthonous Protocontinents in the west and north (Fig. 11.9a).

The Western Khondalite Zone of Ramakrishnan et al. (1998) comprises sillimanite-bearing gneisses, quartzite, calc-silicates, and rare marbles. The mineral associations including cordierite, spinel, and sapphire have been reported in isolated pockets located within this zone. The alkaline complexes of the Eastern Ghats Granulite Belt are apparently restricted to the margin between the Western Charnockite Zone and the Western Khondalite Zone. There are bodies of granite gneisses (showing evidence of partial melting) along with enclaves of charnockites and khondalite, along with some large bodies of garnet-bearing porphyritic granitoids in the Central Migmatite Zone. The Eastern Khondalite Zone is dominated by migmatized sillimanite gneisses with minor quartzite and calc-silicate rocks.

Rickers et al. (2001) based on the Nd model ages distinguished four age groups with distinct regional distribution. The rocks of all the four age groups display older crustal residence ages, which are different from the age of the known metamorphic events. Regionally, the four domains defined by Rickers et al. (2001) coincide partially with those of lithological domains recognized by Ramakrishnan et al. (1998) (Fig. 11.9b). The main conclusions drawn by Rickers et al. (2001) on the age-wise distribution of the precursor (premetamorphic) rocks of the Eastern Ghats Granulite Belt are: (i) The Eastern Ghats Granulite Belt comprises rocks that are derived mostly from the reworked Archaean and early-to-middle Proterozoic crustal material, (ii) there were minor additions of juvenile material during the younger Grenvillian Orogeny at 1 Ga, and (iii) the crustal materials of Domains 2 and 3 (Fig. 11.9b) either are unrelated or represent distinct parts of one orogeny. The boundary between the two domains as conceived by Rickers et al. (2001) is not gradual, and may be of tectonic origin.

Implicit in the above-mentioned divisions of the Eastern Ghats Granulite Belt by Ramakrishnan et al. (1998) and Rickers et al. (2001) is that the lithotectonic units of the belt represented distinctly separate crustal fragments having incongruent evolutionary histories. Similar conclusions have also been highlighted by different authors working independently on metamorphism, regional structural analysis, and geochronological studies in different parts of the Eastern Ghats Granulite Belts. For example, the metamorphic studies indicated divergent P-T-t paths in different parts of the belt leading the interpretations of divergent and restricted tectonic histories (Dasgupta et al., 2013). Similarly, structural analysis in different parts of the Eastern Ghats Granulite Belts revealed complex deformation histories that are unlikely in any tectonically homogenized belt (Bhattacharya 1996). Likewise, the results of recent isotope studies indicated that metamorphic and structural histories across the Eastern Ghats Granulite Belts are not straightforward as previously assumed (Gupta, 2004).

Taking cue from these studies, Chetty and Murthy (1993, 1994) initiated to look into the structural framework of the Eastern Ghats Granulite Belt using Landsat TM data, followed by reconnaissance field surveys. The studies revealed occurrences of a network of major Ductile Shear Zones at the margin and within the entire terrane of Eastern Ghats Granulite Belt. These Ductile Shear Zones enclose several isolated terrane showing divergent tectonic pattern. Chetty (2001) described the Eastern Ghats Granulite Belt as a collage of juxtaposed terrane representing stacks of nappe sheets. The author identified eight disparate domains bounded by Ductile Shear Zones (Fig. 11.9c).

The key to the tectonic model of Chetty (2001) is the recognition of an array of Ductile Shear Zones that bound the domains representing collage of thrust nappes. All the shear zones are characterized by the development of mylonitic foliation associated with grain-size reduction, metamorphic retrogression, development of stretching lineation, and distinct signatures of magmatism of different sorts especially of alkaline, anorthositic, and granitic characters. The main gneissic foliation in all the domains is refolded in near noncoaxial manner generally about the axis subparallel to the elongation of the terrane. The structural history in each terrane is distinct in terms of orientation of stretching lineation, nature of gneissic character, and tectonic make-up.

11.5 CONCLUDING REMARKS

The Eastern Ghats Granulite Belt is a linear high-grade belt that skirts the eastern fringe of the Proto-India astride the belts of Dharwar, Baster, and Singhbhum Protocontinents. The belt is truncated as it is sliced off at the eastern continental margin of the Indian Subcontinent. Traditionally, the belt is regarded as a homogeneous geological entity. Recent studies on different aspects like lithological constitution, age of the constituents, metamorphic history, and polyphased structural deformation provide evidence of strong tectonometamorphic heterogeneity of this accreted terrane. This is evident in the reconnaissance structural and tectonic studies that provide a clear indication that the Eastern Ghats Granulite Belt represents a collage of juxtaposed terrane. This certainly is a move in the right direction that should offer motivations for future studies ensuring detailed structural studies in regional and smaller domains to find out true picture on structure and tectonics. Mere description of the Eastern Ghats Granulite Belt as a mobile belt may not help in proper understanding of the tectonometamorphic character of the heterogeneously deformed belt.

REFERENCES

Bhadra, S., Gupta, S., Banerjee, M., 2004. Structural evolution across the Eastern Ghats Mobile Belt–Bastar craton boundary, India: hot over cold thrusting in an ancient collision zone. J. Struct. Geol. 26, 233–245.

Bhattacharya, S., 1996. Eastern Ghats granulite terrain of India of India: an overview. J. SE Asian Earth Sci. 14, 165–174.

Biswal, T.K., Sinha, S., 2003. Deformation history of the NW salient of the Eastern Ghats Mobile Belt, India. J. Asian Earth Sci. 22, 157–169.

Biswal, T.K., Sanjeevan, O., Nayak, B.P., 1998. Deformational history of Eastern Ghat Mobile Belt around Lathore, Balangir district, Orissa. J. Geol. Soc. India 51, 219–225.

Biswal, T.K., Jena, S.K., Datta, S., Das, S., Khan, K., 2000. Deformation of Terrane boundary shear zone (Lakhna shear zone) between Eastern Ghats Mobile Belt and Bastar Craton in Balangir and Kalahandi Districts, Orissa. J. Geol. Soc. India 55, 367–380.

Biswal, T.K., Ahuja, H., Sahu, H.S., 2004. Emplacement kinematics of nepheline syenites from the Terrane Boundary Shear Zone of the Eastern Ghats Mobile Belt, west of Khariar 5, NW Orissa, evidence from meso- and microstructures. Proc. Indiana Acad. Sci. 113, 785–793.

Chetty, T.R.K., 2001. The Eastern Ghats Mobile Belt, India: a collage of juxtaposed terranes (?). Gondwana Res. 4, 319–328.

Chetty, T.R.K., Murthy, D.S.N., 1993. LANDSAT thematic mapper data applied to structural studies of the eastern Ghats granulite terrain in part of Andhra Pradesh. J. Geol. Soc. India 42, 373–391.

Chetty, T.R.K., Murthy, D.S.N., 1994. Collision tectonics in the Eastern Ghats Mobile Belt: a new interpretation. Workshop on Eastern Ghats Mobile Belt. Geol. Surv. India Spec. Publ. 44, 39–50.

Crowe, W.A., Nash, C.R., Harris, L.B., Leeming, P.M., Rankin, L.R., 2003. The geology of the Rengali Province: implications for the tectonic development of northern Orissa. J. Asian Earth Sci. 21, 697–710.

Dasgupta, S., Bose, S., Das, S., 2013. Tectonic evolution of the Eastern Ghats Belt, India. Precambrian Res. 227, 247–258.

Dobmeier, C., Reith, M., 2003. Crustal architecture and evolution of the Eastern Ghats Belt and adjacent regions of India. In: Yoshida, M., Windley, B.F., Dasgupta, S. (Eds.), Proterozoic East Gondwana: Supercontinent Assembly and Breakup. vol. 206. Geol. Soc., London, pp. 145–168.

Gupta, S., 2004. The Eastern Ghats Belt, India—a new look at an old Orogen. Geol. Surv. India Spec. Publ. 84, 75–100.

Leelanandam, C., 1997. The Kondapalli layered complex, Andhra Pradesh, India: a synoptic overview. Gondwana Res. 1, 95–114.

Neogi, S., Das, N., 2000. Lithotectonic domains and petrotectonic evolution of the boundary zone between the Eastern Ghats mobile belt and the Baster craton, Deobag area. Central India. Geol. Surv. India Spec. Publ. 57, 180–204.

Radhakrishna, B.P., 1989. Suspect tectono-stratigraphic terrane elements in the Indian Subcontinent. J. Geol. Soc. India 34, 1–24.

Ramakrishnan, M., Vaidyanadhan, R., 2008. Geology of India. Geological Society of India, Bangalore556.

Ramakrishnan, M., Nanda, J.K., Augustine, P.F., 1998. Geological evolution of the Proterozoic Eastern Ghats Mobile Belt. Geol. Surv. India Spec. Publ. 44, 1–21.

Ramam, P.K., Murthy, V.N., 1997. Geology of Andhra Pradesh. Geological Society of India, Bangalore, p. 245.

Rao, A.T., Dvakara Rao, V., Yashida, M., Arima, M., 1995. Geochemistry of charnockites from the Eastern Ghats Granulite Belt—evidence for possible linkage between India and Antarctica. In: Yashida, M., Santosh, M. (Eds.), India and Antarctica During the Precambrian. vol. 34. Geol. Soc. India, Mem, pp. 273–291.

Rickers, K., Mezger, K., Raith, M.M., 2001. Evolution of the continental crust in the Proterozoic Eastern Ghats belt, India and new constraints for Rodinia reconstruction: implications from Sm-Nd, Rb-Sr, and Pb-Pb isotopes. Precambrian Res. 112, 183–212.

Roy, A.B., 2014. Indian Subcontinent, Reference Module in Earth Systems and Environmental Sciences. vol. 2014. Elsevier. https://doi.org/10.1016/B978-0-12-409548-9.09372-6.

Sarkar, A., Paul, D.K., 1998. Geochronology of Eastern Ghats Mobile Belt—a review. Geol. Surv. India Spec. Publ. 44, 51–86.

CHAPTER

PRECAMBRIANS OF SHILLONG PLATEAU

12

12.1 INTRODUCTION

The Shillong Plateau, a highland region in the eastern Meghalaya State, northeastern India, represents a rolling tableland that comprises most of Meghalaya. Geologically, it forms the northeastern extension of the Indian Peninsular Shield forming an east-west-trending oblong horst-like block having elevation between 600 and 1800 m above the Bangladesh Plains in the south (Fig. 12.1). The plateau's western, northern, and southern escarpments are known as the Garo, Khasi, and Jaintia Hills, respectively. The Shillong Plateau forms a detached crustal block separated from the Indian Peninsula by the 'Garo-Rajmahal Gap' in the Bengal Basin.

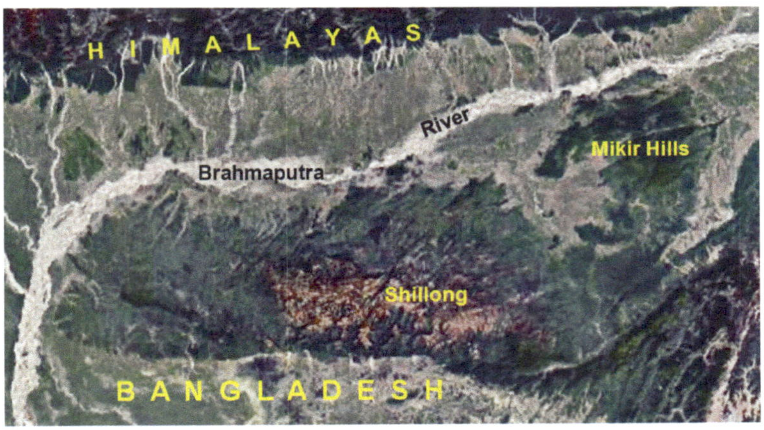

FIG. 12.1

Google satellite Imagery showing broadly rectangular tableland of the Shillong Plateau, including Mikir Hills in the east. The plateau is bounded between Brahmaputra River in the north and west and the Plains of Bangladesh in the south.

The Shillong Plateau including the Mikir Hills in the east is a misnomer term, which has also been described as the Meghalaya Plateau (Valdiya, 1984; Nandy, 2001), the Shillong-Meghalaya Plateau (Chatterjee et al., 2006), and also the Assam-Meghalaya Plateau (Sarma and Dey, 1996). In this chapter, however, we would prefer to describe the dominantly Precambrian crustal block by the pristine name, the Shillong Plateau, for avoiding misunderstandings.

Indian Shield. https://doi.org/10.1016/B978-0-12-809839-4.00012-6

12.2 GEOLOGY AND LITHOSTRATIGRAPHIC RELATIONSHIPS OF THE PRECAMBRIAN ROCKS

The broad lithostratigraphic formations of the Precambrian rocks of the Shillong Plateau include (i) ensembles of granite-gneisses having inclusions of several supracrustal enclaves collectively described as the Gneissic Complex; (ii) the Shillong Group, dominantly comprising quartzite-phyllite sequence with minor calc-silicate rocks; (iii) the Khasi metadolerites; and (iv) a number of intrusive bodies of massive plutonic granitoids (Fig. 12.2).

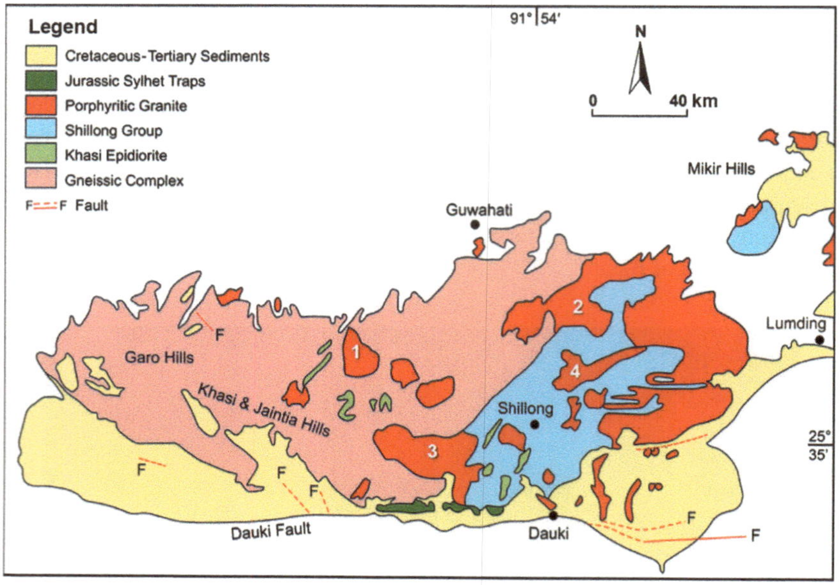

FIG. 12.2

Geological map of the Shillong Plateau showing distribution of major lithological components. The Khasi metadolerite is shown as Khasi Epidiorite in the map. Numbers in the map show outcrops of some important group of porphyritic granites: (1) Myllium, (2) Nongpoh, (3) South Khasi, and (4) Kyrdem Granites.

The Gneissic Complex is well exposed in the western, northern, and northeastern part of the Shillong Plateau. Scattered occurrences of the dominantly gneiss-granite ensembles also occur as isolated 'inselberg-type' outcrops along the north and south bank of the Brahmaputra River in the Goalpara, Kamrup, Darrang, and Nowgaon districts of Assam (Barooah, 2011). Apart from the different types of orthogneisses variously described as biotite gneiss, banded gneiss, grey gneiss, or even tonalite-trondjhemite-granodiorite gneiss, there are some smaller enclaves of greenstone belt rocks that include enclaves of mafic, ultramafic bodies, and some isolated pockets of banded iron formation. The field data as revealed in some outcrop-scale images indicate a complex and protracted evolutionary history that could date back to the oldest known gneissic rocks known from different parts of the Indian Shield. Banded gneisses showing a complex deformation pattern and evolutionary history are illustrated in Fig. 12.3a–d. Some indication about this is pointed out by Bidyananda and Deomurari (2007) while suggesting a correlation of the Shillong Plateau rocks with those of the Dharwar Protocontinent.

FIG. 12.3

Illustrations of some common types of the banded gneisses showing complex deformation and evolutionary history.

Images courtesy, Jonali Medhi.

The banded iron formation that occurs as lenticular bands, varying in shapes and sizes in association with mafic rocks at some localities, constitutes the most significant supracrustal component of the Gneissic Complex. Apart from these, there are calc-schists representing metamorphosed impure carbonate rocks, which generally occur as small bands within the Gneissic Complex. There are a few larger bands that range between 500m and 1km in dimension. Small lenses of siliceous marbles also occur in association with the calc-schists.

Several bodies of granulites (with ill-defined formational boundaries) of different composition and character occur within the vast expanse of gneiss-granite ensembles. These include lenticular bands of hornblende-pyroxene granulite, calc-granulite, sillimanite-quartz schist, sillimanite-corundum rocks, biotite-cordierite granulite, quartz-dumortierite-tourmaline schist, phlogopite-spinel granulite, anthophyllite-spinel-phlogopite granulite, chrondrodite-phlogopite-corundum granulite, and quartz magnetite rocks (Gogoi, 1973).

The massive granulites occur mainly as 'intrusive' masses within the granite-gneiss as well as the metasedimentary bodies in three different forms (Barooah, 2011) (Fig. 12.4): (1) medium- to coarse-grained acid granulites or charnockites, sensu stricto, made up of blue or green quartz, feldspar, and subordinate amount of ferromagnesian minerals including hypersthene; (2) fine-grained basic granulites, occurring as thin bands, lenses, and drawn out boudins; and (3) coarse-grained ultrabasic granulites, occurring as small masses and lenses.

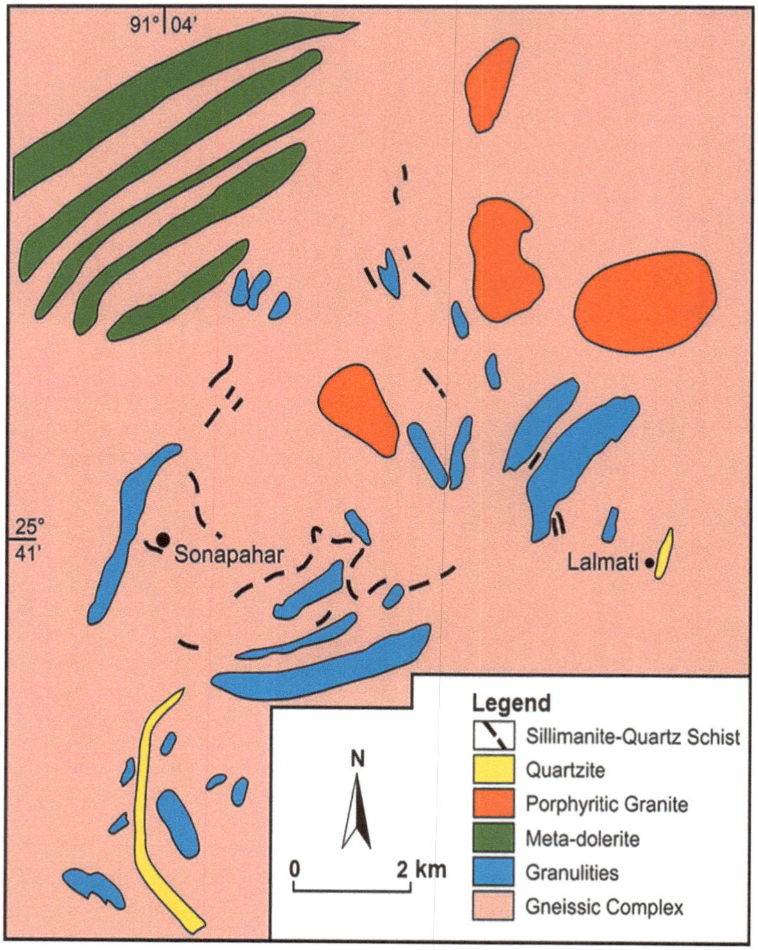

FIG. 12.4

Geological sketch map of Sonapahar area west of Khasi District, Meghalaya. Blue-coloured bands and patches represent emplaced massive granulites of diverse composition.

Compiled from different sources.

A series of scattered occurrences of the sillimanite-quartz schists are noted in the Sonapahar area (25°40′42″: 91°03′42″) (Fig. 12.4), which occur in close association with massive granulites.

The sillimanite/cordierite bearing paragneisses locally include Mg–Al rich components. A number of lenses and pockets of quartz-sillimanite schists occur within the Gneissic Complex especially in the Sonapahar area. There are also several pockets of 'sillimanite-quartz'-bearing high-alumina silicate rocks identified as the metamorphosed Precambrian palaeosols by Golani (1989). In addition, there are several patchy occurrence of Mg–Al-rich schist rocks that comprise variable proportions of phlogopite, spinel, anthophyllite, cordierite, sapphirine, chondrodite, and corundum. The origin of this unusual association of Mg–Al components within the high-aluminous sediments remains an intriguing issue requiring further study.

Overlying the Gneissic Complex occurs the Shillong Group (the term 'Group' replacing 'series' of Medlicott, 1869, 1886) that is an ensemble of metasediments and metavolcanics, showing local presence of basal conglomerate (Fig. 12.5). The presence of the basal conglomerate, coupled with the evidence of abrupt change in the metamorphic grade and the intensity of deformation, provides proof of an unconformity between the Shillong Group and the underlying basement of the Gneissic Complex.

FIG. 12.5

The basal Nongkhya conglomerate occurring at the base of the Shillong Group.

Image courtesy: Jonali Medhi.

The Shillong Group represents a weekly metamorphosed and little deformed succession of Proterozoic rocks having a broad NNE–SSW trend. Following the bipartite division proposed by Medlicott (1869), Barooah and Goswami (1972) divided the Shillong Group into the older Tyrsad Formation (dominantly argillaceous) and the younger Shillong Formation (dominantly arenaceous). The two formations are separated by a conglomerate horizon well exposed around Mawphlang (25°26′26″; 91°43′50″) (Fig. 12.6).

FIG. 12.6

Mawphlang conglomerate that separates the two formations of the Shillong Group.

Image courtesy: Jonali Medhi.

The lower phyllitic succession of the Shillong Group, the Tyrsad Formation, grades from siliceous to carbonaceous phyllite in composition containing quartz, muscovite graphite, and pyrite (Fig. 12.7a). Near the contact of the younger intrusive granitoid plutons, the phyllites and mica schists show occurrence of muscovite, biotite, garnet, chloritoid, and andalusite. The marl varieties of phyllite consist of hornblende, garnet, clinozoisite, tremolite, calcite, microcline, and oligoclase. The upper succession, the Shillong Formation, is represented by Mawphlang Quartzite (Fig. 12.7b) that varies from coarse-to-medium-grained sandstone along with siltstone and minor shale. The sandstone consists of grains of quartz and subordinate kaolinized feldspar with fragments of chert and quartzite set in a siliceous matrix. The siltstones have sericite-dominant matrix.

FIG. 12.7

Lithology of the Shillong Group. (a) Carbonaceous slate/phyllite overlain by younger Cenozoic boulder bed; (b) Mawphlang Quartzite representing the upper formation of the Shillong Group.

Image courtesy: Jonali Medhi.

Several bodies of mafic dykes (and a few sill-like bodies) have intruded into the Shillong Group and in the Gneissic Complex. The smaller bodies of dykes show cross-cutting relationship in two different directions. The northeast-southwest trending bodies are more dominant than those in the northwest-southeast trending intrusives. These structurally undeformed intrusive bodies showing minor surficial alteration of pyroxene grains have changed into hornblende and chlorite attributing pale green colour to the rocks. Because of the metamorphic changes in the mafic intrusive bodies, these intrusive bodies have been described as greenstones. These intrusive bodies occurring in and around Khasi Hills area have been described as the Khasi Epidiorite by Medlicott (1869) (Fig. 12.4).

A number of massive post-tectonic granite plutons occur in the Shillong Plateau, of which the most important ones are (i) Mylliem Granite, (ii) Kyrdem Granite, (iii) Nongpoh Granite, and (iv) South Khasi Granite (Fig. 12.2). There are also several smaller bodies of porphyritic granite. A few of these granite plutons are observed that have intruded into both the basement Gneissic Complex and the overlying the Shillong Group. These younger granite plutons of the Shillong Plateau region are mostly heterogeneous in composition, medium- to coarse-grained, and vary widely in texture from equigranular to porphyritic types commonly preserving xenoliths of the different country rock (Fig. 12.8a and b).

The commonly occurring younger granites show coarse-grained porphyritic character and are represented by grey- and pink-coloured K-feldspar megacryst.

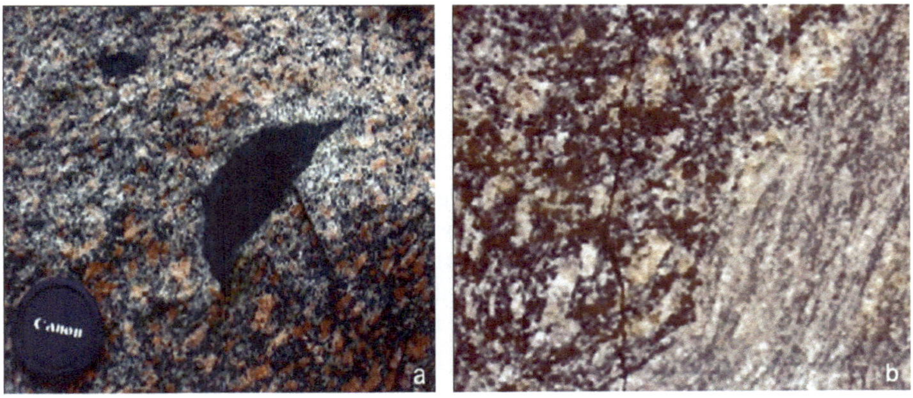

FIG. 12.8

The youngest granitic plutons of the Shillong Plateau region showing massive mostly heterogeneous, medium- to coarse-grained character. (a) Porphyritic granite showing inclusion of mafic xenolith. (b) Porphyritic granite intruding into strongly deformed bodies of Gneissic Complex.

Reproduced from Santosh et al. (2017) with permission.

12.3 SUMMARY OF THE EVOLUTIONARY HISTORY

Conventionally, the Precambrian ensembles of the Shillong Plateau are divided into four lithostratigraphic groups:

1. Porphyritic granites—youngest intrusives affecting all the preexisting rocks
2. Khasi Greenstones—dominantly northeast-southwest-oriented swarms of dykes/sills
3. Shillong Group—mainly metasedimentary sequences with minor mafic volcanics, and
4. Gneissic Complex—basement granite-gneisses and granulites

There are, however, two features that need serious reconsideration while working out lithostratigraphy of the terrain of Shillong Plateau. The first is the inclusion of the granulites into the ambit of the Archaean basement though there is hardly any geological evidence to prove that the basement gneisses have been converted entirely into granulite facies metamorphism during Archaean. Field studies, however, do not convey anything like that. On the other hand, a major part of the supracrustals that includes pelites and associated mafic and ultramafic bodies does show granulite facies metamorphism. Though only a couple of isotope data based on zircon studies indicate late- or end-Archaean age of 2566.4 ± 26.9 Ma, the mere physical character of the gneisses indicates the strong possibility of a long and protracted Archaean evolutionary history that could be as old as 3.5–3.3 Ga as noted in most of the crustal blocks making up the Indian Shield. Two possibilities exist about the fewer reports of isotope ages based on the zircon studies. One is that the older rocks have not been sampled for the purpose of dating of rocks, and the other could be the result of near-complete reconstitution during the end-Archaean time.

Apart from that, an important benchmark event that eluded attention of the workers/authors of the Shillong geology is the information about the occurrence of the metamorphosed high-aluminous sediments representing what has been described as the Precambrian palaeosols by Golani (1989) from within the gneissic terrain near Sonapahar. Golani (1989) has rightly described the rocks as Precambrian and not Archaean looking into the similarity in the development of the sillimanite-quartz layer with the metamorphosed clay beds along the base of the earliest Proterozoic formations in the Aravalli Protocontinent around Udaipur (Roy and Paliwal, 1981; Roy and Kröner, 1996; Sreenivas et al., 2001; Roy and Jakhar, 2002; Wall et al., 2012) (Fig. 9.5a). In the case of the geology around Udaipur, the deposits of high-aluminous (metamorphosed) clay beds mark a prominent unconformity along the peneplained surfaces of the Archaean basement. The only notable difference is that the unconformity surface in the Shillong Plateau region is marked by the metamorphosed 'high-aluminous' clay beds (represented by the sillimanite-quartz-bearing rocks) and shows evidence of being repeatedly folded, unlike the more straight-lying unconformity surface that occur in the Udaipur region. The development of the iron-free high-aluminous clay pockets at the base of the Palaeoproterozoic Aravalli Basin has been interpreted to have formed during preoxidation phase of weathering process that led to the formation of white clay at the base of the Palaeoproterozoic rocks. Based on the similarity in the occurrence of high-aluminous sediment represented by sillimanite-quartz in the Shillong Plateau and the white mica-quartz rocks in the Udaipur region in the Aravalli Protcontinents, we may logically postulate that the zone of occurrence of sillimanite-quartz bands in the Shillong Plateau region truly marks the Archaean-Proterozoic Boundary.

Considering the above and coupled with the critical review of the geological map (Fig. 12.4), we may concluded that the granulites, including all varieties, types, and petrochemistry, postdated the timeline marked by the folded unconformity surface along the 'Archaean-Proterozoic Boundary'. Age wise, therefore, the granulites cannot be Archaean in age as commonly considered. Acceptance of the presence of the Archaean-Proterozoic Boundary implies the occurrence of Palaeoproterozoic rock association above the Archaean basement. There is little record of this post-Archaean lithologic sequence except the reported occurrence of sillimanite-cordierite-dumortierite-bearing alumina-silicate rocks around the Sonapahar region.

Except the occurrence of the youngest porphyritic granite plutons during the latest Precambrian (550 ± 50 Ma), very little geochronological data are available on the older Precambrian rocks. A few U–Pb SHRIMP data on zircon are available from the Shillong Plateau region (Santosh et al., 2017). Based on the data on the inherited zircon core, an age of 2566.4 ± 26.9 Ma has been suggested as the age of recycled Neoarchean Crust in the basement granite-gneisses of the region. However, as it has been pointed out earlier, looking into the physical character of the gneisses, it may be stated with conviction that the Archaean evolutionary history of the region could be quite long and protracted similar to that noted in most of the crustal blocks making up the Indian Shield.

U–Pb SHRIMP data on zircon also indicated Proterozoic ages between 1758.1 ± 54.3 and 1617.1 ± 14 Ma (Santosh et al., 2017) that may be interpreted to mark the late Palaeoproterozoic thermal event linked to the granulite facies metamorphism coinciding with culmination of the Palaeoproterozoic history. It is hard to get the proper time slot for the Shillong Group that is, in general, little deformed and virtually unmetamorphosed. Santosh et al. (2017) have also recorded some zircon ages of 1430.4 ± 9.6 Ma from the granite-gneiss of the Mikir Hills, which they interpreted as the crystallization age. This early Mesoproterozoic age could represent the thermal event marking the emplacement of dolerite bodies described as the Khasi Epidiorite ensemble. The Shillong Group could have evolved (meaning opening and inversion of the basin) during the time slot between youngest Palaeoproterozoic granulite facies metamorphism and the earliest Mesoproterozoic thermal event when the Khasi mafic dykes (Khasi Epidiorite) were emplaced. The youngest thermal event is recorded between 506 and 535 Ma when the Proterozoic 'plutonic' granites formed intruding into all the earlier formed Precambrian rocks of the Shillong Plateau.

REFERENCES

Barooah, B.C., 2011. Precambrian Rocks of Northeast India. North East Geo-Resource Consultancy Service, Guwahati. 108 p.

Barooah, B.C., Goswami, I.D., 1972. Precambrian stratigraphy of the Assam Plateau. J. Mines Met. Fuel 20, 368–373.

Bidyananda, M., Deomurari, M.P., 2007. Geochronological constraints on the evolution of Meghalaya massif, northeastern India: an ionprobe study. Curr. Sci. 93, 1620–1623.

Chatterjee, N., Mazumdar, A.C., Bhattacharya, A., Saikia, R.R., 2006. Mesoproterozoic granulites of the Shillong-Meghalaya Plateau: evidence of westward continuation of the Prydze Bay Pan-African suture into Northeastern India. Precambrian Res. 125, 1–26.

Gogoi, K., 1973. The geology of the Precambrian rocks in the north-western parts of the Khasi and Jaintia Hills, Meghalaya. Geol. Surv. India Misc. Publ. 23 (1), 37–48.

Golani, P.R., 1989. Sillimanite-corundum deposits of Sonapahar, Meghalaya, India: a metamorphosed Precambrian palaeosol. Precambrian Res. 43, 175–189.

Medlicott, H.B., 1869. Geological sketch of the Shillong Plateau in North-Eastern Bengal. Mem. Geol. Soc. India 7 (1), 197–207.

Medlicott, H.B., 1886. The Precambrian framework of part of the Khasi Hills, Meghalaya. Rec. Geol. Surv. India 117 (2), 1–59.

Nandy, D.R., 2001. Geodynamics of Northeastern India and the Adjoining Region. ABC Publication, Kolkata. 209 p.

Roy, A.B., Jakhar, S.R., 2002. Geology of Rajasthan (Northwest India)—Precambrian to Recent. Scientific Publishers India, Jodhpur. 421 p.

Roy, A.B., Kröner, A., 1996. Single zircon evaporation ages constraining growth of the Aravalli craton, northwestern Indian shield. Geol. Mag. 133, 333–342.

Roy, A.B., Paliwal, B.S., 1981. Evolution of lower Proterozoic epicontinental deposits: stromatolite-bearing Aravalli rocks of Udaipur, Rajasthan, India. Precambrian Res. 14, 49–74.

Santosh, K., Rino, V., Yasutaka, H., Kimura, K., Raju, S., Terada, K., Pathak, M., 2017. Contribution of Columbia and Gondwana Supercontinent assembly- and growth-related magmatism in the evolution of the Meghalaya Plateau and the Mikir Hills, Northeast India: constraints from U-Pb SHRIMP zircon geochronology and geochemistry. Lithos 277, 356–375.

Sarma, K.P., Dey, T., 1996. Re-look on Shillong Plateau. Bull. Pure Appl. Sci. 15F (2), 51–54.

Sreenivas, B., Roy, A.B., Srinivasan, R., 2001. Geochemistry of sericite deposits at the base of the Palaeoproterozoic Aravalli Supergroup, Rajasthan, India: evidence for metamorphosed and metasomatised Precambrian paleosol. Proc. Indian Acad. Sci. (Earth Planet. Sci.) 110, 39–61.

Valdiya, K.S., 1984. Aspects of Tectonics: Focus on South and Central Asia. Tata McGraw Hills, New Delhi. 319 p.

Wall, H.d., Pandit, M.K., Chauhan, N.K., 2012. Paleosol occurrences along the Archean–Proterozoic contact in the Aravalli craton, NW India. Precambrian Res. 216–219, 120–131.

PART

3

EARLY PHANEROZOIC GEOLOGY OF INDIA

CHAPTER

PALAEOZOIC GEOLOGICAL HISTORY

13

13.1 INTRODUCTION

The geological history of the Indian Shield can be broadly divided into two different phases of evolution. The first involves the entire Precambrian covering the phases of growth and evolution of the continental Crust. During the second, the post-Cambrian phase, the pristine Indian Shield underwent transformation, described as the 'Phanerozoic reconstitution' (Roy, 2004), which brought about geomorphotectonic changes in the crustal block that ultimately changed to what is now known as the Indian Subcontinent. Several basins, both continental and oceanic, had also developed during the process of post-Precambrian reconstitution of the Indian Shield.

The chapter deals with the major part of the Palaeozoic history of the Indian Shield from the Cambrian to Devonian focusing on the crustal evolutionary history on aspects of palaeogeographic and palaeoclimatic changes, as well as on the Palaeozoic faunal evolution. For the sake of continuity, a brief discussion is added in the last part dealing with the marine Permo-Carboniferous beds that formed in the northern and northwestern shelf as also in the continental interior. Discussion on the essentially continental Gondwana geology will follow in the next two successive chapters.

Barring some granite intrusions during the Pan-African Crust building history centring around 500 ± 50 Ma, there scanty records of geological activity in the Indian Shield for over 180 million years between the early Ordovician and late Carboniferous. Several continental rift basins had developed during the late Palaeozoic Permo-Carboniferous that were filled up with the thick sediments constituting the Gondwana Supergroup. There are a few references in the literature about the continuity of sedimentation in the shelf sea region in some isolated northern parts of the Indian Shield during the early Palaeozoic that ultimately became a part of the 'Tethys Himalayas' at a later stage.

13.2 EVIDENCE OF THE PRECAMBRIAN-CAMBRIAN BOUNDARY

The transition across the Precambrian-Cambrian Boundary is noted in a few places along the northern margin of the Indian Subcontinent; much of the originally formed outcrops have been obliterated due to the Himalayan collision-related destruction in the northern part of the Indian crustal block concurrently with the rise of the Himalayas.

Noticeable transformation in respect of palaeoclimatic changes and the evolution in life forms is recorded in different formations deposited during this phase of the geological history in parts of the Indian Shield. Excessive precipitation of Ca/Mg and other anions with the gradual diminution of the Tethys Sea is thought to be the possible cause for transformation of soft-bodied life forms into

Indian Shield. https://doi.org/10.1016/B978-0-12-809839-4.00014-X

skeleton-bearing fauna. Metabolic changes in the early life cycle were possibly triggered due to excessive availability of anions under evaporitic palaeoclimatic conditions. The Precambrian-Cambrian Boundary is sharply marked in the deposits in different basins denoting such variations.

The basal boundary of the Cambrian (~543 Ma) has been biostratigraphically marked in the Lesser Himalaya, which is detectable on the basis of the occurrence of Ediacaran carbonate-rich beds bearing signatures of the organic-walled tubular *Shaanxilithes ningqiangensis* (Hughes, 2016a) (Fig. 13.1) that underlies the basal beds of the Cambrian (~543 Ma).

FIG. 13.1

Shaanxilithes ningqiangensis indicating Lower Cambrian Boundary from the lowermost Tal Group in the Lesser Himalaya. Scale bar represents half-millimetre interval.

Reproduced from Hughes (2016a) with permission.

BOX 13.1

The Ediacaran (or Vendian) biota consisted of enigmatic tubular and frond-shaped, mostly sessile organisms that lived during the Ediacaran Period (635–542 Ma). Trace fossils of these organisms have been found worldwide and represent the earliest known complex multicellular organisms. The Ediacaran biota radiated at ~575 Ma, after the Earth had thawed from the Cryogenian period's extensive glaciation. The biota largely disappeared with the rapid increase in biodiversity known as the Cambrian Explosion.

13.3 DEVELOPMENT OF THE PALAEOZOIC BASINS

The earliest Palaeozoic basins developed along a few pockets, mainly in the foreland region of the Indian Shield marked by the incidence of magmatic activities. These include the Baldwin Volcanics on the eastern edge of the India-Myanmar boundary, the Manaslu Volcanics in Nepal, the Mansehra Volcanics in Pakistan, and the Mangsu La Volcanic tuffs in the Kinnaur region (Bassi and Chopra, 1984). This phase of magmatism though remains undated is correlated with the widely known ~500 ± 50 Ma Pan-African magmatic activity in different parts of the southern Gondwanaland.

Amongst the Cambrian basins, the most studied one is that which developed in the Salt Range hills in the Punjab province of Pakistan (Fig. 13.2). The extension of the Cambrian basin is traceable up to Kashmir, Spiti, Zanskar, Kinnaur, and Uttarakhand regions. Occurrence of Cambrian basins is also reported from Peshawar and Attock-Cherat Ranges in the Lesser Himalaya.

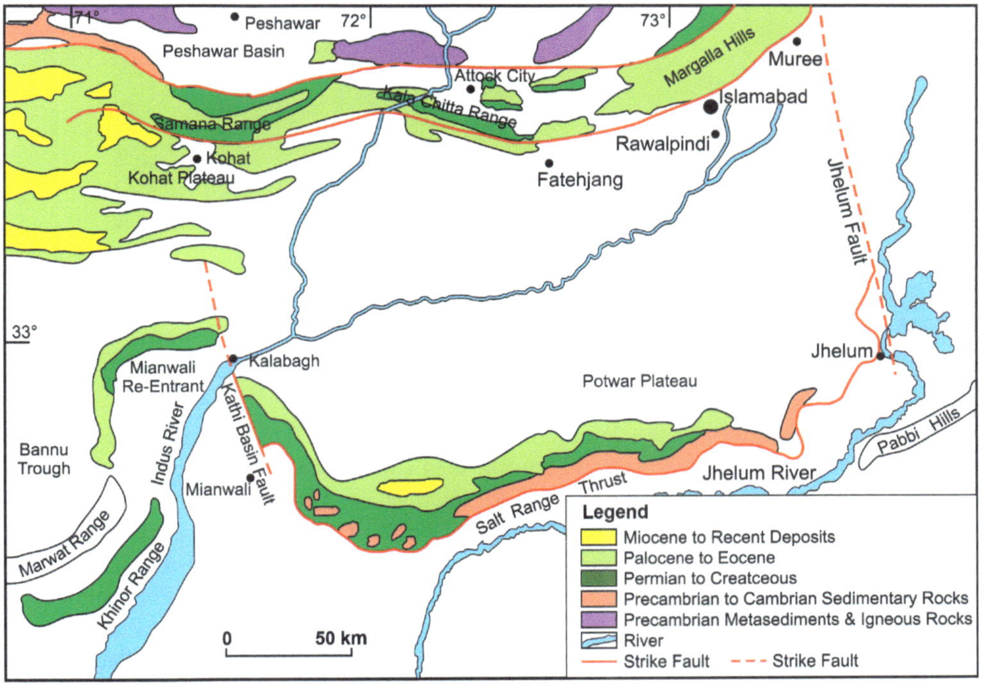

FIG. 13.2

Geological map of the northwestern Pakistan. Outcrops of the Salt Range formations occur south of the Potwar Plateau.

Figure courtesy, S.A. Sameeni.

The Cambrian sequence in the Salt Range hills overlies the basement rocks known as the Salt Range Formation, which comprises mainly of gypsum and salt deposits. Earlier described as the Saline Series, the assemblage of the Salt Range Formation can be correlated with the Hanseran Evaporite Group of the Marwar Supergroup developed in western Rajasthan. Unconformable relationship between the Salt Range Formation and the immediately overlying Kherwa Sandstone of early Cambrian age is prominently exposed in the Kherwa Gorge section (Fig. 13.3) in the Potwar Plateau region.

The Cambrian succession includes, besides Kherwa Sandstone (Fig. 13.4a), three other formations, such as the Kussak, Jutana, and Baghanwala Formations. The Cambrian fossil-bearing slate and greywacke also occur at several places. These Cambrian formations contain, apart from other fossils, a wealth of trace fossils including trails of trilobites and worm tracks. The Baghanwala Formation, which contains salt pseudomorphs (Fig. 13.4b), is well exposed along the roadside and displays a sharp unconformable upper contact with the overlying Permian glacial deposits (Fig. 13.5).

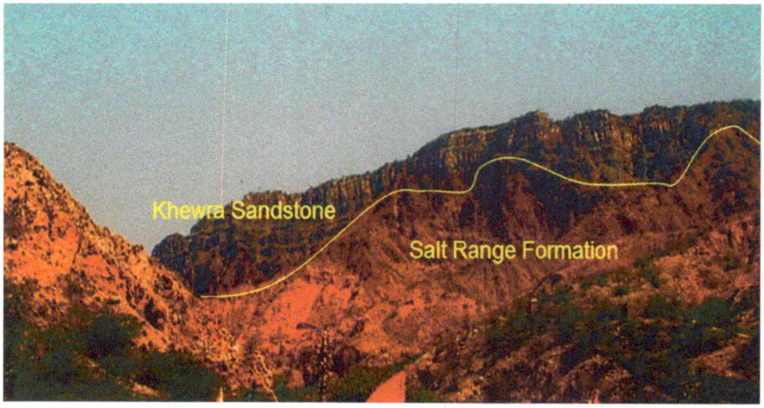

FIG. 13.3

The nature of unconformity between the Precambrian Salt Range Formation overlain by Kherwa Sandstone of Cambrian age.

Image courtesy: S.A. Sameeni.

Along the central margins of the Himalayan foreland, the Inner Tal Group of early Cambrian age occurs at several places like Mussoorie and Lansdowne in Uttarakhand, and Korgai and Nigali Dhar in the Himachal Pradesh. Re-Os ages of ~552 Ma old Small Shelly Fauna (SSF) from black shales and chert of Tal Formation confirm its Lower Cambrian age (Singh, et al., 1999). The Cambrians of the Spiti region is classified under the Haimanta Group (Srikantia, 1981; Bhargava and Bassi, 1998) and divisible into the Early to Middle Cambrian Kunzam La (Parahio) Formation and the Precambrian to Early Cambrian Batal Formation (Singh et al., 2014). The early to middle Cambrian sequence is more than 700 m thick and rests over the Precambrian-Early Cambrian Batal Formation in the Parahio Valley of Spiti (Fig. 13.6). Apart from these, there are also reports of Cambrian fossil-bearing lithoformations from Nampa and Nanda Devi Peaks in Nepal. The uppermost Cambrian is absent throughout the Himalaya.

FIG. 13.4

(a) Kherwa Sandstone forming the basal Cambrian formation, Salt Range hills, Pakistan. (b) Salt pseudomorph structures in the Baghanwala Formation.

Images courtesy: S.A. Sameeni.

FIG. 13.5

Outcrops of the youngest Baghanwala Formation of Cambrian age unconformably underlying the Permian glacial deposits in a roadside outcrop in the Salt Range.

Image courtesy: S.A. Sameeni.

FIG. 13.6

Photograph showing a section on the Parahio River in the Parahio Valley, Spiti, exposing Cambrian beds.

Reproduced from Singh et al. (2016) with permission.

The Ordovician conglomerate-sandstone designated as the Thango Formation rests along a pronounced angular unconformity over the Cambrian rocks in Spiti, Zanskar, and Kinnaur. In Kashmir, the Ordovician is known as Hallamulla Formation. The Ralam Conglomerate in Uttarakhand, which was earlier assigned a Precambrian age, has been found to be in strike continuity of the Thango Formation. In Nepal, Damgad Formation, with conglomerate in the basal part, represents the Ordovician. In the Peshawar Basin, the Ordovician is represented by Misri Banda Quartzite, Hisartang Formation, and Panjpir Formation.

In the Spiti, Zanskar, and Kinnaur, the Thango Formation is conformably overlain by the Takche Formation of Late Ordovician to Middle Silurian age (Ashgill, Wenlock). The Silurian carbonate-shale beds belonging to the Yong, Shiala, and Garbyang Formations occur in the northeastern Kumaon and Garhwal regions. In Nepal, the Silurian is known as the Dark Band Formation from which Upper Llandovery graptolites have been reported. Small to medium size coral-algal-stromatoporoid build-ups are reported from Kashmir, Spiti, Zanskar, Kinnaur, and Garhwal areas.

The development of Ordovician, Devonian, and Silurian basins is reported from the Anantnag, Baramulla, Poonch, and Doda districts in Jammu and Kashmir. The Muth Quartzites of Devonian age occurs in the Liddar, Sind, Naubagh and the Warwan Valleys in Jammu and Kashmir. These are followed upwards by the Ordovician and Silurian fossil-bearing beds represented by the Thango and Takche Formations in the Chorgal and Spiti Valleys and in the Kinnaur area in Jammu and Kashmir. The Muth Quartzite is the most characteristic formation representing part of Devonian age. In Nepal, the Tilicho Pass Formation and equivalent represent the Devonian (Eifelian-Frasnian). In Spiti and Zanskar, a sedimentologic break is recorded between the Takche and the Muth formations.

On the western foreland of the Indian Shield, that is, along the northwestern border of Rajasthan, the outcrops of the Cambrian lithoformations occur overlying the Neoproterozoic Malani Group. Apart from that, patches of Permo-Carboniferous sediments represented by Bap and Badhura Formations occur unconformably overlying the Early Palaeozoic rocks over the Precambrian basement rocks (Roy and Jakhar, 2002).

Evidences of marine incursions are noted both from the western and the north-central frontiers of the Peninsular India. In the western foreland margins, the marine water invaded the mainland through the Narmada-Son Lineament zone, while in the north-central foreland, the marine incursion occurred from the presently represented by the Nepal Himalayas that transgressed up to the central Indian mainland where fossil-bearing marine beds are reported at Mahendragarh and Umaria near Jabalpur.

13.4 PALAEOCLIMATIC AND PALAEOGEOGRAPHIC CONTROLS ON THE SEDIMENTATION PATTERN

The nature of sediments deposited in different basins during the Palaeozoic provided evidence of the depositional condition (palaeogeographic and palaeoclimatic). A few of which are listed below:

(a) Occurrence of phosphate beds indicating warm tropical climate is recorded during the Precambrian-Cambrian transition in the Tal Formation of the Krol Belt and also further west in the Hazara Basin of Pakistan (Mazumdar and Banerjee, 2015).

(b) The occurrence of boulder beds and associated features like drop stones and striated pavements along the western flanks indicates glacial condition during the Permo-Carboniferous (Fig. 13.7a and b).

(c) Occurrence of effusive rocks in Myanmar, Nepal, and western margins in Pakistan suggests volcanic contributions in the early Palaeozoic formations.

(d) Occurrence of profuse marine fossils in the Silurian, Ordovician, and Devonian strata preserved in quartzite, shale (at places black shale), and carbonate rocks confirm marine depositional environment of these rocks.

All the Palaeozoic strata represent euxinic palaeoclimatic conditions. The presence of oolite beds in some formations like the Krol Formation also denotes evaporitic conditions of deposition. Marl beds possibly represent terminal desiccation phase of the inland sea basins.

FIG. 13.7

(a) The outcrops of Bap Boulder Spread of Permo-Carboniferous age, west of Jaisalmer. (b) Well-bedded subhorizontal Badhura beds of Permo-Carboniferous age.

Image courtesy, Surja Ram Jakhar.

13.5 EVOLUTION OF LIFE FORMS

Most conspicuous evolutionary trend in the life forms on the Earth took place during the Precambrian-Cambrian transition. In the context of the Indian Shield, the Palaeozoic is characterized by the evolution of number of life forms that had a widespread geographical extension. Records of exceptionally well-preserved fossils in the Cambrian formations in the different parts of the Indian Shield demonstrate a range of fossilization processes for a variety of soft tissues, especially in the processes of phosphatization and pyritization and in some cases recording a level of anatomical details.

Life form changes from the soft-bodied late-Precambrian Ediacaran-type biota into hard-skeleton-bearing fossils recorded in the 'Korgai Syncline' in Himachal Pradesh where the Krol Formation shows well-preserved soft-bodied acritarchs in the overlying Tal Formation replacing the earliest spinet-form fossils as the possible precursors of trilobites. Wide-scale diversification in life forms is also observed in the development of corals, crinoids, bryozoans, brachiopods, molluscs, ostracodes, primitive vertebrates, and plant life. An important feature noted in the Palaeozoic stratigraphy is the significant hiatus noted in some basins, especially in the western foreland basins where Ordovician rocks are directly overlain by the younger Carboniferous beds. This is particularly observed along the western foreland basins as in western Rajasthan and in the Salt Range hills.

The stromatolites (cyanobacterial fossils) constitute the most noticeable earliest life forms representing nonerratic transition from Precambrian to Cambrian. The occurrence of stromatolites is known from widely scattered regions in the Lesser Himalayas.

The trace fossils bear significance especially in the nonfossiliferous zones, and in the present context, it is quite common in the Cambrian. The changeover in the morphology of ichnofossils comprising early metazoan life from the simplest to the compound type marks an evolutionary trend. The Ediacaran trace fossils like *Taphrhelminthopsis circularis* are reported from southeast Kashmir Valley denoting Cambrian age (Bhargava and Srikantia, 1982). The ichnotaxon *Archaeonassa* is described from the lower part of the Banded Siltstone Member of the Sankholi Formation (Tal Group), which indicates the extent of molluscan animals during early Cambrian (Singh et al., 2015).

The Tommotian period shows noticeable occurrence of Annelida, Mollusca, and Arthropoda. Trace fossils of trilobites (*Cruziana*, *Rusophycus*, *Chondrite*, and *Phycodes*) from Lower Cambrian have been established quite importantly in the global presence (Shah et al., 1998). Cambrian-age trace fossils are reported from the Lolab Formation in the Liddar Valley in Jammu and Kashmir represented by *Planolites* spp., *Skolithos*, *Taphrhelminthopsis* spp., and *Bergaueria*. *Planolites* and *Arenicolites* trace fossils have also been reported from the Muth Quartzites (Srivastava and Kumar, 1992). In Spiti, some ichnofossils like *Palaeophycus*, *Planolites*, *Skolithos*, and xiphosurid markings (Bhargava and Bassi, 1998) are reported.

Ornamented Metazoan eggs have been reported from the early Cambrian Chert Member of Deo-ka-Tibba Formation, Tal Group in Uttarakhand region of the Lesser Himalaya. Hughes (2016b) ascertained the presence of first soft-bodied metazoans from the Kunzam La (Parahio) Formation of the Cambrian of the Spiti Valley.

Trilobites and graptolites are some of the earliest skeleton fossils from the Palaeozoic besides Brachiopods. Trilobites are common in the Spiti Valley. *Opsidiscus*, an important trilobite, has been reported from Parahio Valley in the Spiti Basin (Pandey and Parcha, 2013). Cambrian succession in Zanskar Basin in the Parahio Valley, Spiti, is dominantly represented by *Cruziana* ichnofacies (Virmani et al., 2015). Singh et al. (2014, 2016, 2017) have been able to differentiate Cambrian 2 and Cambrian 3 series globally using key taxa *Yuehsienszella*, *Oryctocephalus indicus*, and *Pagetia significans* from Kunzam La Formation of Spiti Valley (Fig. 13.8a–c). Ostracodes of Lower Cambrian age are also reported from Krol Formation in the Kumaon region in Uttarakhand (Das et al., 1990).

Ordovician and Silurian periods are characterized by diminished fossil occurrences compared to the Cambrian, and no sharp boundary can be drawn between the two based on the fossil assemblages. The exceptional presence of *Chitinozoan* fossil assemblage from the Yong Limestone of the Garhwal region of Uttarakhand helps to distinguish the Ordovician-Silurian Boundary (Khanna et al., 1985).

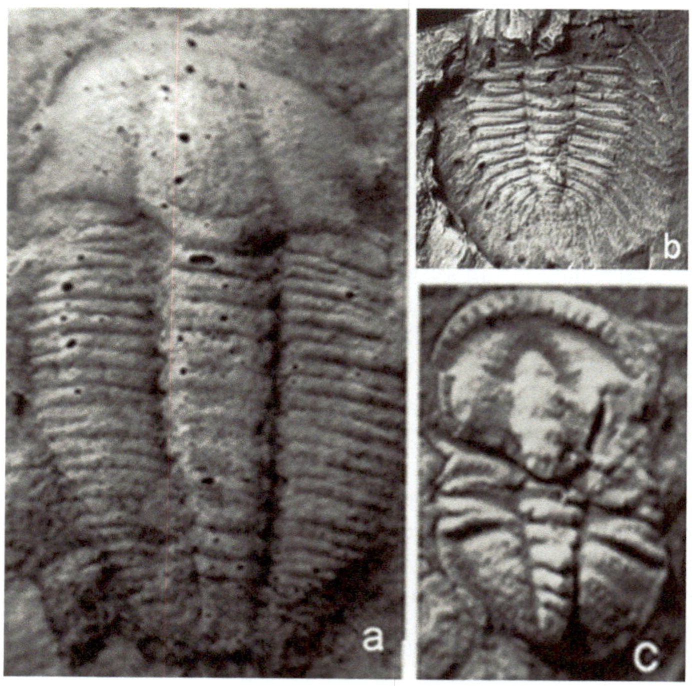

FIG. 13.8

Trilobite species (a) Yuehsienszella, (b) Oryctocephalus indicus, and (c) Pagetia significans from the Parahio River section of the Kunzam La Formation in the Spiti Valley.

Reproduced from (a) Singh et al. (2014). (b) Singh et al. (2016) and (c) Singh et al. (2017) with permission.

The full-bodied fossil occurrence increased with the onset of Silurian like Trilobite (*Calymene*), Coral (*Coenites* and *Favosites*), and Brachiopod (*Orthis* and *Atrypa*), which also showed impactful presence. Devonian plant fossils included *Protolepidodendron* and *Taeniocrada,* besides few other fragmentary plant fossils. The Devonian Fauna is characterized by *Spirifer*, *Phacops*, *Megalodus*, *Atrypa*, *Orthis*, *Favosites*, *Orthoceras*, and *Athyris*. Conodonts and crinoids are important Devonian fossils reported from the different parts of Ladakh, Kashmir, Uttarakhand, and Himachal Pradesh. Trace fossils of *Planolites* and *Skolithos* are reported from Devonian Muth Quartzite (Parcha and Pandey, 2011).

13.6 EDIACARAN EXPLOSION AND EXTINCTION OF BIOTA: INDIAN CONTEXT

A lot has been said and published recently on the extinction of the Ediacaran fossils that is considered as a bench-mark event demarcating the Precambrian-Cambrian Boundary. The explosion observed in the Cambrian strata extends back into the Neoproterozoic Ediacaran (635–542 Ma). This makes the Vendian-early Cambrian transition a global phenomenon even in the Indian context.

The predominating phase of biological evolution took place during the Cambrian Evolutionary Radiation of life forms that started to evolve at around 580 Ma. Represented by the enigmatic Ediacaran fauna, this was followed by the Cambrian Explosion of the modern phyla between ~540 and 520 Ma. Extensive peneplanation of the mountainous regions supplied massive quantity of nutrients, and because of that, there was an explosive enhancement of the Cambrian Evolutionary Radiation. Prior to that, the bacterial mat grounds and associated communities of Ediacaran organisms dominated the shallow seafloor (primarily the tidal flats). Near the end of the Ediacaran Period at ~543 Ma, several soft-bodied biota suffered extinction. During the Precambrian-Cambrian transition, vertically burrowing animals altered the ecology of the seafloor, and the biomineralization became a commonplace feature.

The presence of Ediacaran fossils is found to be decisive in discriminating the age of rocks in different parts of the world including India. An ironically noticeable feature associated with carbonates lying above the Ediacaran carbonate beds is depletion in heavy carbon isotope values, which has been reported from the carbonate beds of Birmania Formation (Hughes et al., 2015).

BOX 13.2

Three major events demarcate the Precambrian-Cambrian Boundary: (A) terminal Ediacaran extinction, (B) the Cambrian 'substrate revolution' (i.e. mass availability of nutrients to enhanced palaeoweathering), and (C) the diversification of biomineralizing organisms, which is linked with the reversal frequency of the Earth's magnetic field.

13.7 EVIDENCE MARINE PERMO-CARBONIFEROUS FORMATIONS

The Permo-Carboniferous is globally witnessed as the Hercynian (or Variscan) Orogeny and henceforth transitions of mass scale occurred on the Earth. The transition period evidenced fresh cycle of sedimentation that invariably continued till Cretaceous. The sedimentary cycle commenced with deposition of boulder beds marking unconformity in several newly formed basins. Spiti Valley and Kinnaur in Himachal Pradesh, Kumaon in Uttarakhand, Hazara-Kashmir section, and the Salt Range in the west were the areas showing sedimentary beds with fossil assemblages. Calcareous sandstone, limestone, and shales were predominantly deposited in these basins.

The formation of the earliest beds of charcoal has been reported by Jasper et al. (2016) in the form of fragments of tracheids that show homogenized cell walls, a characteristic feature of charcoal. The charcoal formation is reported to be due to the incidence of forest fires during the Late Permian transitional deposition in Zewan beds in Jammu and Kashmir.

BOX 13.3

Tracheids are elongated cells in the xylem of vascular plants that serve in the transport of water and mineral salts. Tracheids are one of two types of tracheary elements, vessel elements being the other. Tracheids, unlike vessel elements, do not have perforation plates.

REFERENCES

Bassi, U.K., Chopra, S., 1984. Magnetite tuff from the Himalaya India. J. Geol. Soc. India 25, 67–69.

Bhargava, O.N., Bassi, U.K., 1998. Geology of Spiti-Kinnaur Himachal Himalaya. Mem. Geol. Surv. India 124, 1–210.

Bhargava, O.N., Srikantia, S.V., 1982. Taphrhelminthopsis circularis from? Cambrian sediments of southeast Kashmir valley. J. Geol. Soc. India 23, 406–407.

Das, D.P., Raha, P.K., Achharay, P.K., 1990. Record of earliest (Lower Cambrian) Ostracoda from the Krol Formation, Nainital area, Kumaun Himalaya. J. Geol. Soc. India 36, 523–529.

Hughes, N.C., 2016a. The Cambrian palaeontological record of the Indian subcontinent. Earth Sci. Rev. 159, 428–461.

Hughes, N.C., 2016b. Soft-bodied fossil preservation in the Cambrian Parahio Formation of Spiti. Curr. Sci. 110, 774–775.

Hughes, C.N., Myrow, P.M., McKenzie, N.R., Xiao, S., Banerjee, D.M., Stockli, D.F., Tang, Q., 2015. Age and implications of the phosphatic Birmania Formation, Rajasthan, India. Precambrian Res. 173, 164–173.

Jasper, A., Uhl, D., Agnihotri, D., Tewari, R., Pandita, S.K., Benício, J.R.W., Pires, E.F., Da Rosa, A.A.S., Bhat, G.D., Pillai, S.S.K., 2016. Evidence of wildfires in the Late Permian (Changsinghian) Zewan Formation of Kashmir, India. Curr. Sci. 110, 419–423.

Khanna, A.K., Sinha, A.K., Saha, S.C.D., 1985. Yong Limestone of Tethys Himalaya—its stratigraphic status and palynological fossils. J. Geol. Soc. India 26, 191–198.

Mazumdar, A., Banerjee, D.M., 2015. Genesis of Early Cambrian phosphorite of Krol Belt, Lesser Himalaya. Curr. Sci. 108, 1247–1252.

Pandey, S., Parcha, S.K., 2013. Systematics, biometry of the species Opsidiscus from the middle Cambrian succession of the Spiti basin, India. J. Geol. Soc. India 82, 330–338.

Parcha, S.K., Pandey, S., 2011. Devonian ichnofossils from the Farakah Muth Section of the Pin Valley, Spiti Himalaya. J. Geol. Soc. India 78, 263–270.

Roy, A.B., 2004. The Phanerozoic reconstitution of Indian shield as the aftermath of break-up of the Gondwanaland. Gondw. Res. 7 (2), 387–406.

Roy, A.B., Jakhar, S.R., 2002. Geology of Rajasthan: Precambrian to Recent. Scientific Publishers, Jodhpur, India 421 p.

Sameeni, S.A., 2009. The Salt Range: Pakistan's unique field museum of geology and paleontology. In: Lipps, J.H., Granier, B.R.C. (Eds.), Paleo-Parks—The Protection and Conservation of Fossil Sites Worldwide. Carnets de Géologie/Notebooks on Geology, Brest. Book 2009/03, Chapter 6 (CG2009_BOOK_03/06).

Shah, S.K., Kumar, A., Sudan, C.S., 1998. Trace fossils from the Cambrian sequence of Zanskar (Ladakh Himalaya). J. Geol. Soc. India 51, 777–784.

Singh, S.K., Trivedi, J.R., Krishnaswamy, S., 1999. In: Re-Os chronology of black shales from the Tal and Shali Formations, Lesser Himalaya, India. ABSTRACT, 14th Himalayan-Karakoram-Tibet Workshop, Kloster Ettal, Germany, pp. 143–144.

Singh, B.P., Virmani, N., Bhargava, O.N., Kishore, N., Gill, A., 2014. Yuehsienszella (Cambrian Series 2) trilobite from the Parahio Valley, Spiti region (Zanskar-Spiti Sub-basin), India and its biostratigraphic significance. J. Paleontol. Soc. India 59, 81–88.

Singh, B.P., Bharagava, O.N., Chaubey, R.S., 2015. Early Cambrian trail Archaeonassa from the Sankholi Formation (Tal Group), Nigali Dhar syncline (Sirmur district), Himachal Pradesh. J. Geol. Soc. India 85, 717–721.

Singh, B.P., Virmani, N., Bhargava, O.N., Negi, R.S., Kishore, N., Gill, A., 2016. Trilobite fauna of basal Cambrian Series 3 (Stage 5) from the Parahio Valley (Spiti), Northwest Himalaya India and its biostratigraphic significance. Ann. Palaeontol. 102, 59–67.

Singh, B.P., Chaubey, R.S., Bhargava, O.N., Prasad, S.K., Negi, R.S., 2017. The Cambrian trilobite fauna from the Shian (Saybang) section, Pin Valley (Spiti) and its biostratigraphic significance. Palaeoworld 26, 25–36.

Srikantia, S.V., 1981. The lithostratigraphy, sedimentation and structure of Proterozoic-Phanerozoic formations of Spiti basin in the higher Himalaya of Himachal Pradesh, India. In: Sinha, A.K., Nautiyal, S.P. (Eds.), Contemporary Geo-Scientific Researches in India. Mahendra Pal Singh Publisher, Dehradun, India, pp. 31–48.

Srivastava, A.K., Kumar, S., 1992. Trace fossils from the Muth Quartzite of Malla Johar area, Tethys Kumaon Himalaya, India. J. Geol. Soc. India 40, 43–47.

Virmani, N., Singh, B.P., Gill, A., 2015. Integrated Litho-ichnofacies and ichnofacbric analysis of the lowermost part of the Kunzam La Formation along the Khemnagar Khad and the Pariho Valley sections, Spiti region (Zanskar-Spiti-Kinnaur basin), northwest Himalaya, India. J. Geol. Soc. India 85, 557–566.

CHAPTER

GEOLOGY OF THE GONDWANA SUPERGROUP

14

14.1 INTRODUCTION: GONDWANA FORMATIONS IN THE GLOBAL CONTEXT

The Gondwana Supergroup constitutes one of the significant geological ensembles deposited during the early Phanerozoic Eon in the Indian Subcontinent constituting a part of the pristine Indian Shield. The social importance of the Gondwana Supergroup is vested in the fact that more than 99% of the total coal resource of the Subcontinent is hosted in the Gondwana basins. This post-Precambrian geological formation evolved after a break in sedimentation for more than 180 million years between the Ordovician and late Carboniferous. Development of the Gondwana basins started with the deposition of land-derived sediments, with minor marine input at the initial stage. The Gondwana deposition in the Indian Subcontinent like most other places began with the formation of tillites and glacial boulder beds in close association with marine beds. This was followed by the deposition of fluvial and fluvio-lacustrine sediments intercalated with enormous deposits of plant remains that ultimately turned into coal seams in linear intracontinental rift basins. These sedimentary successions with coal-bearing beds (called seams) constituted the Gondwana Supergroup.

Precise definition of the Gondwana formations as formal stratigraphic units has suffered because of the inclusion of rocks deposited in diverse geological conditions into its ambit and also because of the overemphasis placed on the floral evidence in the stratigraphic correlation of different formations. In view of that, the Gondwana Supergroup has been redefined to include dominantly continental rift-basin deposits formed between the late Carbonaceous and the time of the Gondwana break-up during the early/middle Jurassic.

The Gondwana sequences of the Indian Subcontinent were part of the Gondwana Supercontinent that also included Antarctica, Australia, Madagascar, Southern Africa, and South America. The eastern Gondwanaland of which India was also a constituent included Antarctica, Australia, and Madagascar, while the rest comprised western Gondwanaland. The global correlation between these constituent landmasses of the Gondwanaland is principally based on the basis of floral and faunal fossils and palaeoenvironmental conditions depicted from the sediments deposited in each of these.

BOX 14.1

It was H.B. Medlicott who first coined the term Gondwana in 1872 for the coal-bearing formations of India. The term was derived from the ancient 'Gond' tribes of central India, and the name Gondwana is derived by combining two words, 'Gond' (land of the Gond tribes) and 'wana' (a wood, forest, or grove).

Indian Shield. https://doi.org/10.1016/B978-0-12-809839-4.00015-1

14.2 DISTRIBUTION, CLASSIFICATION, AND CORRELATION OF GONDWANA FORMATIONS

The Gondwana Supergroup in the Indian Subcontinent comprises continental rift-basin deposits formed between late Carbonaceous and middle or late Cretaceous with minor marine inputs. Some Gondwana-type coal basins are also known from different parts of the eastern Himalayas. We are omitting these highly dismembered and tectonized Gondwana basins in the Himalayas from the present discussion, restricting only to the continental rift basins formed in the Peninsular India for the sake of simplicity (Fig. 14.1). Also excluded are the contemporaneous (Permo-Carboniferous) marine shelf deposits in the northwestern Rajasthan (Bap Boulder Spreads and Badhaura Formation) and the much younger marine shelf deposits along the eastern coast.

Three different rift-basin-type occurrences of the Peninsular Gondwana basins include the following:

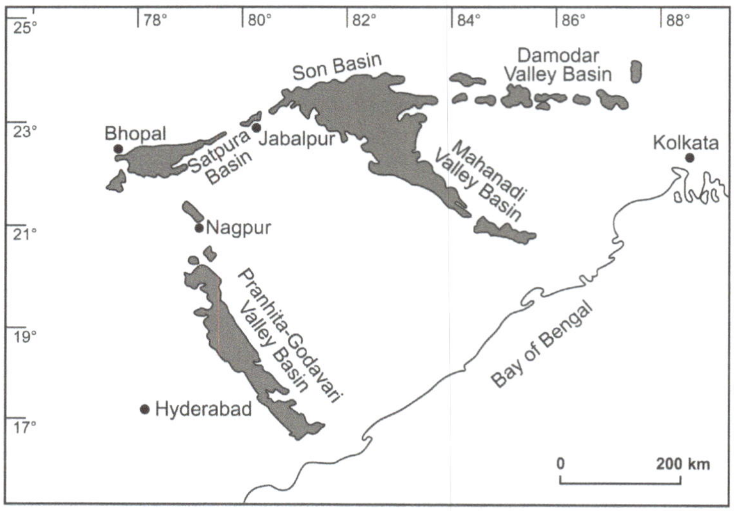

FIG. 14.1

Distribution of the Gondwana Basins along the major peninsular river valleys.

(1) East–west trending Damodar Valley Basins. Included in this belt is the narrow subparallel belt of subsidiary basins occurring north of the main belt. One such is also described as the Rajmahal Basin.

(2) Northwest–southeast trending Son-Mahanadi Valley Basins. The belt widens and thickens in a northwesterly direction, meeting the extension of the Damodar Valley Basins. The northernmost part is also described as Son Basin.

(3) Northwest–southeast trending Pranhita-Godavari Valley Basins including the Satpura Basin in the north.

Stratigraphic classification of the Gondwana formations developed in the three belts, suggested by Shastri et al. (1977), is shown in Table 14.1. The basal Talchir unit includes glacial tillites with shale and shale-siltstone rhythmites which have been interpreted as the glacial varve deposits. All the other units comprise sandstone with shale as the dominant lithology. A number of coal seams (along with carbonaceous shale) occur in the Karharbari, Barakar, and Raniganj Formations. The Barren Measure, which occurs between the Barakar and the Raniganj, as the name implies, is devoid of any significant coal seam. The Panchet and the Supra-Panchet are also coal-free sequences. The Supra-Panchet is also known as the Mahadeva Formation.

Table 14.1 Generalized Classification of the Gondwana Formations in Three Different Basins of the Indian Subcontinent

Age	Damodar Valley	Son-Mahanadi Valley	Pranhita-Godavari Valley
Jurassic			
Lower			Kota
Triassic			
Upper	Supra-Panchet	Parsora-Mahadeva	Dharmavaram Maleri
Middle		Pali	
Lower	Panchet		Bhimavaram Yerapalli Mangali
Permian			
Upper	Raniganj/Barren Measures	Kamthi	Kamthi Motur Barakar
Lower	Barakar Karharbari Talchir	Barakar Karharbari Umaria-Talchir	Talchir

The stratigraphic correlation of the Gondwana formations in three different rift basins along the Damodar, Son-Mahanadi, and Pranhita-Godavari as indicated in Table 14.1 is not free from any disagreement and debate. On the other hand, serious controversy persists on the stratigraphic relationship between different formations, especially in the Son-Mahanadi and Pranhita-Godavari Basins. In fact, the controversy regarding the stratigraphic interpretation of the Gondwana succession in the three Peninsular river valley basins has continued to worry the Gondwana workers for more than a century. The precise determination of formal stratigraphic units in different basins has suffered mainly because of the overemphasis given to the floral evidence in stratigraphic correlation.

Addressing this problem of Indian Gondwana stratigraphy, Mukhopadhyay et al. (2010) suggested the use of some unique tectonic and depositional features and palaeoclimatic events that are also recognized in other parts of the Gondwanaland, to classify and correlate the intra- and interbasinal Gondwana formations in the Indian Subcontinent. Using features like of marine flooding surfaces, tectonic uplift and sagging of basin floors, and evidence of major change in depositional environment, Mukhopadhyay et al. (2010) prepared an interbasin correlation chart (Table 14.2) of the different

Table 14.2 Correlation of Gondwana Formations Occurring in Different Gondwana Basins

			Damodar-Koel	Southern Mahanadi	Son	Satpura	Godavari
Cretaceous	Lower	Albian Aptian			Bansa bed	Jabalpur	Chikiala/ Gangapur
Cretaceous	Lower						
Jurassic	Upper	Oxfordian					
Jurassic	Middle						Kota /
Jurassic	Lower	Pliensbachian	Suprapanchet	Up. Kamthi	Bandhavgarh /Parsora	Up. Panchmarhi / Up. Bagra	Up. Kamthi
Triassic	Upper	Rhaetian				Lr. Bagra	Dharmaram
Triassic	Upper	Norian Carnian			Tiki	Denwa	Maleri
Triassic	Middle	Ladinian Anisian					Bhimaram
Triassic	Lower		Panchet	Lr. Kamthi	Pali	Lr. Panchmarhi	Yerapalli/ Middle Kamthi
Permian	Lopingian		Raniganj	Raniganj	Raniganj	Bijuri	Lr. Kamthi
Permian	Guadalupian		Barren Measures	Barren Measures	Barren Measures	Motur	Barren Measures
Permian	Cisuralian	Kungurian	Up. Barakar	Up. Barakar	Up. Barakar	Up. Barakar	Up. Barakar
Permian	Cisuralian	Artinskian	Lr. Barakar/ Karharbari	Lr. Barakar	Lr. Barakar	Lr. Barakar	Lr. Barakar
Permian	Cisuralian	Sakmarian Asselian	Talchir	Talchir	Talchir	Talchir	Talchir
Carboniferous		Gzhelian					

Nondeposition/erosion

After Mukhopadhyay et al. (2010) with permission.

formations of the Gondwana Supergroup, assigning the age wherever available. The most significant part of the correlation is the identification of seven time zones for the deposition of the Gondwana sediments.

Dutta (2002) had earlier highlighted the importance of lithostratigraphy in the interpretation of the Peninsular Gondwana basins based on the physical criteria such as lithological association, their petrographical characteristics, and determination of the order of superposition.

14.3 BASIN FORMATION, LITHOSTRATIGRAPHY, AND SEDIMENTATION HISTORY

The initial sedimentation in all the basins of the Gondwanaland took place coincided in time with the phase of melting of glaciers. In the earliest radiometric age, determined from juvenile magmatic zircon associated with interglacial marine mudstone of Dwyka Group from Kalahari Basin, Africa, is

302±3 Ma (Bangert et al., 1999). Veevers (2006) correlated the 302 Ma (Gzhelian) age as the time of inception of the Gondwana basins over the preexisting Precambrian basement concurrently with the phases of multiple deglaciation.

In the Indian Subcontinent, the Gondwana deposition was initiated during the widespread transgression inundating almost all the Gondwana basins that had formed along the Protocontinent boundaries during Early Sakmarian at around 290 Ma (Mukhopadhyay et al., 2010). The Tethyan marine front, according to these authors, had entered into the interior area from the north, east, and west.

The evidence of glacial regime during the deposition of Talchir, the oldest formation of the Gondwana Supergroup is indicated by the presence of diamictites, polished and striated pavements, bedding plane striation, faceted/striated gravels, bullet-shaped clasts, dropstone, cryoturbation, and some other features (Fig. 14.2) (Bhattacharya and Bhattacharya, 2015; Chakraborty and Bhattacharya, 2008). The Talchir Formation rarely exceeds a thickness of 100–120 m, except in some parts of the Damodar Valley Basin, where it attains a maximum thickness of 300 m. The glacial origin, lack of mineralogical maturity, and the overall greenish colour of sediments along with the unconformity at the base make the unit easily identifiable in the field.

FIG. 14.2

(a) Glacial tillites overlying sandstones of Talchir Formation, Talchir, Odisha. (b) Striated gravels indicate glacial abrasion. (c) Uniformly fine rhythmic layering representing glacial varve deposits of Talchir beds, Talchir coalfield, and Odisha. (d) Bullet-shaped clasts within the chaotic matrix-supported conglomerate, Talchir Formation, Raniganj Basin, and West Bengal.

(a) Image courtesy, Sukanta Dey. (b) Reproduced from Chakraborty and Bhattacharya (2008) with permission. (c) Image courtesy, H.N. Bhattacharya. (d) Reproduced from Bhattacharya and Bhattacharya (2015) with permission.

Overlying the Talchir occur three successive formations: Barakar, Barren Measures, and Raniganj, which developed in the Damodar Valley Basins. These formations together measure a total thickness of nearly 2500 m. The Karharbari Formation that also includes some coal-bearing beds occurs between the Barakar and the Talchir, mainly in the Damodar Valley Basins. The unit is generally made up of sandstone, grey micaceous shale, and black carbonaceous shale.

The Barakar Formation comprises conglomerate, sandstone, shale, siltstone, fire clay, and most importantly the commercially significant coal seams (Fig. 14.3). The thickness of the unit varies between 700 and nearly 1100 m. The lithological composition and characteristics of the rocks of the Barakar Formation indicate deposition in humid climate in restricted swampy basins.

FIG. 14.3

An operational coal mine at Chasnala, Jharia Coalfield, Jharkhand, showing interlayering of coal seams within beds of sandstone.

Image courtesy: Sukanta Dey.

Overlying the Barakar occurs the Barren Measure Formation having gradational contact with the former. The Barren Measure Formation comprises sandstone, grey micaceous shale, and black carbonaceous shale. As the name suggests, this unit is mostly barren of coal seams. The thickness of this unit is extremely variable. In the type area of the Damodar Valley, it is nearly 100 m, while in the Pranhita-Godavari Valley, the thickness exceeds 500 m. The unit overlies the Barakar Formation conformably having a gradational contact.

Stratigraphically above the Barakar occurs the Raniganj Formation having a similar lithological association including coal seams. The associated sediments include shale (including carbonaceous types), siltstone, and sandstone. Like most of the coal-bearing Gondwana basins, the Raniganj Coalfield is riddled with postdepositional faults (Fig. 14.4). The sandstones are locally calcareous. Economically exploitable coals are best developed in the easternmost Raniganj Basin, and their importance dwindles towards west in the Damodar Valley and in rest of the Peninsular area. The thickness of this formation varies from 900 m in the east to 700 m in the Pranhita-Godavari Basin. The Raniganj Formation shows gradational contact with the Barren Measures that it underlies.

The Panchet Formation is made of alternate sandstone and shale. The brightly coloured red shale in this unit in an otherwise grey background of coal-bearing sediments makes the unit easily identifiable. The Panchet Formation is known as the Maleri in the Pranhita-Godavari Basin, as the Pali-Tiki in the Son Basin north of the Son-Mahanadi Basin, and as the Pachmarhi-Denwa in the Satpura Basin in the north of the Pranhita-Godavari Basin.

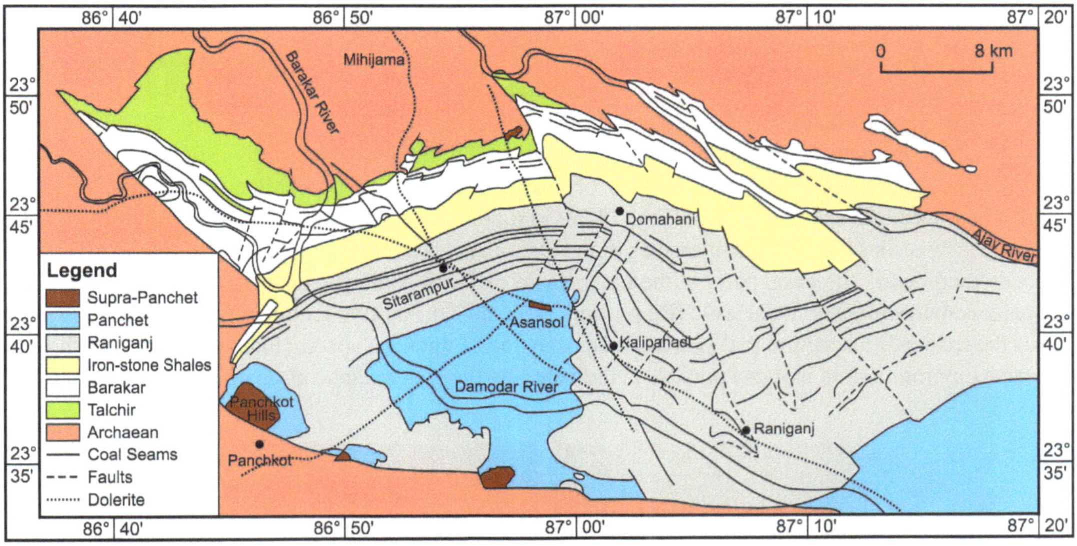

FIG. 14.4

Geological map of the Raniganj Coalfield. The outcrop pattern is disrupted by the large number of postdepositional faults. Compiled from different sources.

The Mahadeva Formation overlying the Panchet is the youngest Gondwana formation in the central India. The rocks of this unit form majestic hill ranges in an otherwise gently undulating, monotonous country. The equivalents of the Mahadeva Formation are variously named in different Gondwana basins. It is known as Supra-Panchet in the Damodar Valley Basin; as the Kota in the Pranhita-Godavari Basin; as the Parsora in the Son Basin, north of the Son-Mahanadi Basin; and as the Bagra in the Satpura Basin, north of the Pranhita-Godavari Basin. The younger Panchet and the Supra-Panchet Formations show only a few coal seams occurring in between sandstones.

The stratigraphic correlation chart of Mukhopadhyay et al. (2010) (Table 14.2) shows a pronounced depositional break after the deposition of the Supra-Panchet and the other equivalent formations in almost all the Gondwana basins. The break that began tentatively during 170–180 Ma (Toarcian-Aalenian) is also the time when a couple of Jurassic basins started to emerge in the Indian Subcontinent particularly in the Kachchh and Jaisalmer region in northern Gujarat and northwestern Rajasthan, respectively (Roy et al., 2017). Incidentally, this was also the time when the break-up of the Gondwanaland had begun. Tectonically, therefore, the formation of any new basins having links with the Gondwana appears unlikely. In other words, the Gondwana basin forming episode must have culminated with the deposition of the Supra-Panchet and its equivalent formations in the different Gondwana basins.

14.4 PALAEOCLIMATIC INFLUENCE ON GONDWANA SEDIMENTATION

The Gondwana formations indicate varied depositional characteristics that correspond to different phases of palaeoclimatic conditions like glacial, fluvial, warm, and humid phase, detailed below:

14.4.1 GLACIAL PHASE

The Gondwana sedimentation was initiated with the deposition of the Talchir Formation under glacial palaeoclimate as evidenced by the occurrence of features like striated pavements, till fabrics, tillites, conglomerates, and rhythmites of glacial origin (also described as glacial varves) (Fig. 14.2c). The subsequent progressive warming up in the palaeoclimatic conditions had significantly influenced the process of sedimentation leading to the development of proglacial lacustrine-type palaeoenvironment on local scale. Dropstones like exotic boulders (Fig. 14.5a) in diamictite-type matrix are witnessed in the Talchir equivalent formations that are considered as indicators of such a climatic condition. Similar features have also been recorded in the basal parts of the Palar Basin of Tamil Nadu suggesting a proglacial lacustrine-type palaeoenvironment on local scale (Ramasamy et al., 2000). Another feature reported from the Talchir beds is the ice-wedge structure that also indicates thawing of thick ice sheets (Fig. 14.5b). The fresh water lacustrine environment in isolated pockets is indicated by the syngenetic carbonate nodules in the tillites.

FIG. 14.5

(a) A gravel-sized dropstone in a fine-grained matrix, Talchir Formation. (b) Ice-wedge structure indicating thawing of thick ice sheets preserved in the Talchir siltstone.

Reproduced from Chakraborty and Ghosh (2008) with permission.

Abrupt changeover from glaciomarine to terrestrial fluviolacustrine depositional environment is noted across the Talchir and the Karharbari Formations. The evidence for this comes from the occurrence of the storm-influenced sediments (known as tempestites) in the Karharbari Formation showing hummocky cross stratification overlying the glaciogenic sediments of Talchir Formation (Bhattacharya et al., 2005). The occurrences of marine beds at Umaria and Manendragarh, near Jabalpur, are indicators of marine transgression event. The till fabric indicates northerly palaeoslope with southerly provenance.

14.4.2 FLUVIAL PHASE OF SEDIMENTATION

The fluvial sediment sequences overlie the glacial beds that ultimately resulted in the formation of flood basin complex comprising either anastomosed or braided river systems that interchanged with the lacustrine conditions. Braided river systems are characterized by the amalgamated nature of the multistoreyed, multilateral sand bodies of high stacking density (Tewari, 1996). Sediments are coarse-grained confirming uniformity in deposition, whereas the presence of intermittent fine clastics suggests fluctuation in depositional energy condition. The development of asymmetrical cycles together with abundance of very coarse to coarse sandstone is attributed to rapid lateral migration of braided channels, whereas the deposition of symmetrical cycles with subequal amount of coarse to medium sandstone and fine clastics including coal correspond to slow and gradual overstepping of meandering stream channels and associated subenvironments across the alluvial plain (Tewari, 1997). Cycles of fining-upward sequence are commonly observed in the fluvial succession of the Permian Barakar Formation (Fig. 14.6). Each cycle starts with a channel facies with coarse, trough cross bedded sandstone (light coloured), deposited near the bottom of sequence over a scoured surface on the underlying shale bed. The sandstone becomes finer towards top with planar and climbing-ripple cross laminations. This is topped by dark layers of shale and coal beds (seams). Fluvial-coal swamp complex facies are also reported from the Chintalapudi subbasin of the Pranhita-Godavari Valley (Laxminaryana, 1995).

FIG. 14.6

Raniganj Coal Mine, Damodar Valley Basin, showing fining-upward sequence in the Barakar sandstone beds.

Image courtesy: Sukanta Dey.

14.4.3 WARM AND HUMID PHASE

The Barakar and the Raniganj Formations that host most of the Gondwana coal deposits were deposited under the warm and humid climatic condition. The environmental condition favoured proliferation of floral growth that also led to rise in surface temperature. Enormous vegetation was dumped by meandering streams into swamps and oxbow lakes to form coal seams. Basins repeatedly kept on sinking with the increase sedimentation and basin extension. Sediments were deposited in alternating high- and low-energy regimes as the overbank/levee deposits. Overbank facies were linked by laterally connected channel-fill sand bodies and the presence of thick mudstone.

The deposition of the Barren Measures marks a break in the palaeoclimatic condition when the temperature had risen concurrently with the development of an arid condition. The prevailing situation led to obliteration of the thick vegetable cover of the vegetation, and a dry spell was evident during the deposition of the Barren Measures that lacked plant fossils.

The return of the warm and humid conditions during the Raniganj Formation was extended further up to the deposition of the Panchet and Supra-Panchet formations where flora of wide variety of species flourished. However, occurrence of ferruginous sandstones with thin layers of haematite denotes arid conditions. The red colouration observed in the floodplain sediments fines, and the intermittent calcareous paleosol horizons indicate a semiarid climate with seasonal rainfall during the deposition (Retallack et al., 1996).

The post-Supra-Panchet time had been a period of basin inversion ushering in the onset of high rates of palaeoweathering resulting in the formation of lateritic conglomerates, red clayey, and haematitic palaeosols. The mottled clays with intercalated lime beds are also seen along with red sandstone deposited in shallow waters. The post inversion changes in lithofacies types suggest a fresh episode of tectonic uplift, ushering in the closure of the Gondwana basins.

14.5 GONDWANA FOSSILS

The Gondwana Supergroup includes a host of plant fossils occurring in the different lithoformations. Floral varieties found during the Gondwana period were mostly climate-dependent. Moist flora luxuriantly flourished during Permian when coal forming climatic condition prevailed. Some of the varieties that were transitional in existence were slowly replaced with the change in the climatic condition as shown diagrammatically in Fig. 14.7.

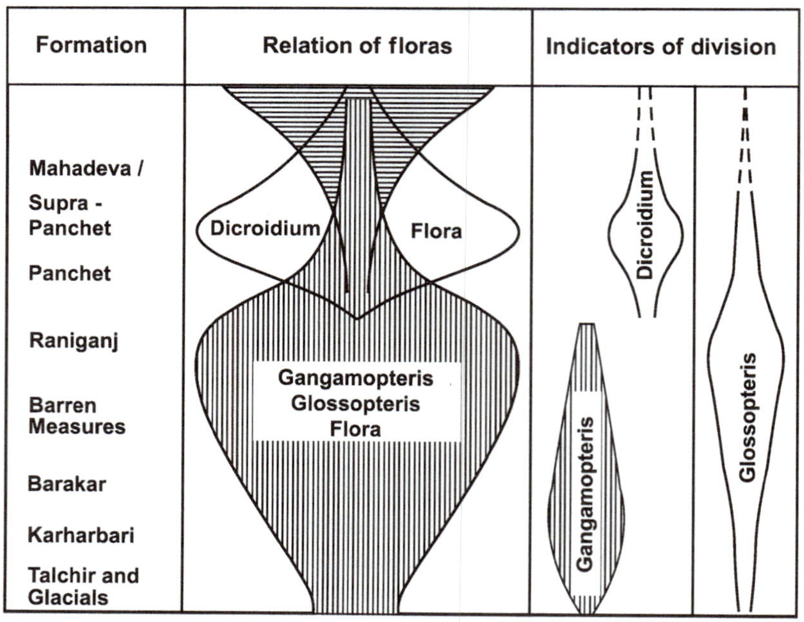

FIG. 14.7

Schematic representation of the relation between the floras of the Gondwana Supergroup showing vertical distribution pattern of *Glossopteris* (as contrasted with *Gangmopteris*) with time.

From Lele (1964).

Preservation potential of pollens and spores (besides algal oncolites) was most prolific during the Gondwana time, which helps to understand the evolutionary trend. The pollen-type sporopollenin is unique and completely manifests the genetic character of the plant and at the same time helps in understanding the palaeoclimatic condition. Fossil spores of various varieties have been reported from by Tewari et al. (2009). Recent studies by Jha et al. (2011a,b) have reported a number of occurrences of palynomorphs of early Triassic age from the Pranhita-Godavari Valley by Mahesh et al. (2014),

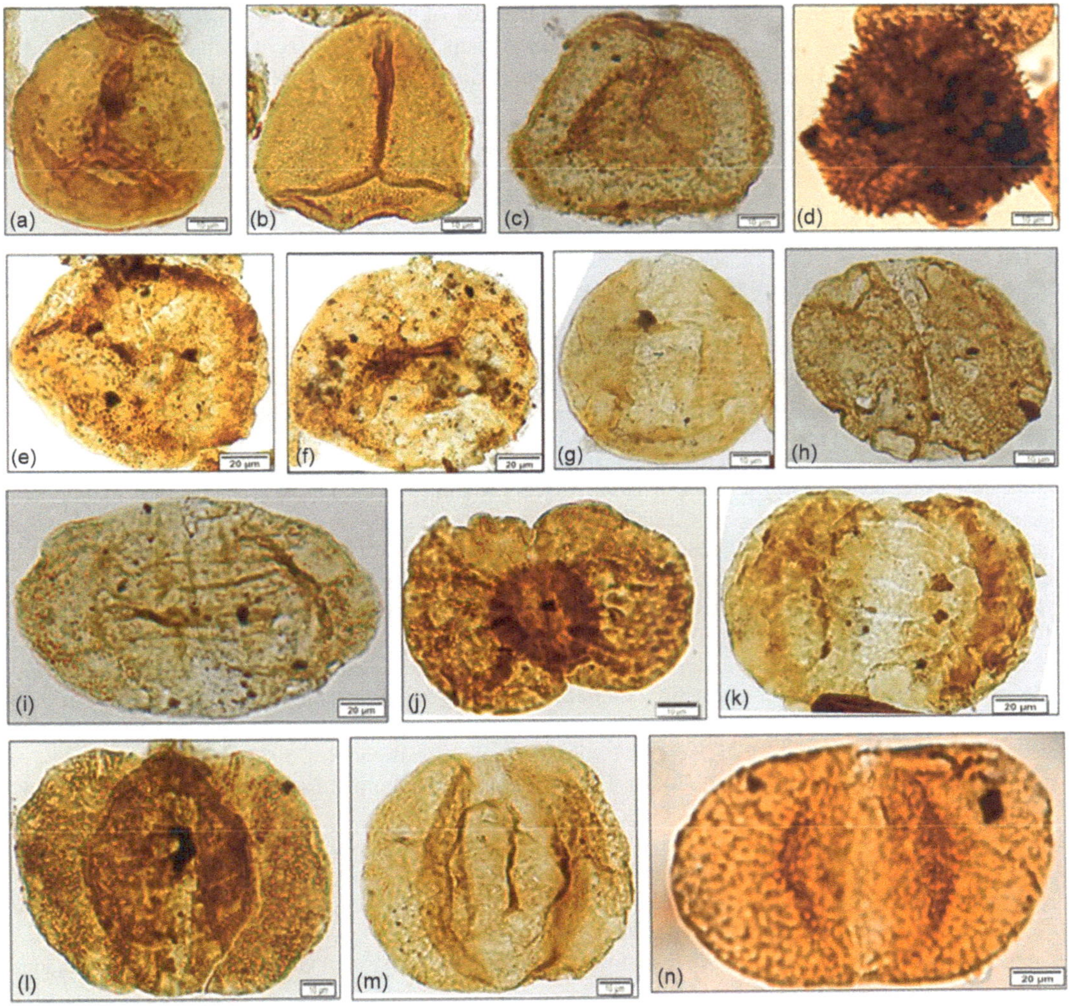

FIG. 14.8

Representative photomicrographs of palynomorphs recovered from borehole in Raniganj Coalfield: (a) *Microfoveolatispora raniganjensis*, (b) *Microbaculispora villosa*, (c) *Gondisporites raniganjensis*, (d) *Brevitriletes unicus*, (e) *Densipollenites magnicorpus*, (f) *Densipollenites invisus*, (g) *Striasulcites* sp., (h) *Scheuringipollenites barakarensis*, (i) *Faunipollenites varius*, (j) *Rhizomaspora indica*, (k) *Striatopodocarpites subcircularis*, (l) *Striatites varius*, (m) *Crescentipollenites crasus*, and (n) *Alisporites plicatus*.

Reproduced from Mahesh et al. (2016) with permission.

who reported that the peat-forming vegetation was mainly composed of palynomorphs of gymnosperms represented by glossopterids, conifers, cordaites, and sphenopsids from the Wardha basin. The two studies on palynomorphs indicated correlatable Lower Permian (Artinskian) age for Wardha and Pranhita-Godavari Basins. The dominating presence of bisaccates in the Raniganj coalfields and the presence of monosaccates reflect a peat-forming community composed mainly of gymnosperms

(Mahesh et al., 2016) (Fig. 14.8). Scattered occurrences of vertebrate and invertebrates fossils are found in the different formations of Gondwana Supergroup. Vertebrates include fishes, reptiles (turtle, *Lystrosaurus*, a mammal-like reptile), mammals, and dinosaurs. Invertebrates include brachiopods, gastropods, bivalves, and bryozoa.

REFERENCES

Bangert, B., Stolhofen, H., Geiger, M., Lorenz, V., 1999. Fossil record and high-resolution tephrostratigraphy of Carboniferous glaciomarine mudstones, Dwyka Group, southern Namibia. Commun. Geol. Surv. Namibia 12, 235–245.

Bhattacharya, H.N., Bhattacharya, B., 2015. Lithofacies architecture and palaeogeography of the Late Paleozoic glaciomarine Talchir Formation, Raniganj Basin, India. J. Palaeogeogr. 4, 269–283.

Bhattacharya, H.N., Bhattacharya, B., Chakraborty, A., 2009. Architecture of Storm Event Beds from Late Paleozoic Glaciomarine Talchir Formation, Talchir Coalfield, Orissa. Vistas in Geological Research, Utkal University. (Spec. Publ. in Geology 8), pp. 1–10.

Bhattacharya, H.N., Chakraborty, A., Bhattacharya, B., 2005. Significance of transition between Talchir Formation and Karharbari Formation in Lower Gondwana basin evolution: a study of west Bokaro coal basin, Jharkhand, India. J. Earth Syst. Sci. 114, 275–286.

Chakraborty, C., Bhattacharya, S., 2008. Pattern of sedimentation during the Late Paleozoic, Gondwanaland glaciation: an example from the Talchir Formation, Satpura Gondwana basin, central India. J. Earth Syst. Sci. 117, 499–519.

Chakraborty, C., Ghosh, S.K., 2008. Pattern of sedimentation during the Late Paleozoic, Gondwanaland glaciation: an example from the Talchir Formation, Satpura Gondwana basin, central India. J. Earth Syst. Sci. 117, 499–519.

Dutta, P., 2002. Gondwana Lithostratigraphy of Peninsular India. Gondwana Res. 5, 540–553.

Jha, N., Pauline, S.K., Mehrotra, N.C., 2011a. Early Triassic palynomorphs from Manuguru area of Godavari Graben, Andhra Pradesh. J. Geol. Soc. India 78, 167–174.

Jha, N., Pauline, S.K., Tewari, R., Mehrotra, N.C., 2011b. Palynological dating and correlation of surface and subsurface sediments from Wardha Valley Coalfield, Maharashtra. J. Geol. Soc. India 78, 137–148.

Lele, K.M., 1964. In: The problem of Middle Gondwana in India. Proceedings of 22nd International Geological Congress, New Delhi, Section 9, pp. 181–202.

Laxminaryana, G., 1995. Gondwana sedimentation in Chintalapudi sub-basin of Godavari Valley, Andhra Pradesh. J. Geol. Soc. India 46, 375–383.

Mahesh, S., Sabina, K.P., Bilwa, L.M., 2014. Palynology and depositional facet of Lower Permian (Artinskian) sediments from New Majri open caste mine, Wardha Basin India. J. Geol. Soc. India 83, 697–708.

Mahesh, S., Murthy, S., Sabina, K.P., Saran, S., Singh, V.P., 2016. Organic matter characterization of carbonaceous shales from Raniganj coalfields and its implications on depositional conditions: a palynofacies and petrographical overview. J. Geol. Soc. India 87, 132–144.

Mukhopadhyay, G., Mukhopadhyay, S.K., Roychowdhury, M., Parui, P.K., 2010. Stratigraphic correlation between different Gondwana basins of India. J. Geol. Soc. India 76, 251–266.

Ramasamy, S., Madhavraju, J., Armstrong, J., 2000. Dropstones in Talchir sediments of Palar Basin, Tamil Nadu—implications on depositional conditions and palaeoclimate. J. Geol. Soc. India 56, 47–52.

Retallack, G.J., Veevers, V.J., Morante, R., 1996. Global coal gap between Permian-Triassic extinction and Middle Triassic recovery of peat forming plants. Geol. Soc. Am. Bull. 108, 195–207.

Roy, A.B., Chatterjee, A.K., Chauhan, N.K., 2017. Geological evolution of Kachchh: an epitome of successive Phanerozoic events. Curr. Sci. 112, 1051–1055.

Tewari, R.S., 1996. Palynoevent stratigraphy in Gondwana sequence of India. In: Gondwana Nine. vol. 1. Ninth International Gondwana Symposium, Oxford-IBH Publishing Pvt. Ltd., New Delhi, pp. 3–19.

Tewari, R.C., 1997. Numerical classification of coal-bearing cycles of early Permian Barakar coal measures of eastern-central Indian Gondwana basins using Q-mode cluster analysis. J. Geol. Soc. India 50, 593–599.

Tewari, R., Mehrotra, N.C., Meena, K.L., Pillai, S.S.K., 2009. Permian megaspores from Kuraloi area, Ib-river coalfield, Mahanadi Basin, Orissa. J. Geol. Soc. India 74, 669–678.

Veevers, J.J., 2006. Updated Gondwana (Permian-Cretaceous) Earth history of Australia. Gondwana Res. 9, 231–260.

CHAPTER

GONDWANA SUPERGROUP: ORIGIN AND EVOLUTION

15

15.1 INTRODUCTION

The Phanerozoic reconstitution of the Indian Shield across the Precambrian-Cambrian Boundary witnessed significant geologic changes that ultimately resulted in the geomorphotectonic transformation of the Precambrian crustal block. The culmination of Precambrian witnessed the inversion (upliftment) of the Vindhyan and the associated 'Purana basins' in the mainland of the Indian Shield. Globally, the end of the Precambrian marked a gradual change over from the configuration of Rodinia to that of Pangea. The precise timing and the process of growth and evolution of Pangea are still debated. However, it may not be very irrational to assume that the Pangean configuration was in the process of emergence by the end of Precambrian.

The newly emerged Supercontinent Pangea included two distinctive continental associations, the Laurasia in the north and Gondwanaland in the south, which were linked due to proximal connectivity of major landmasses. This alignment is well interpreted in several studies, and logical placements of the interlocking boundaries of the continents have been discussed on the basis of rock, mineral, and fossil affinities. The Indian crustal block (presently known as Indian Subcontinent) was placed northeast of the erstwhile Gondwanaland (Fig. 15.1).

BOX 15.1

There is a prevailing confusion about the timing of the growth of Gondwanaland and the formation of Gondwana basins. Both the features could not have evolved concurrently. The conventionally accepted time of the growth of the Gondwana basins was around Late Carbonaceous or Early Permian. The basin formations could not have taken place before or even simultaneously with the growth of the landmass called Gondwanaland.

The generally held view is that there was a pre-Pangea Supercontinent called Pannotia that broke into several smaller fragments by about 550 Ma, that is, precisely at the stroke of Precambrian-Cambrian Boundary. The larger piece containing part of China, India, Africa, South America, and Antarctica is called Gondwanaland.

15.2 ORIGIN AND EVOLUTION OF GONDWANA BASINS

Barring the Early Palaeozoic granitic activities, there is virtually no record of any geological activity in the Indian Shield for about 180 million years between the Early Ordovician and Early Carboniferous. It was during the Late Carboniferous when a number of continental rift basins started to evolve as the receptacle of land-worn sediments. Marine incursions noted only locally in the central Indian region during the extension of arms of the sea from the western and northern part of the Indian crustal block.

Indian Shield. https://doi.org/10.1016/B978-0-12-809839-4.00016-3

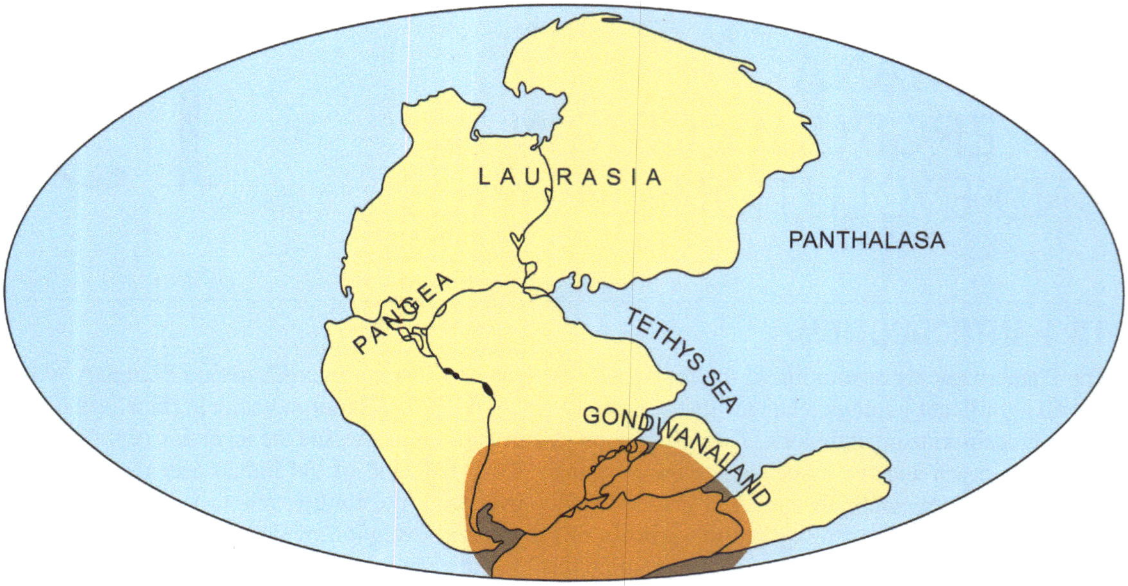

FIG. 15.1

The framework of the Pangean Supercontinent comprising Laurasia in the north and the Gondwanaland in the south. IND stands for the continental block that constituted the Indian Subcontinent at a later stage. The dark-brown part in the supercontinent framework represents the zone of south polar glaciation.

Very little is known about the causative factors that led to the formation of the rift basins for the deposition of the Gondwana sediments. Timewise, there is a connection between the Permo-Carboniferous glaciation and the formation of the Gondwana rift basins. But the link could be a mere coincidence between two disparate features. One possibility is that it was only during the Late Carboniferous/Permian time that the major part of the Gondwanaland, including the Indian crustal block, entered into the Circle of the South Pole (Fig. 15.1).

An interesting feature about the Permo-Carboniferous glaciation in the Gondwanaland is the movement directions of the glacial sheets that developed in the South Polar Region. The reconstructed flow directions of the ice sheets indicated a radiating pattern diverging from a central topographic high (Fig. 15.2). The development of rift basins as the depository of Gondwana sediments almost simultaneously with the phase of glaciation suggests the possibility of the development of a subcrustal thermal dome close to the South Polar Region, which triggered the geomorphic elevation on the surface. At the same time, the thermal doming could also be responsible for the opening of the rift basins along some structural weak zones. In short, the prevailing thermodynamic situation could have acted as the driving force not only for the opening of rift basins but also for the movement of different crustal blocks. The possibility of thermal doming as the motive force for the opening of the Gondwana rift basins is indirectly proved by the reports of widespread volcanism and the granitic magmatism during Permian (Artiskian) along the northern margin of the existing Indian landmass.

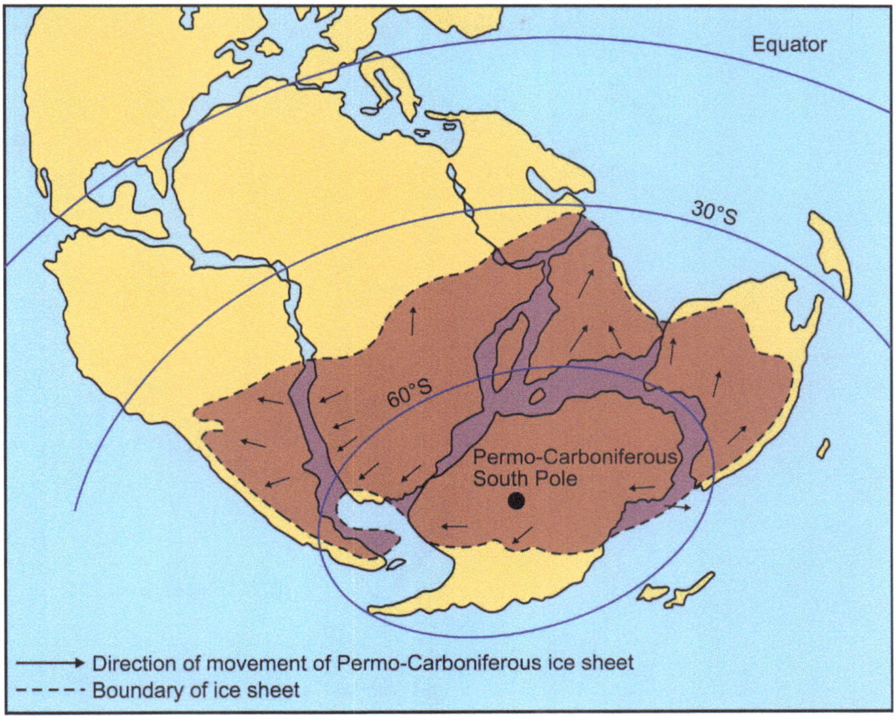

FIG. 15.2

Illustration of the direction of movement of the South Polar ice sheets during the Permo-Carboniferous.

In the context of the Indian Subcontinent, a number of continental rift basins evolved as the receptacle of land-worn sediments during Late Carboniferous. As mentioned earlier, the marine incursions noted only locally in the central Indian region as extension of the arms of the sea either from the western or from the northern part of the Indian landmass.

All the coal-bearing Gondwana basins in the Indian Subcontinent evolved along the contacts of the Protocontinents, barring some exceptions (Roy, 2012). The classic examples are the Godavari (or Pranhita-Godavari), Mahanadi, and Damodar Valley Basins that occur south of the Son-Narmada Lineament (Fig. 15.3).

A notable feature in the development of Gondwana basins in the Indian Shield is the total absence of any coal-bearing Gondwana (rift) basin in the entire belt of the Aravalli-Bundelkhand Protocontinents that occur north of the tectonic margin represented by the Narmada-Son Lineament and its northeastern continuation (Fig. 15.3). Roy (2004) offered a cogent tectonic explanation for this preferential development of the Gondwana basins in the Indian Shield. The development of rift zones (grabens), according to Roy (2004), is easier where the extensional stresses act at high angles to the weak structural grains such as the contact zones of protocontinental margins (described as 'Join' by Rogers, 1986). The non-development of any such rift zone or basin would imply that the operative stresses were either parallel or at low angles to the structural grains in the Aravalli-Bundelkhand cratonic blocks (Protocontinents).

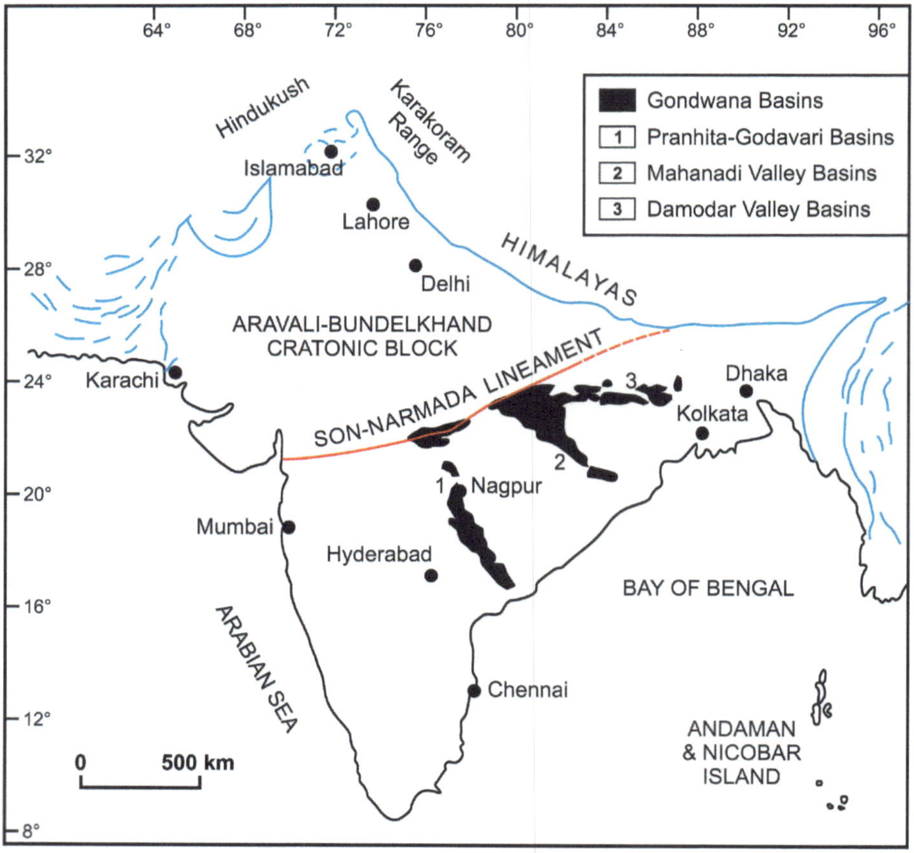

FIG. 15.3

Schematic map showing disposition of the Gondwana rift basins, south of the Son-Narmada Lineament.

Since the prevailing structural grains in the Aravalli-Bundelkhand cratonic blocks or Protocontinents are dominantly NE-SW, it has been inferred that the stress vectors that caused the opening of the Gondwana basins were also NE-SW directed (*all the directions refer to the present-day position of the Indian Subcontinent*). Such a stress system would help in opening all the earlier grains/rifts in the southern block, which were oriented at high angles to it but not in the northern block where the structural grains were subparallel to it. The same stress system might also have helped opening a 'chasm' along the Son-Narmada Lineament zone causing marine incursion.

Manifestations of magmatism in the Gondwana basins are not very common. Occurrences of some ultrapotassic and other alkaline intrusive rocks are reported from different parts of the Damodar Valley Gondwana sedimentary basins. Available field and palaeomagnetic and geochronological data for most of the studied dykes suggest their emplacement during early Cretaceous period (between 110 and 117 Ma) (Agarwal and Rama, 1976; Kent et al., 1997, 2002; Srivastava et al., 2014). These ages are closer to Kerguelen Plume-related early Cretaceous igneous activities, and therefore cannot be responsible in anyway in the development of the Gondwana basins that evolved much earlier.

15.3 BASIN TECTONICS AND SEDIMENTATION PATTERN

The Gondwana basins are typically bounded by faults that developed along ancient Precambrian grains as mentioned earlier. Chakraborty et al. (2003) discussed the importance of fault development (both basinal and basin margin faults) in the Gondwana basin especially on the nature of sedimentation in different Gondwana basins of the Peninsular India (Fig. 15.4). The deposition of sediments was affected by intrabasinal faults indicating fault-controlled *syn*-sedimentary subsidence. The patterns of the intrabasinal faults and their relationships with the respective basin-bounding faults represent both extensional and strike-slip regimes (Chakraborty et al., 2003).

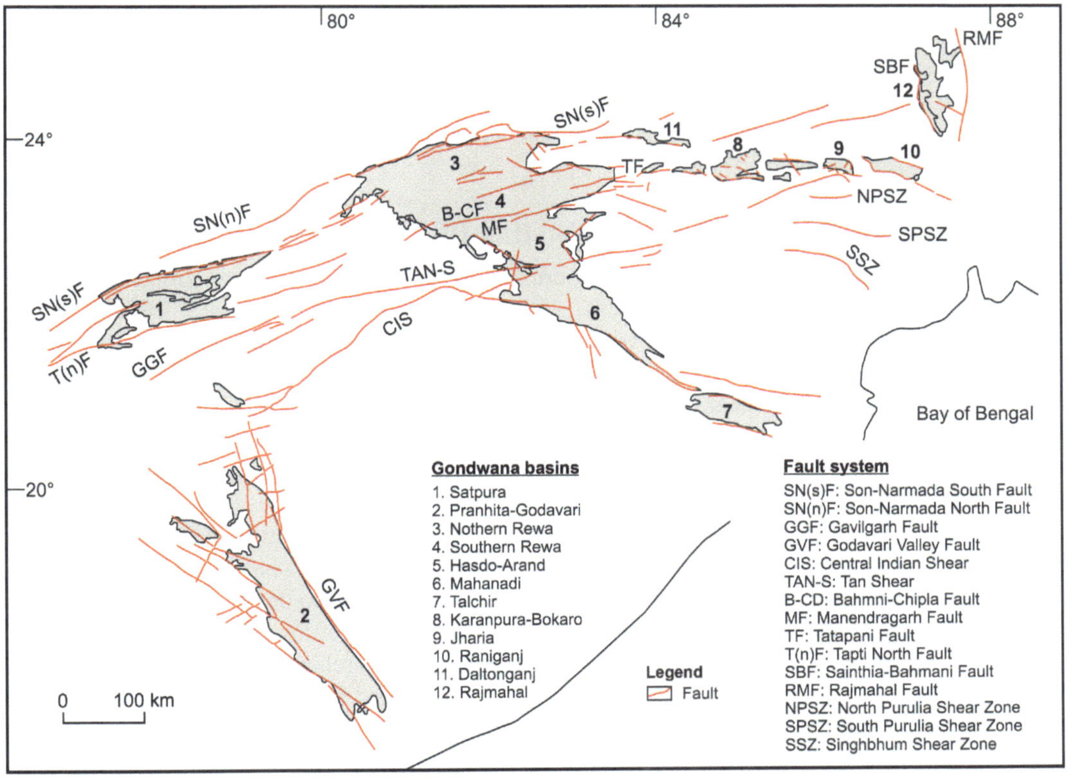

FIG. 15.4

Layout of the major faults and lineaments associated with different Gondwana basins of Peninsular India.

Reproduced from Chakraborty et al. (2003) with permission.

The earliest phase of Gondwana sedimentation (especially during the Talchir) took place in the eroded topographic depressions. Similarities existed between several sedimentary successions that occurred as new isolated basins. The glacial detritus was dumped in the earlier formed basins with interplay of unidirectional currents and oscillating waves during storms (Bhattacharya et al., 2009). Palaeodrainage systems at the initial (incipient) stages led to the development of pockets of

Palaeosols. The palaeosols comprise red-brown clay with ferruginous concretions developed under poorly drained to well-drained groundwater conditions. Even where the palaeosols are not present, the bedrock displays evidences of deep weathering. These probably represent the earlier topographic positive areas affected by physical weathering and intense groundwater activity (Dutt and Mukhopadhyay, 2000).

The main phases of sedimentation in the Gondwana basins took place along several linear continental rift basins comprising sags and depressions, which at a later stage developed into half or full grabens. Topographic relief and faulting during the postdepositional phase resulted in the varying morphology of the individual basins. The sedimentation in various Gondwana basins was controlled by the progressive and repetitive movements along block faulted underlying basement slabs. The overall basin geometry was generated in three stages of faulting during and postdating the sedimentation (Chakraborty et al., 2003).

There were longer intervals of rapid subsidence when fluvial sediments were deposited. These alternated with shorter intervals of relative tectonic quiescence or slower rate of subsidence, favourable for the development of the regional coal seams (Mukhopadhyay, 1984).

15.4 GEOPHYSICAL CHARACTERIZATION OF THE GONDWANA BASINS

There are several studies on the estimations of crustal thickness based on the gravity data below the different Gondwana basins (Mishra et al., 1987; Narula et al., 2000; Qureshy et al., 1967). In fact, the Gondwana basins with their complex tectonics and high hydrocarbon potential had become targets of serious geophysical investigations in the recent years, although the nature and thickness of sedimentary basins and the associated basement structures are yet to be precisely ascertained. The limited studies on the Bouguer gravity anomalies of Gondwana basins have been modeled to infer the thickness and structure of the basins, especially beneath the region under the Deccan Trap cover. These studies coupled with the inferences made from the study of the deep seismic sounding profiles (Kaila, 1982; Kaila and Krishna, 1992; Kaila et al., 1990) failed to confirm any significant contribution of the crustal thinning process in the development of the horst-graben structures in the Gondwana basins (Mahadevan, 1994). Similarly, the available heat flow data from Gondwana basin regions also do not suggest any noteworthy thermal perturbation. Higher heat flow values observed in some of the central and eastern Indian coalfields are attributed to either the Deccan or the Rajmahal Plume-related magmatism (Rao and Rao, 1982). The Precambrian antiquity of the rift zones is demonstrated by the development of the linear Proterozoic basins and emplacement of granulites along the flanks of the river valleys, as confirmed by the geophysical data (Mishra and Prajapati, 2003).

REFERENCES

Agarwal, J.K., Rama, S.N.I., 1976. Chronology of Mesozoic volcanics of India. Proc. Indian Acad. Sci. 84A, 157–179.

Bhattacharya, H.N., Bhattacharya, B., Chakraborty, A., 2009. Architecture of Storm Event Beds from Late Paleozoic Glaciomarine Talchir Formation, Talchir Coalfield, Orissa. Vistas in Geological Research, Utkal University. (Spl. Publ. in Geology 8), pp. 1–10.

Chakraborty, C., Mandal, N., Ghosh, S.K., 2003. Kinematics of the Gondwana basins of peninsular India. Tectonophysics 377, 299–324.

Dutt, A.B., Mukhopadhyay, A., 2000. Ancient denudational surfaces in the Gondwana sequences of Peninsular India. J. Geol. Soc. India 56 (11), 505–516.

Kaila, K.L., 1982. Deep Seismic sounding studies in India. Geophys. Res. Bull. 20 (Spec. Issue), 309–328.

Kaila, K.L., Krishna, V.G., 1992. Deep seismic sounding studies in India and major discoveries. In: Gupta, H.K. (Guest Ed.), Special Issue on Seismology in India—An Overview. Curr. Sci. 62, 117–154.

Kaila, K.L., Murthy, P.R.K., Rao, V.K., Venkateswarlu, N., 1990. Deep seismic sounding in the Godavari graben and Godavari (coastal) basin, India. Tectonophysics 173, 307–317.

Kent, R.W., Saunders, A.D., Kempton, P.D., Ghose, N.C., 1997. Rajmahal basalts, Eastern India: mantle sources and melt distribution at a volcanic rifted margin, large igneous provinces: continental, oceanic and planetary volcanism. Geophysical Monograph. Am. Geophys. Union 100, 145–182.

Kent, R.W., Pringle, M.S., Muller, R.D., Saunders, A.D., Ghose, N.C., 2002. $^{40}Ar/^{39}Ar$ Geochronology of the Rajmahal basalts, India and their relationship to the Kerguelen Plateau. J. Petrol. 43, 1141–1153.

Mahadevan, T.M., 1994. Deep continental structures of India: a review. Geol. Soc. India 28, 569. (Memoir).

Mishra, D.C., Prajapati, S.K., 2003. A plausible model for evolution of schist belt and granite plutons of Dharwar craton, India and Madagascar during 3.0–2.5 Ga: insight from gravity modelling constrained in part from seismic studies. Gondwana Res. 6 (3), 501–511.

Mishra, D.C., Gupta, S.B., Rao, M.B.S.V., Venkatarayudu, M., Laxman, G., 1987. Godavari basin. A geophysical study. J. Geol. Soc. India 30, 469–476.

Mukhopadhyay, A., 1984. Interrelationship of tectonism and sedimentation in the Jharia basin. J. Geol. Soc. India 25 (7), 445–454.

Narula, P.L., Acharyya, S.K., Banerjee, J., 2000. Seismotectonic atlas of India and its environs. Geol. Surv. India 87.

Qureshy, M.N., Aravamadhu, P.S., Bhatia, S.C., 1967. A review of the geophysical constraints in modelling the Gondwana crust in India. Tectonophysics 212, 141–151.

Rao, R.U.M., Rao, G.V., 1982. Heat flow in Indian Gondwana basins and heat production of their basement rocks. Tectonophysics 91, 105–117.

Rogers, J.J., 1986. Dharwar craton and the assembly of Peninsular India. J. Geol. 94, 129–143.

Roy, A.B., 2004. The Phanerozoic reconstitution of Indian shield as the aftermath of break-up of the Gondwanaland. Gondwana Res. 7 (2), 387–406.

Roy, A.B., 2012. Indian Shield: insight into pristine size shape and tectonic framework. Indian J. Geosci. 66, 181–192.

Srivastava, R.K., Kumar, S., Sinha, A.K., Chalapathi Rao, N.V., 2014. Petrology and geochemistry of high-titanium and low-titanium mafic dykes from the Damodar valley, Chhotanagpur Gneissic Terrain, eastern India and their relation to Cretaceous mantle plume(s). J. Asian Earth Sci. 84, 34–50.

PART

4

BETWEEN JURASSIC BREAKUP AND THE HIMALAYAN COLLISION

CHAPTER

FROM JURASSIC BREAK-UP TO HIMALAYAN COLLISION

16

16.1 BREAK-UP OF GONDWANALAND: THE FIRST UPSHOT

Major marine incursions occurred along newly formed rift zones at Jaisalmer and Kachchh region on the western India during the Middle Jurassic, based on faunal evidence corresponding to the Bajocian-Callovian time span. The timing helps to correlate the event with the break-up of Gondwanaland and the subsequent northward drifting of the Indian continental block, which was initiated at around 165 Ma or a little earlier (Roy et al., 2017).

The Jurassic Break-up of the Gondwanaland had a pronounced effect on the Indian Lithosphere, mainly through the development of sets of new fracture systems concurrently with the reactivation of some ancient tectonic grains. Some of the fracture surfaces were introduced later during the rise of thermal Plumes under the Indian Lithosphere.

The newly developed fracture systems along with the reactivated older tectonic grains helped to chip off the Indian Shield into a number of rigid crustal blocks. Geomorphologically expressed as Lineaments, these fracture systems had significantly affected the tectonic and seismic character of the Indian Crust during the subsequent geological history.

BOX 16.1

Lineaments are tectonically induced linear features that appear as lines in aerial photographs or space imagery. Most of these features represent surface expression of deep crustal tectonic dislocation. Though it appears quite similar sounding, the term lineation is used to describe linear structures in rocks formed during deformation associated with metamorphism.

Roy (2004) divided different Lineaments into two broad categories: (i) Lineaments that were constrained almost entirely within the Precambrian rocks and (ii) Lineament that intersects all the rock formations from the Precambrian to the youngest Cenozoic. Barring a few, almost all the first-category Lineaments appear as vertical fracture surfaces that help to identify them as very young features and at the same time recognize these as reactivated Precambrian grains. The second group of Lineaments, which transect almost all the geological formations including the youngest Phanerozoic rocks, has persistent azimuth orientations corresponding approximately to N35°W-S35°E and N65°E-S65°W trends. Out of these, the NW-SE-trending Lineaments are quite rectilinear in their trends compared with the NE-SW-oriented ones. The latter set of Lineaments partly imitates the Precambrian trends (Roy, 2004). Apart from the two sets of Lineaments, there is a third type as shown in Fig. 16.1, which has different orientations in different parts of the Subcontinent. These Lineaments are linked with Plume outbursts that had affected the region at different times.

Indian Shield. https://doi.org/10.1016/B978-0-12-809839-4.00017-5

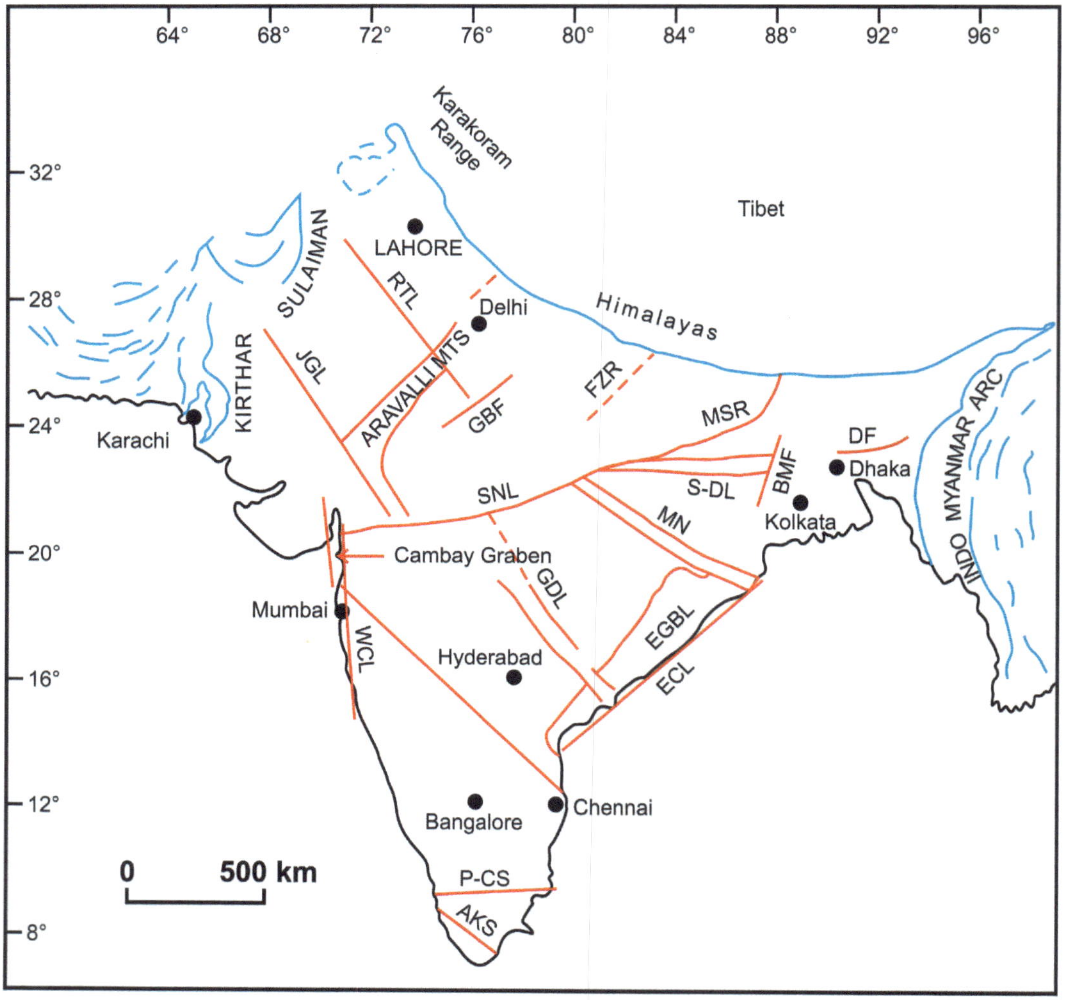

FIG. 16.1

Generalized map showing the distribution of major Lineament fractures in the Indian Subcontinent. Legend: *AKS*, Achankovil Shear Zone; *CNL*, Chennai-Nasik Lineament; *DF*, Dauki Fault; *ECL*, East-Coast Lineament; *GDL*, Godavari Lineament; *GBF*, Great Boundary Fault; *FZR*, Faizabad Ridge; *JGL*, Jaisalmer-Barwani Lineament; *MBS*, Moyar-Bhavani Shear Zone; *MNL*, Mahanadi Lineament; *MSR*, Munger-Saharsa Lineament; *P-CS*, Palghat-Cauvery Shear Zone; *RMF*, Rajmahal Fault; *SNL*, Narmada-Son Lineament; *S-DL*, Son-Damodar Lineament; *RTL*, Raisinghgarh-Tonk Lineament; *WCL*, West Coast Lineament.

Reproduced from Roy (2004) with permission.

16.2 PLUME OUTBURSTS: IMPLICATIONS ON PRUNING OF 'GONDWANIC' INDIAN CRUST

Subsequent to the dismemberment of Gondwanaland at ~165 Ma, the 'Gondwanic' India along with Madagascar and Seychelles and Antarctica remaining attached to it had started to move northward from the southern latitudes. On the way to the northward journey, the 'Indian Crust' passed over several Plume heads like the Kerguelen, Marion, Crozet, and Reunion at different times before colliding with the Eurasian Plate (Fig. 16.2).

BOX 16.2

(i) The term Plume or Mantle Plume is used for the 'molten' hot rock material (magma) that nucleates at the Core-Mantle Boundary and rises through the Earth's Mantle becoming a diapir in the Earth's Crust. By definition, diapir is a type of geological intrusion in which the hot magma is forced into the brittle overlying rocks.

(ii) The centres of the currently active zones or spots of volcanic activity are known as hotspots. Barren Island in the Andaman-Nicobar chain of islands is an example of hotspot in the north-eastern part of the Indian Ocean (Fig. 16.3).

The most significant consequence of the Plume impingements was the chipping of the Gondwanic Indian Crust with simultaneous opening of a new ocean. The earliest stripping event was during the impingement of Kerguelen Plume that triggered separation of Antarctica from the Indian crustal block during the early Cretaceous. Initially, a narrow intracontinental marine basin had formed in the Tiruchirappalli region (*nee* 'Trichinapalli') that was later aborted. The drifting of Antarctica towards southeast was initiated during 116–119 Ma (Baksi et al., 1987; Ghatak and Basu, 2011) with the consequent opening of the Bay of Bengal in the Indian Ocean. The separation of Antarctica was also marked by the extensive Plume-related volcanism over a wide area in the eastern part of the (present) Indian Subcontinent and outside (Fig. 16.4).

Close to the separation of Antarctica, a new hotspot activity was initiated at ~80 Ma south of Bhubaneswar in the eastern part of the Indian Subcontinent, which is thought to be continuous with the northern extension of the 85 degree East Ridge (Michael and Krishna, 2011). The Bengal Basin in the eastern part of the Indian Subcontinent, according to Roy and Chatterjee (2015), evolved coevally with the 85 degree East Ridge in the Bay of Bengal that is linked with the Afanasy-Nitkitin Ridge, a Late Cretaceous product of the Crozet Hotspot (Mahoney et al., 1996). The linking of Afanasy-Nikitin Ridge with the Crozet Plume may appear a debatable issue.

The separation of the Madagascar from the Indian continental block took place during the Marion Plume activity at 88–90 Ma (Torsvik et al., 1998). The evidence for this comes from the occurrence of acid and mafic volcanic rocks of similar age in different parts of central and northern Kerala, St. Mary's Island off the Karnataka coast, and also from Madagascar (Anil Kumar et al., 2001; Torsvik et al., 1998). Further evidence of Marion Plume-related activity is marked by the event of basin closing in the Kachchh region in western India much to the north of dismembered zone in the south, about 84 million years ago, during Upper Cretaceous (Santonian, 83–86 Ma) (Roy et al., 2017). The time of basin inversion coincides with the event of impingement of the Marion Plume under the Indian Lithosphere as it moved over the Plume head around that time. Significantly, this was also the time of opening of a new basin in the Kachchh region in northwestern Gujarat and also in the Bagh Basin in central India, much to the east of Kachchh (Roy et al., 2017).

FIG. 16.2

Map of Indian Ocean showing locations of four Plume heads at Reunion, Crozet, Kerguelen, and Marion.

FIG. 16.3

A classic example of "Strombolian"-type volcanic eruption on 24 December, 2010, in the Barren Island of the Andaman-Nicober Island Chain. The summit of the cone-shaped crater is about 300 m above the solidified lava flows in the foreground.

Image courtesy Hetu Sheth.

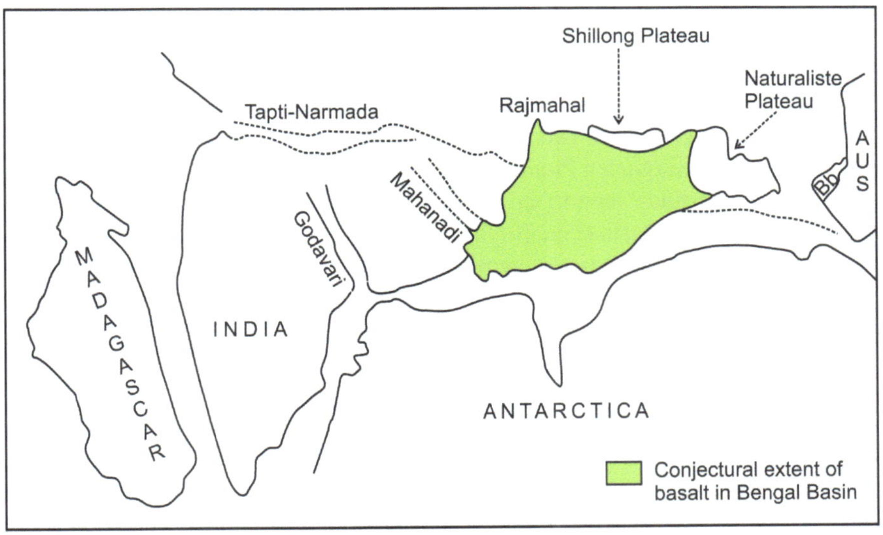

FIG. 16.4

The East Gondwana palaeoreconstruction showing extent of Rajmahal/Sylhet Trap basalt resulting from the Kerguelen Plume-related magmatism. AUS, Australia.

Reproduced from Kent (1991) with permission.

Further trimming of the Indian Lithosphere took place during the youngest Reunion Plume during the period of transition from Cretaceous to Paleogene. The main effect of the Reunion Plume outburst was the further modifications of the west-coast line with the consequent vertical rise of the Western Ghats. The rise of the mountainous Western Ghats coincided with tilting of the (present day) Peninsular India to the east with simultaneous easterly flowing drainage pattern. In addition to these tectonogeomorphic changes, significant geomorphologic transformations took place in the Kachchh region in northwestern India. In short, the different Plume outbursts caused pruning of the Gondwanic crustal block turning it to the shape of the Greater India (a conceptual model of the pre-Himalayan India) imagined by Emile Argand (1924).

Apart from changing the overall geographic shape of the Precambrian Indian Shield, the underplating-related Plume outbursts caused widespread magmatism in different parts of India. Some changes in the geophysical character of the terrane from the pristine Precambrian traits are noted in some parts, mainly from the triangular peninsular part of the pristine Indian Shield. These include development of localized gravity 'highs' and zones of high heat flow and the occurrences of magnetic and telluromagnetic anomalies over the entire Peninsular region (Qureshy, 1982; Qureshy and Midha, 1986).

The Bouguer gravity anomaly colour-image map (Fig. 16.5) prepared by Murthy (1999) appears quite illustrative in this regard. The interactions of the Indian Lithosphere with the Plume heads and also the related underplating events during the Plume outbursts are quite clearly expressed in the form of significant gravity highs fields that appear in the western and eastern areas of the Indian Shield. The appearance of two linear gravity highs in the N-S (between Kolkata, nee Calcutta, and Patna) and E-W (south of the Shillong Plateau region of eastern India) directions can be correlated with the Kerguelen Plume-related magmatism in the region.

The development of rift fractures subparallel to the directions of the two linear gravity highs has presumably been induced by the tensional stresses generated during the Plume outbursts. The two rift fractures are (1) N-S-trending basin-margin fault bordering the western margin of the Bengal Basin in the west and (2) E-W-trending Dauki Fault bounding the southern margin of the Shillong Plateau (Fig. 16.1). Both the fracture systems are associated with basin evolution in the respective regions. It is, however, difficult to categorize the occurrence of patches of the gravity highs along the eastern Indian coastline. Based on the presence of two volcanic layers of 107 and 114–116 Ma (Subrahmanyam, 2000), the extension of the Kerguelen Plume activity further south of the Bhubaneshwar area has been suggested. The suggestion finds support in the occurrences of patches of gravity high shown in the coastal areas of the Mahanadi Basin (Fig. 16.5).

The available geological information suggests the possibility of linking the gravity highs that occur near Mangalore and north of Thiruvanathapuram (*nee* Trivandrum) in the southwest of the Peninsular India with the Marion Plume-related underplating event.

The Bouguer anomaly contour map of northern Gujarat and western Rajasthan shows presence of large patches of high-gravity anomaly fields. This picture of scattered but very apparent clustered high-gravity anomaly fields has been linked with magmatism associated with the Reunion Plume outburst (Fig. 16.6) (Roy, 2003).

The important surface geological signatures that resulted from the Reunion Plume outbursts are (1) N-S and E-W sedimentary basins (of latest Cretaceous and the earliest Cenozoic age) in western Rajasthan and northern Gujarat, (2) Lineaments having roughly E-W and N-S fracture systems (lineaments) and reactivation of early formed lineaments, and (3) intrusions of alkaline and mafic bodies and bimodal lava flows. The history of evolution of rift basins contemporary with the Reunion Plume outbursts is very well-documented in western Rajasthan (Roy, 2003). The stratigraphic records from the Jaisalmer Basin suggest that the sedimentation, which had earlier closed at around 85 Ma, resumed by the late Cretaceous. The two other basins, which formed simultaneously with the Jaisalmer Basin, are the N-S-oriented Barmer Basin and the almost E-W-oriented Bikaner-Nagaur Basin. The evidences suggest a close interaction between the basin opening and marine transgression in the region with the underplating events related to the Indian Lithosphere-Reunion Plume.

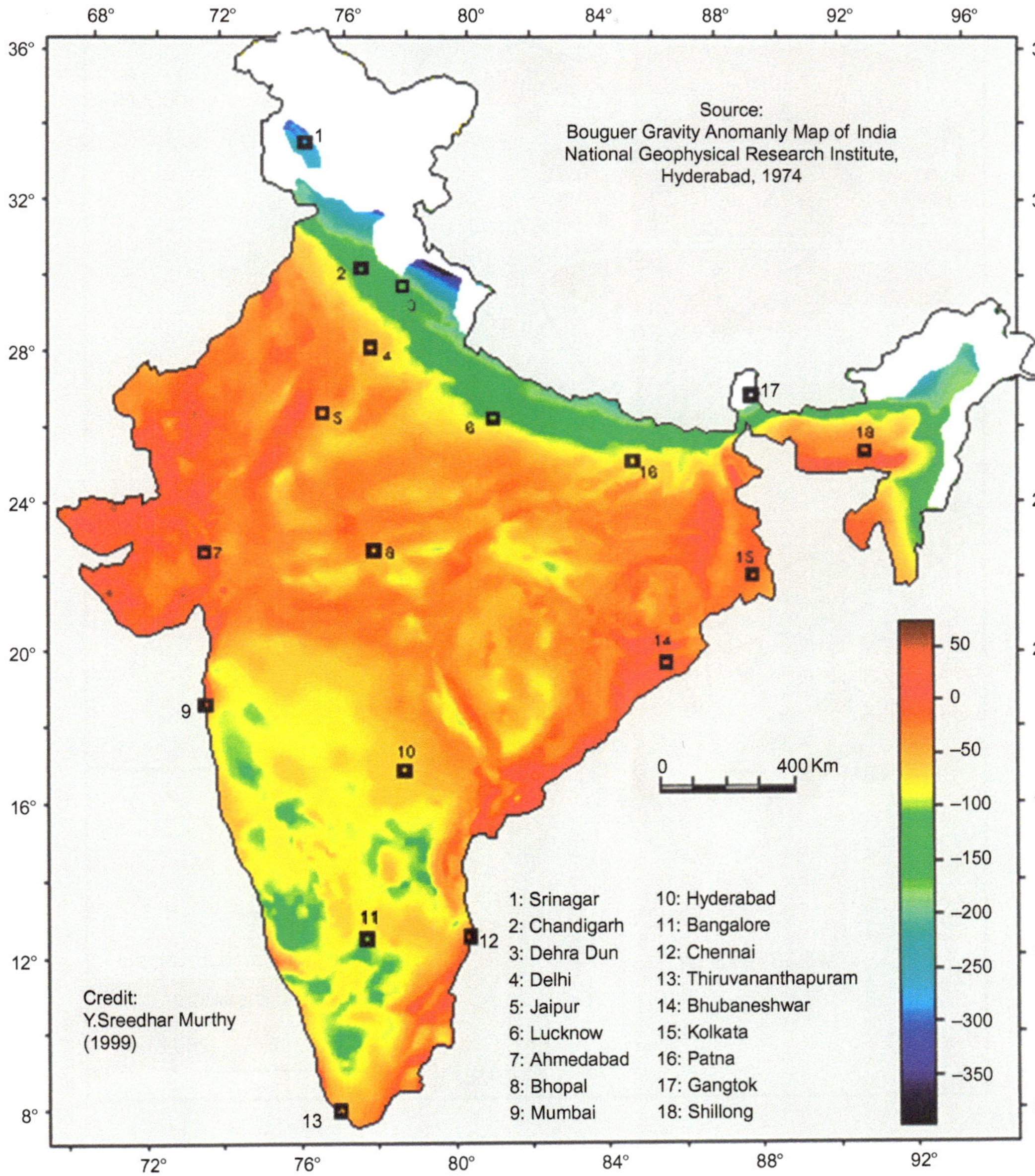

FIG. 16.5

Bouger gravity anomaly colour-image map of India.

Reproduced from Myrthy (1999) with permission.

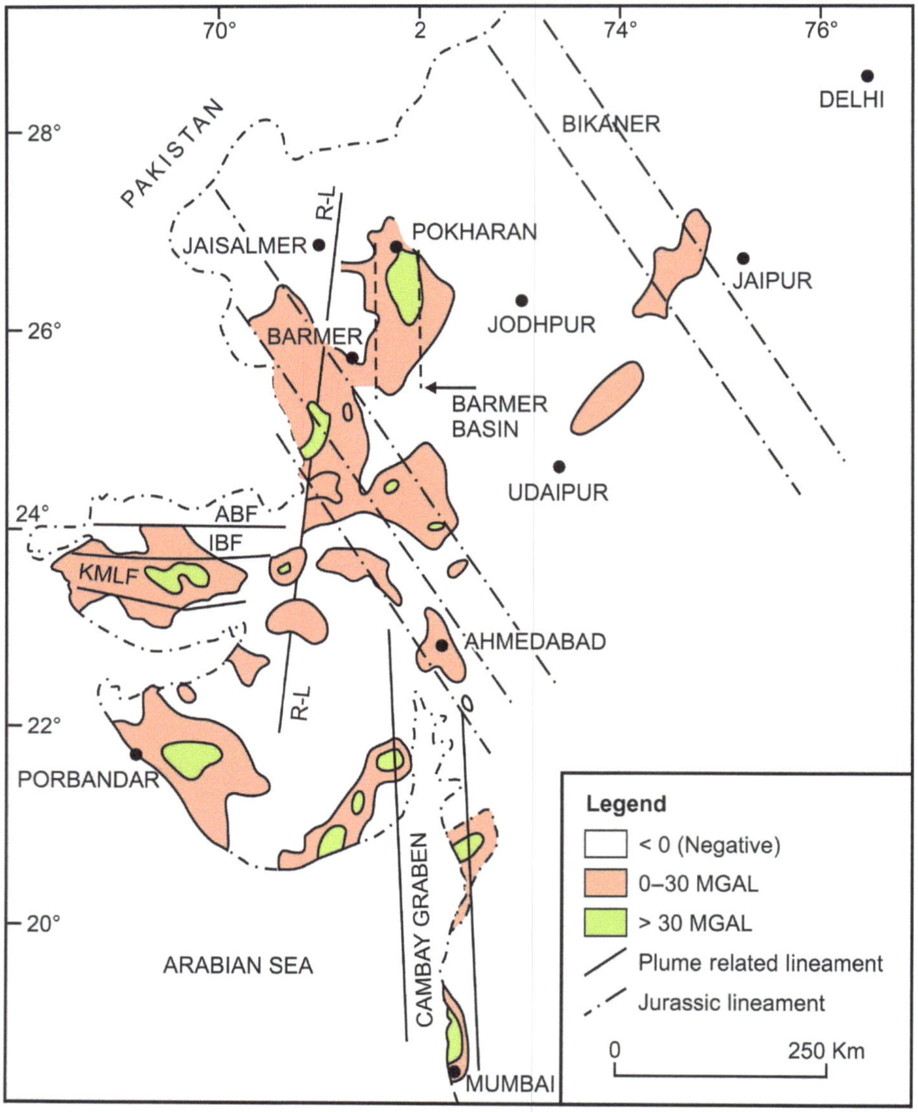

FIG. 16.6

Bouger gravity anomaly map of western India showing two systems of Lineament.

A very significant information is provided by the discovery of a linear Low-Velocity Zone (LVZ) at about 100 km depth in the subcrustal Upper Mantle region that underlies the N-S-oriented Barmer Basin. Based on the P-wave seismic tomographic study, Kennet and Widiyantoro (1999) could locate a cylindrical LVZ of about 200 km across in the Upper Mantle that underlies the region between the north of Gulf of Cambay and the northwest of Barmer. The striking coincidence in the extent of the

LVZ in the Upper Mantle (described as the trace of the Reunion Plume) with that of the N-S-orientated Cenozoic Basin of Barmer possibly suggests a cause-and-effect relationship between the underplating event related to Reunion Plume outburst and the formation of the contemporary Barmer Basin.

In the Rann of Kachchh region of northern Gujarat, there are a series of E-W-trending tectonic domes constituted of Jurassic rocks that have been formed by the diapiric intrusions (Fig. 16.7a and b) of basalts from below (Kanjilal, 1978). The intrusive relationship between the basalts exposed in the central parts of the eroded domes and the highly fossiliferous Jurassic Sandstones clearly proves the younger age of the basaltic rocks that occupy the core regions. The domes, which occur as chains of islands, are bordered in the north by a major E-W-running fault Lineament.

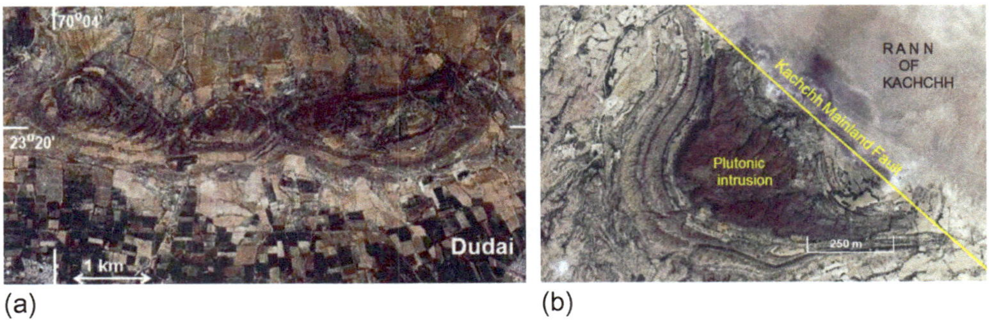

FIG. 16.7

(a) Satellite imagery showing closed outlines of a series of antiformal domes at Dubai near Bhuj, Gujarat. (b) Plutonic intrusion of alkali basalt in the core of a truncated dome. (Inset) Outcrops of mafic rocks in the core region of the dome.

Reproduced from Roy et al. (2017) with permission.

Several isolated plutonic bodies, plugs, and minor intrusive bodies of alkaline mafic and syenite suites associated with comagmatic alkaline dykes of different composition along with a number of suites of volcanic rocks ranging in composition from basic to acid occur at Mundwara, Sarnu-Dandali, and Tavidar regions in southwestern Rajasthan. The geochemical characteristics of the different rocks of Mundwara, Sarnu-Dandali, and Tavidar indicate that these are the products of partial melting of rocks situated at different levels in the Upper Mantle and in the Crust. The available isotope ages from the rocks of the region suggest Deccan affinity (Basu et al., 1993). The occurrences of a number of Mundwara-type volcano-plutonic complexes are known from the Saurashtra and Kachchh regions of northern Gujarat that include the Girnar Hills and the Alec, Osam, and Chamardi Hills in the different parts of northern Gujarat. In addition to these, innumerable volcanic plugs are present in the Kachchh region in Gujarat (Roy, 2003). There are several dyke-like bodies of limbergite, nepheline-syenite, monchiquite, and camptonite that intruded into olivine gabbro constituting the base (Sukheswala, 1981). In spite of the close regional association, there is no apparent correlation between the zones and patches of gravity high and the occurrences of the igneous complexes that occur as plutons and/or plugs of various dimensions. The only exception is the high-gravity anomaly field (+50 mGal) marked over Alec Hills near Porbandar, which is a volcano-plutonic basaltic complex. The presence of high-gravity Bouguer anomaly over the Alec Hills region indicates the presence of high-density rocks (representing a possible mafic magma pouch) underlying the region (Roy, 2003). It is interesting to note that there is virtually no such high-gravity anomaly field in the entire Deccan Trap volcanic province outside the Kachchh-Saurashtra and the western coast region. In conformity with the interpretation of Takin (1966), it may be suggested that the patches of positive Bouguer gravity anomaly of over +30 mGal values in the aforesaid region are due to the mass of high-density rocks (such as the olivine gabbro) that were emplaced in the Crust as the manifestation of Plume-Lithosphere interaction in the region. Based on the occurrences of alkaline rocks at Sarnu-Dandali and Mundwara in Rajasthan, Kadi, and Netrang in Gujarat and Jawahar and Salsette in Maharashtra, a north-south corridor has been conceived for the southward trace of the Plume outburst as the Indian Shield moved northward over the Reunion hotspot (Basu et al., 1993). The corridor reportedly passes through the Laccadive-Maldive 'oceanic island chain' in the south. The important geomorphotectonic features associated with the passage of the Plume outburst are the formation of the Cambay Graben and the straight running west coast of the Indian Peninsula. There are also several fault lines, which divide the coast-shelf zone into a number of parallel N-S-trending horst-graben-like rift blocks (Roy, 2003). This fault system is possibly linked with the Reunion Plume outburst-related tectonics. The occurrence of linear, high-gravity Bouguer anomaly fields that virtually criss-cross the Indian Shield, especially in the northern part, needs some explanation. Most of these linear geophysical features show parallelism with the zones of important tectonic belts, represented by the Proterozoic orogenic belts and the Proterozoic-Palaeozoic rift basins. In the case of the belt that lies over the Aravalli Mountains, west of Jaipur showing markedly high values of the Bouguer gravity anomaly, there is reported evidence of Mesozoic-Cenozoic underplating event (Roy, 2001; Rao et al., 2002; Sivaraman and Raval, 1995). Extending this interpretation to the other geophysically defined Lineaments (Murthy, 1999), it may be suggested that these (the gravity highs) are the manifestations of Plume-induced underplating events.

REFERENCES

Argand, E., 1924. La tectonique de L'Asie. Proc. Int. Geol. Cong. 7, 171–372.

Baksi, A.K., Marman, T.R., Paul, D.K., Farrar, E., 1987. Widespread early Cretaceous flood basalt volcanism in eastern India: geochemical data from the Rajmahal-Bengal Sylhet Traps. Chem. Geol. 63, 133–141.

Basu, A.R., Renne, P.R., Dasgupta, D.K., Teichmann, E., Poreda, R.J., 1993. Early and late alkali igneous pulses and a high-^{3}He Plume origin for Deccan Flood Basalts. Science 261, 902–906.

Ghatak, A., Basu, A.R., 2011. Vestiges of the Kerguelen plume in the Sylhet Traps, northeastern India. Earth Planet. Sci. Lett. 308, 52–64.

Kanjilal, S.K., 1978. Geology and stratigraphy of the Jurassic rocks of Habo hill, district Kutch (Gujarat). Proc. Indian Natl. Sci. Acad. 44, 1–15.

Kennet, B.L.N., Widiyantoro, S., 1999. A low seismic wave speed anomaly beneath northwestern India: a seismic signature of the Deccan Plume. Earth Planet. Sci. Lett. 165, 145–155.

Kent, R., 1991. Lithospheric uplift in eastern Gondwana: evidence for a long-lived mantle system. Geology 19, 19–23.

Kumar, A., Pande, K., Venkatesan, T.R., 2001. The Karnataka Late Cretaceous dykes as products of the Marion hot spot at the Madagascar-India breakup event-evidence from ^{40}Ar-^{39}Ar geochronology and geochemistry. Geophys. Res. Lett. 28, 2715–2718.

Mahoney, J.J., White, W.M., Upton, B.G.J., Neal, C.R., Scrutton, R.A., 1996. Beyond EM-1: lavas from Afanasy-Nikitin Rise and the Crozet Archipelago, Indian Ocean. Geology 24 (7), 615–618.

Michael, L., Krishna, K.S., 2011. Dating of the 85°E ridge (northeastern Indian Ocean) using marine magnetic anomalies. Curr. Sci. 100, 1314–1322.

Murthy, Y.S., 1999. Images of the gravity field of India their salient features. J. Geol. Soc. India 54, 221–235.

Qureshy, M.N., 1982. Geophysical and landsat lineament mapping—an approach illustrated from west central and South India. Photogrammetria 37, 161–184.

Qureshy, M.N., Midha, R.K., 1986. Deep crustal signature in India and contiguous regions from satellite and ground geophysical data. In: Barazangi, M., Brown, L. (Eds.), Reflection Seismology, the Continental Crust. Geodyn. Ser.Vol. 14. American Geophys. Union, pp. 77–94.

Rao, G.V.S.P., Kataria, P., Roy, A.B., Laksmi, K.J.P., 2002. Reconstitution of Precambrian Aravalli crust at K-T Boundary: evidence from south-central Rajasthan. J. Appl. Geochem. 4, 15–125.

Roy, A.B., 2001. Neoproterozoic crustal evolution of northwestern Indian Shield: implications on break up and assembly of supercontinents. Gondwana Res. 4, 289–306.

Roy, A.B., 2003. Geological and geophysical manifestations of the reunion plume-Indian lithosphere interactions-evidence from Northwest India. Gondwana Res. 6, 487–500.

Roy, A.B., 2004. The phanerozoic reconstitution of Indian shield as the aftermath of break-up of the Gondwanaland. Gondwana Res. 7, 387–406.

Roy, A.B., Chatterjee, A., 2015. Tectonic framework and evolutionary history of the Bengal Basin in the Indian subcontinent. Curr. Sci. 109, 271–279.

Roy, A.B., Chatterjee, A., Chauhan, N.K., 2017. Geological evolution of Kachchh: an epitome of successive Phanerozoic events. Curr. Sci. 112, 1051–1056.

Sivaraman, T.V., Raval, U., 1995. U-Pb isotopic study of zircons from a few granitoids of Delhi-Aravalli belt. J. Geol. Soc. India 46, 461–475.

Subrahmanyam, C., 2000. In: Plume Tectonics in eastern India and the Bay of Bengal: the case of the 85°E Ridge. Proc. Plume Tectonics Workshop: 131h-141h June, 2000, National Geophysical Research Institute, Hyderabad, pp. 54–55.

Sukheswala, R.N., 1981. Deccan basalt volcanism. In: Subbarao, K.V., Sukheswala, R.N. (Eds.), Deccan Volcanism and Related Provinces in Other Parts of the World. Vol. 3. Geol. Soc. India, Bangalore, Mem, pp. 327–339.

Takin, M., 1966. An interpretation of the positive gravity anomaly over Bombay on the west coast of India. Geophys. J. R. Astron. Soc. 11, 527–533.

Torsvik, T.H., Tucker, R.D., Ashwald, L.D., Eidea, E.A., Rakotosolofod, N.A., de Wite, M.J., 1998. Late Cretaceous magmatism in Madagascar: palaeomagnetic evidence for a stationary Marion hotspot. Earth Planet. Sci. Lett. 164, 221–232.

PART

5

GEOLOGY AND EVOLUTION OF THE HIMALAYAS

CHAPTER

THE HIMALAYAS: EVOLUTION THROUGH COLLISION

17

17.1 GEOMORPHOTECTONIC AND GEOPHYSICAL CHARACTERISTICS OF THE HIMALAYAS

The Himalayas constitutes an imposing crescent-shaped mountain range extending for over 2500 km from the south of the Indus Valley beyond Nanga Parbat in the west to Namcha Barwa in the east (Fig. 17.1). The range varies in width from 350 km in the west to 150 km in the east. The majestic mountain chain showing prominent southward convexity stands like a wall bordering the entire northern margin of the Indian Subcontinent.

Geomorphologically, the most unique feature of the Himalayas is the soaring height it attained, crowned with many of the Earth's premier snow-covered peaks, including 10 of the 14 over 8000 m peaks in the world (Fig. 17.2).

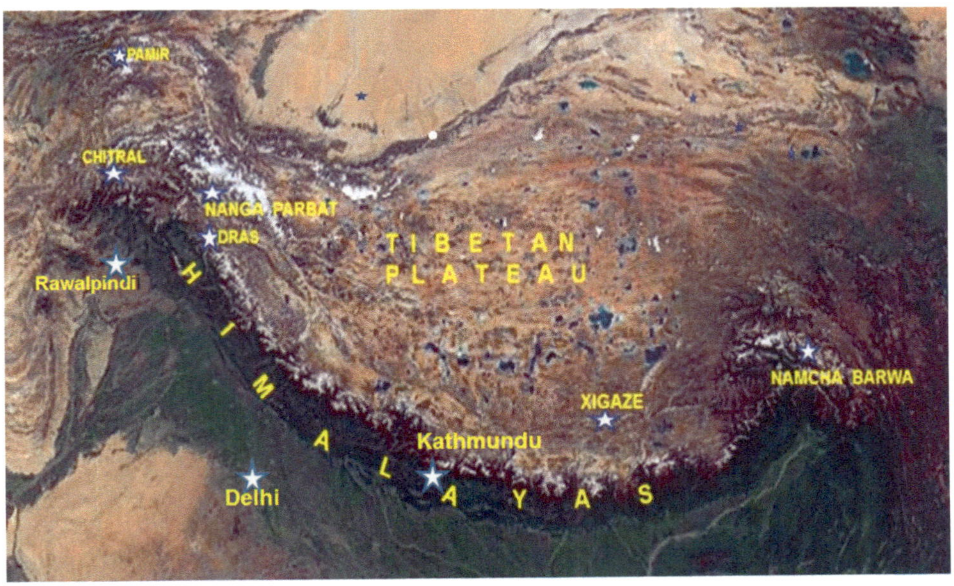

FIG. 17.1

Google Satellite Imagery showing geomorphic traits of the Himalayas and Tibetan Plateau region.

Indian Shield. https://doi.org/10.1016/B978-0-12-809839-4.00018-7

FIG. 17.2

Kanchenjungha, the third highest mountain peak in the world, is one of the crowning glories of the Himalayas.

An important geomorphotectonic character of the Himalayas is the sharp bending of the Himalayas and the associated mountain ranges, which at the western end join with the Sulaiman and Kirthar Ranges in the southwestern end. There is a similar sharp bending at the eastern end, where the mountain range joins the north-trending Indo-Myanmar Range, represented by the Naga Hills and the Arakan Yoma. These two sharp bends on either side of the mountain range are popularly known as the 'Syntaxial bends' of the Himalayan Mountain Range. The most of the high (over 8000m) peaks in the world are located in different parts of the Himalayas but mostly in the central part.

The geophysical characteristics of the Himalayas are as unique as the geomorphotectonic traits of the mountain range. The most distinctive feature of the Himalayas is the crustal thickness, which increases from about 35 to 40 km in the Indus-Ganga-Brahmaputra Plains to between 65 and 80 km over the Higher Himalayas (Qureshy, 1969). The thickness of the continental Crust underlying the Himalayas is reflected in the strong negative gravity anomaly pattern between >−150 and >−350 mGal along the entire length of the mountain belt (Qureshy and Kumar, 1992). The increasing crustal thickness is reflected in the dip of the MOHO, which is estimated to be 7–8°N under the Sub-Himalayas to over 15° further north. A complex architecture of the MOHO with dipping and overlapping segments indicates lithospheric imbrication.

17.2 HISTORY OF GONDWANA BREAK-UP AND NORTHWARD DRIFT OF INDIA

After the break-up of Gondwanaland by around 165 Ma, the Indian lithospheric Crust along with other constituents of the Eastern Gondwana started its northward journey. On the way to the long, almost 7000 km voyage, the 'pristine' Indian terrane passed over four Plume heads: the Kerguelen, Crozet, Marion, and Reunion Plume within a short time span between 117 and 64 Ma. The Plume eruptions resulted in massive flow of mafic lavas and some plutonic intrusions apart from causing the fragmentation and splitting the Eastern Gondwana Precambrian crustal blocks that were moving together. The splitting of Eastern Gondwana involved separation of Australia and Antarctica on the northeastern and eastern side, and Madagascar and Seychelles on the western side of Indian terrane. The separated crustal blocks started moving in different directions opening up the new Indian Ocean that now is situated in between these continental blocks (Roy, 2004).

Records of the earliest Reunion Plume activity, the youngest of all the Plume eruption events, come from the Kohistan Complex that had already become a part of the Karakoram Range in the northeast. This is considered the oldest and the northernmost evidence of impingement of the Reunion Plume (Roy, 2003). The isotope data indicated occurrence of 73.4–72 Ma old alkali pillow basalt in the Kohistan Complex, which shows geochemical similarity to the recent lavas of the Reunion Island in the Indian Mainland (Mahoney et al., 2002). The isotope data from the Reunion Plume-related magmatism in the Kohistan Island Complex indicate that the process of accretion between the Indian crustal block (the Indian Plate) and the Eurasian Plate (assuming Kohistan and Karakoram had already been accreted to it) started before ca. 72 Ma.

17.3 PRECOLLISIONAL ACCRETION OF ISLAND-ARC COMPLEXES WITH THE EURASIAN PLATE

The northward journey of the 'Greater India' continued till its northwest part came close to the margins of Eurasian Plate at around 90 Ma. Much before the end of the northward journey of the Indian Plate, there was development of a chain of Island-Arc Complexes of Kohistan and Ladakh (Fig. 17.3). The arcuate Suture Zone is known as the Shyok Suture Zone (also described as the Northern Suture in the western side).

BOX 17.1

The Indian Precambrian Crustal block, prior to the collision with the Asian Plate, was thought to have an extension in the north between 950 and 550 km from the southernmost base line of the present Himalayas. This pristine Indian Plate had been described as 'Greater India' by Argand (1924), and the extended part now constitutes the terrane of the Himalayas.

The Dras Arc, which formed the link between the Kohistan and Ladakh volcanic Arcs, was also invaded by the granitic magma on a large scale through wide-spread deep-gorging fissures and weak planes that developed along the southern part of the Kohistan-Dras-Ladakh chain of Island Arcs. These volcanic-plutonic chains, which are likely to have evolved between 100 and 70 Ma in the northwest front of the Indian Plate, were added to the Eurasian Plate before the initiation of the major collision between the Eurasian and the Indian Plate (Valdiya, 1998).

The Shyok Suture Zone forming northern terrane boundary of the Island-Arc Complex comprises dominantly olistolithic basalts in the north and LREE-depleted andesites in the south (Rolland et al., 2000). Locally, rhyolite-ignimbrite association has also been reported in the Suture Zone rocks (Fig. 17.4). Based on the combined structural, geochemical, and geochronological study of the Shyok Suture Zone in the Nubra Valley region (Fig. 17.5) in Ladakh, Borneman et al. (2015) suggested evolution of these rocks during 92–85 Ma.

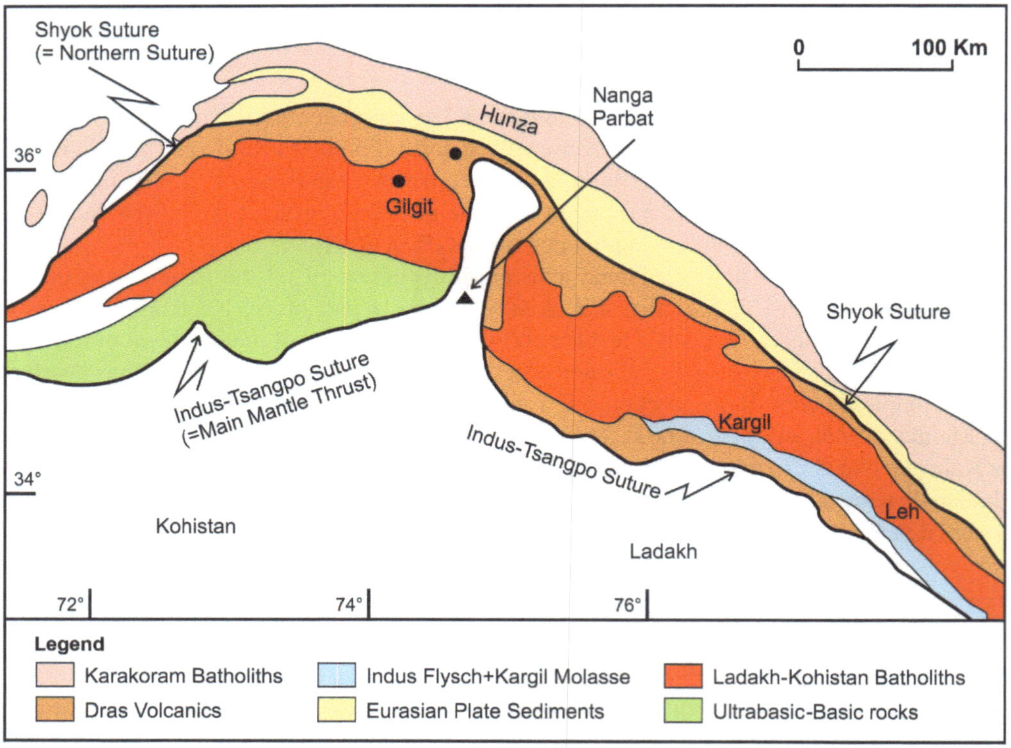

FIG. 17.3

Kohistan and Ladakh terranes, representing Island-Arc Complexes and penetrated by granite batholiths.

Reproduced from Robertson and Collins (2002) with permission.

FIG. 17.4

Field photograph showing occurrence of rhyolite-ignimbrite bearing volcanics from the Shyok Suture Zone at Khardung in Ladakh.

Reproduced from Bhutani et al. (2009) with permission.

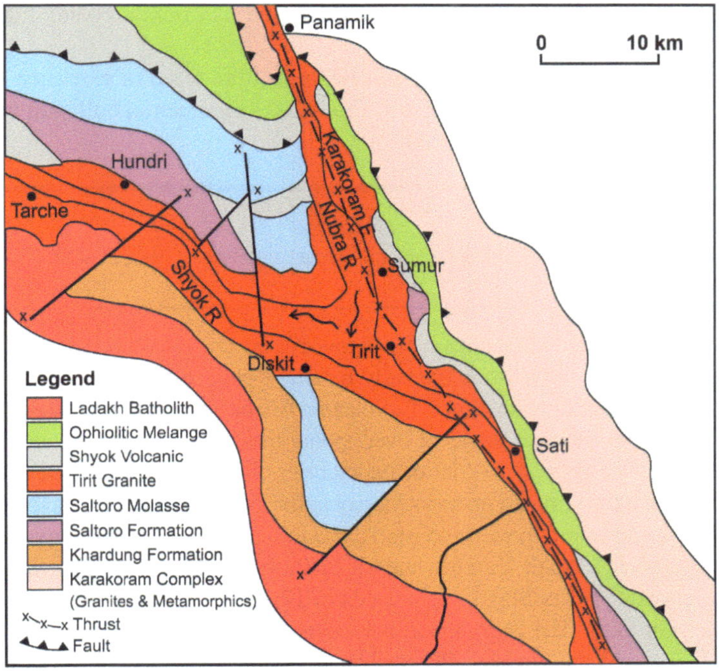

FIG. 17.5

Geological sketch map of the Shyok Suture Zone in northeastern Ladakh.

Reproduced with modification from Upadhyay et al. (1999) with permission.

17.4 INDIA-EURASIA COLLISION: SUTURING OF DISPARATE CRUSTAL BLOCKS

The Karakoram-Kohistan-Afghan terrains in the northwest Himalayas witnessed the initial collision between the Indian and the Eurasian Plates. According to Treloar et al. (1989), the first phase of collision resulted in unobtrusive docking of the Indian Plate along the southern part of the Eurasian Plate between 72 and 65 Ma. A 'land-bridge'-type link was formed between the two Plates that allowed large variety of animals and amphibians including frogs (which were the normal residents of the present-day central Asia) to enter for the first time in the Indian Territory. Sahni (1984) estimated the time of migration of the central Asian animals and amphibians as the Maastrichtian (72–66 Ma).

The duration of collision was, however, long extended. It had initiated in the northwest and ended up in the northeast Himalayas denoting anticlockwise rotation of the Indian Plate. Gibbons et al. (2015) elucidated a multistaged collision history between the Indian and Eurasian Plates:

1. The collision of the Greater India with the Tethyan intraoceanic Arc in Paleocene-Eocene time that led to the final closing of the Tethyan seaway during the middle to late Eocene time. Geological evidence from the collision zone indicates an age of initial Arc-Continent collision at ~52 Ma. The suggested age is consistent with the age of the youngest marine deposits found between India and Eurasia.
2. This was followed by the 'soft' (initial) continent-continent collision between India and Eurasia by ~44 ± 2 Ma. This timing is supported by the marine geophysical data, where the spreading

centres in the Indian Ocean record a drastic decline in the seafloor spreading rates and changes in spreading directions.

3. The abandonment of spreading in the Wharton Basin and the onset of extrusion tectonics in Asia by ~36 Ma are likely indicators of 'hard' (complete) continental collision and highlight the multistage collisional history of this margin.

BOX 17.2

Wharton Basin is the marine area of the northeast quarter of the Indian Ocean. It lies east of the Ninety East Ridge and west of Western Australia.

The continued shoving of the Indian crustal block towards north led to concentration of strain along the zone of collision leading to the stage when the early formed Kohistan-Dras-Ladakh volcanic island complexes together with the deposited sediments of the volcanic Arc basins and the deep seafloor materials were strongly folded and split into multiple stacks of overlapping thrust slices. These intensely deformed zones have been squeezed and pushed out on the continental margins of the colliding blocks. The ocean floor rocks called ophiolites along with the jumbled up of rock association called ophiolite-melange constituted a linear belt that is traceable from Mingora in west Pakistan close to the Afghan border, through Jilal and Sapat in Kohistan, Zilat and Shergol in Ladakh, Darchan and Mt. Kailash (also known as Kangrinboqe '冈仁波齐峰' in Chinese), Xigaze (pronounced as 'Shigatse') in southern Tibet, and finally to Tutting-Tidding belt in eastern Arunachal Pradesh in India. The entire linear belt covering more than 2500 km marks the zone of collision between the Indian and the Eurasian Plates. The ophiolite belt known as the Indus-Tsangpo Suture Zone evolved during the lower Eocene, more precisely between Ypresion and Lutetian when the northward motion of India slowed down from 10 to 5 cm/year. The compression to which Island-Arc Complexes were subjected to gave rise to the development of large isoclinal folding and attendant thrusting. After the collision and the resultant amalgamation of continents, there was no scope of the further northward movement of the Indian Plate. The colliding Indian Crust being less dense than the underlying Lower Crust and/or the Upper Mantle had experienced resistance to slide down under the Eurasian continental collision zone (Valdiya, 1987). The upwarp is discernible all along the length between Nimaling in Ladakh, through Gurla-Mandhata in Kailas-Mansarovar region, and to Laghoi-Kangri belt in Nepal (Fig. 17.6).

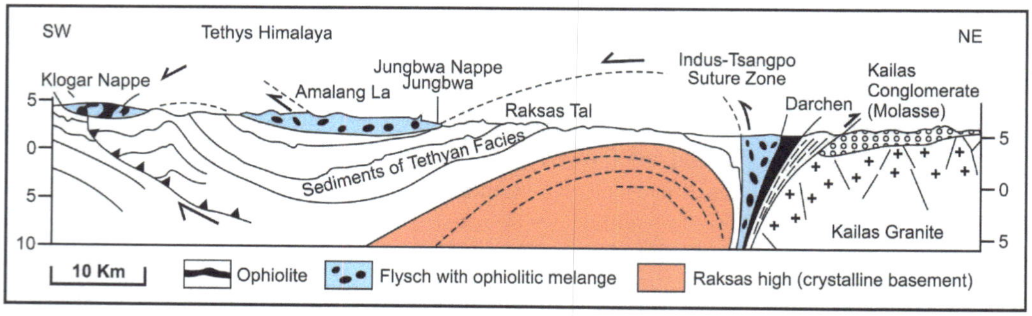

FIG. 17.6

Formation of dome-shaped structure along the leading edge of the Indian Plate near the contact of the Suture Zone. The southward moving thrust sheets that travelled long distances, and formed klippe-like structures.

Based on Gansser (1964).

The impact of the collision of continents in some sectors was so severe that the squeezed-out obducted ophiolites along with the ophiolite melanges of the Indus-Tsangpo Suture Zone were thrust southward between 30 and 80 km away from the roots. The thrust masses made of discrete sheets and nappes piled up one over other forming stacks of klippes (Fig. 17.6).

Significant structural development took place at Kohistan and Ladakh, which were initially separated from each other by a prominent N–S trending promontory of the landmass of the northerly moving Indian Plate. The promontory constituted of Nanga Parbat-Haramosh massif formed the pivot around which the newly accreted Indus-Tsangpo Suture Zone was bent leading to the development of the syntaxial bending of the constituent blocks. The deformation along the syntaxial bending resulted in the development of a complexly deformed tectonized belt around the Nanga Parbat-Haramosh massif region (Fig. 17.7).

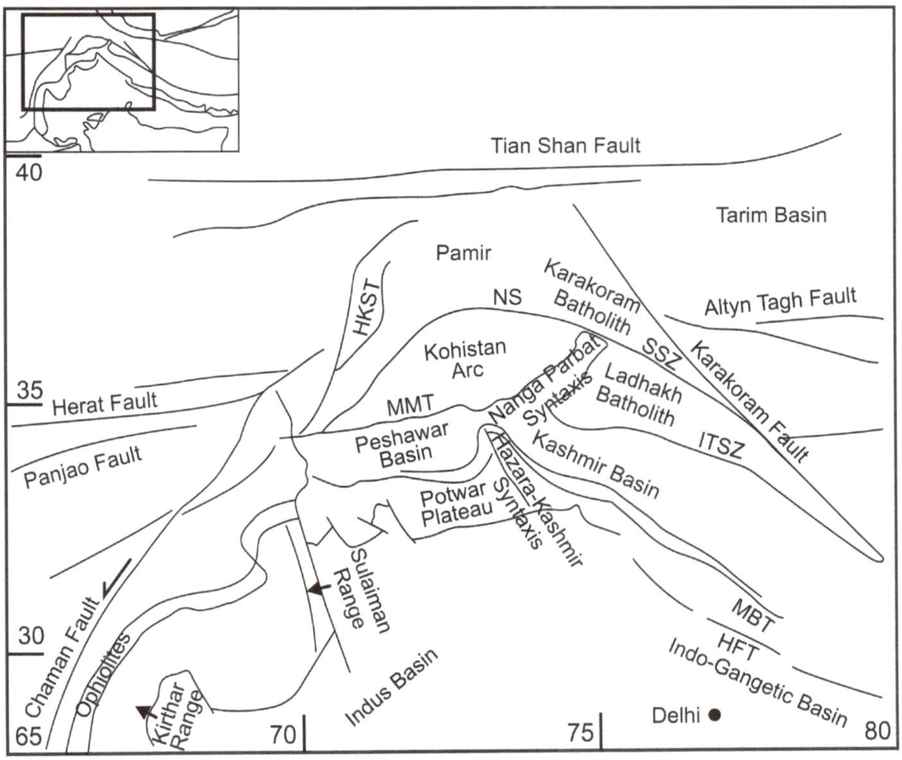

FIG. 17.7

A complexly deformed tectonized belt sandwiched between the Eurasian Plate in the north and the Indian Plate in the south. *HFT*, Himalayan Frontal Thrust; *MBT*, Main Boundary Thrust; *ITSZ*, Indus-Tsangpo Suture Zone; *MMT*, Main Mantle Thrust; *SSZ*, Shyok Suture Zone; *NS*, North Shyok Suture Zone; *HKST*, Hindukush-Kashmir Suture Zone.

17.4.1 SAGGING OF INDUS-TSANGPO SUTURE ZONE

Even as the northern edge of the Indian Plate bulged up, the narrow belt along Indus-Tsangpo Suture Zone sagged down by the late Eocene time between 34 and 30 Ma (Fig. 17.6). The linear belt of depression received sediments comprising alteration of clays, calcareous and siliceous oozes, cherty limestones and greywackes. The entire rock association is described as the Indus Flysch (Sharma and Kumar, 1978; Brookefield and Andrews-Speed, 1984). The flysch comprises, amongst other things, limestone containing Eocene marine foraminifera (Mathur, 1983). The flysch deposits passed on to a thick succession of molasse deposits comprising conglomerate, arkosic sandstone, siltstone and mudstone, deposited in channels as overbank and floodplain deposits (Brookefield and Andrews-Speed, 1984). The depression, which was between 60 and 100 km wide, had formed all along the Suture Zone, which is today occupied by the valleys of the Indus and the Tsangpo Rivers.

BOX 17.3

The depression, between 60 and 100 km wide, that had formed all along the zone of Suture is today occupied by the valleys of the Indus and the Tsangpo Rivers. One reason for sagging in the Indus-Tsangpo Suture Zone (enclosed between lower density rocks on either side of the belt) could be due to shortening of the crustal thickness for attaining isotatic equilibrium.

17.5 GROWTH AND EVOLUTION OF THE MAJOR LITHOTECTONIC DIVISIONS OF THE HIMALAYAS

Conventionally, the following five tectonic divisions are recognized in the Himalayas, which can be followed for over 2500 km (Fig. 17.8) distance between the western Pakistan and the northwestern Myanmar. These, from south to north, are the following:

1. The Sub-Himalayas
2. The Lesser Himalayas

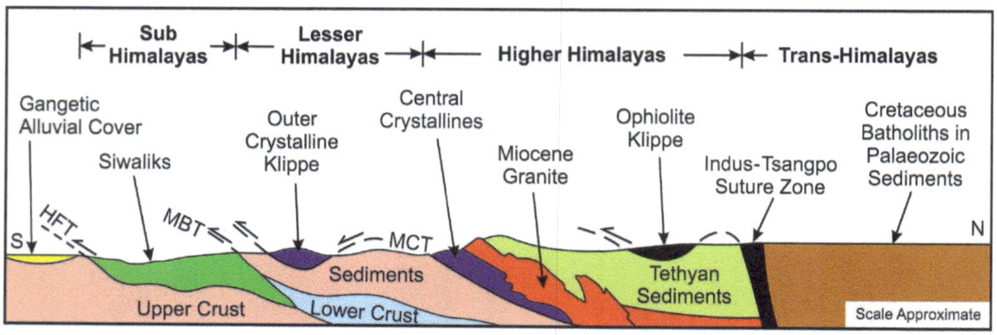

FIG. 17.8

Schematic geological section across the Central Himalayas in Nepal and southern Tibet Plateau. *HFT*, Himalayan Thrust; *MBT*, Main Boundary Thrust; *MCT*, Main Central Thrust.

Reproduced from Windley (1983) with permission.

3. The Higher Himalayas
4. The Indus-Tsangpo Suture Zone
5. The Trans-Himalaya

17.5.1 THE SUB-HIMALAYAS

The Sub-Himalayas form the foothills of the Himalayan Range, and are essentially composed of Miocene to Pleistocene molassic sediments derived from the erosion of the Himalayas. These molasse deposits, known as the Muree and Siwaliks Formations, are internally folded and imbricated. The Sub-Himalayas are thrust along the Main Frontal Thrust over the Quaternary alluvium deposited in the foreland basins. These younger deposits provide evidence of synkinematic deposition features, suggesting that the Himalayas are still an active orogen. The Siwalik sediments were deposited in sag basins that were originated after decoupling along the Main Boundary Thrust. The time of Siwalik Basin formation, deposition, and lithification is dated to be between 18 and 0.2 Ma (Johnson et al., 1983; Ranga Rao et al., 1988). The meandering and encroaching rivers migrated laterally in the flood plains during the fluvial cycle that interspersed with the dry spells (Kumar and Nanda, 1989; Tandon, 1991). The meandering river systems of Lower Siwalik transformed into braided river systems during the Upper Siwalik due to the steepening of slopes because of rise in the altitudes. This also caused significant rise in the rate of sedimentation alongside with the burial rates (Johnson et al., 1983). Siwalik sediments include large number of fossils of different forms and types in almost all stratigraphic horizons.

17.5.2 THE LESSER HIMALAYAS

The Lesser Himalaya bounded in the north by the Main Central Thrust and to the south by the Main Boundary Thrust, mainly comprises rocks of Upper Proterozoic to Lower Cambrian sediments derived from the passive Indian margin intercalated with some granites and acid volcanics of Palaeoproterozoic age (1840 ± 70 Ma). These sediments are thrust over the Sub-Himalayas along the Main Boundary Thrust. The Lesser Himalaya often appears in tectonic windows as at Kishtwar or Larji-Kulu-Rampur windows, within the High Himalaya Crystalline Sequence. The process of formation of this thrust-bounded zone of the Himalayas was initiated during the late Miocene time, when the southern thrust front propagated southward into the foreland region. The Lesser Himalaya comprises dominantly of low-grade metasedimentary rock assemblages of about 20 km thick, and represents an imbricate thrust pile of Mesoproterozoic to Cenozoic sequences originally deposited essentially on the Indian continental Crust. The oldest Phanerozoic rocks that include Cambrian stromatolitic dolomites bear records of epicontinental marine transgression on the Precambrian gneisses. The lowly metamorphosed metasediments show lithological similarity with those of the Higher Himalaya. Much of the Phanerozoic sequence includes sediments of shale-sand-carbonate association deposited in the shallow water shelf-environment. Geochronological studies suggest that most of the Lower Himalaya rocks were located either beneath a palaeo-foreland basin or beneath the Main Central Thrust-related bodies of thrust nappes. Most of the Himalayan thrusting system in the Lower Himalaya developed during the last 15–10 Ma.

17.5.3 THE HIGHER HIMALAYA

The Higher Himalaya also described as the Central Himalayan Domain is considered to form the backbone of the Himalayan 'orogeny' encompassing the areas with the highest topographic relief in the world. It is commonly separated into four zones:

(1) The Higher Himalayan Crystalline Complex (approximately 30 different names exist in the literature to describe this unit; the most frequently found equivalents are Greater Himalayan Sequence, Tibetan Slab, and High Himalayan Crystalline). It is a 30km-thick, medium- to high-grade metamorphic sequence of dominantly of metasedimentary rocks, which is intruded in many places by granitoid of 500 ± 50 Ma and ca. 22 Ma age. It is now generally accepted that the metasediments of the Higher Himalayan Crystallines represent the metamorphic equivalents of the sedimentary series, which forms the base of the overlying Tethys Himalaya. The Higher Himalayan Crystallines form a major nappe that is thrust over the Lesser Himalaya along the Main Central Thrust.

(2) The Tethys Himalaya is an approximately 100 km-wide synclinorium formed by strongly folded and imbricated, weakly metamorphosed sedimentary succession. Several nappes, termed North Himalayan Nappes, have also been described within this unit. An almost complete stratigraphic record ranging from the Upper Proterozoic to the Eocene is preserved within the sediments of the Tethys Himalaya. The transition between the generally low-grade sediments of the Tethys Himalaya and the underlying low- to high-grade rocks of the Higher Himalayan Crystalline Sequence is usually progressive. But in many places along the Himalayan belt, this transition zone is marked by a major tectonic break along the Central Himalayan Detachment System (also known as South Tibetan Detachment System).

(3) The Nyimaling-Tso Morari Metamorphic Dome. In the Ladakh region, the Tethys Himalaya synclinorium passes gradually onto a large dome of greenschist to eclogite facies metamorphic rocks in the north. As with the Higher Himalayan Crystallines, the rocks constituting Nyimaling-Tso Morari Metamorphic Dome represent the metamorphic equivalent of the sediments, which occur at the base of the Tethys Himalaya. The dome is intruded by several bodies of 500 ± 50 Ma old granitoids.

(4) The Lamayuru and Markha Formations constituting a part of the Higher Himalayas comprise flysch and deposits in a turbiditic environment, on the northern part of the Indian continental slope and in the adjoining ocean basin. The age of these sediments ranges from the Late Permian to the Eocene.

The Higher Himalayan Crystallines (also known as the Central Crystallines) comprise two major tectonic units. The upper unit is composed mainly of high-grade psammo-pelitic assemblage associated with anatectic granite, and the lower unit is made up of Precambrian granites, gneisses, amphibolites, metasediments, and their mylonitic equivalents. The entire sequence is recognized as the Munsiari Formation (Roy and Valdiya, 1988). The upper unit of the Higher Himalayan Crystallines makes a huge lithotectonic slab bounded by moderately inclined Main Central Thrust in the south and the steeply dipping Trans-Himalayan Thrust in the north. In small scale, the Higher Himalayan Crystallines show extremely intricate internal deformation resulting from polyphased folding and repeated transposition of foliation planes (Fig. 17.9a and b). The bulk noncoaxial strain in the rocks is attributable to the variation in the case of slip on the shear planes during the thrust propagation. The period of thrusting is coeval with the main phase of progressive metamorphism, anatexis, and attendant granite emplacement during the late Eocene to Lower Miocene. Later deformation and retrogressive metamorphism are related to the uplift during the Middle and Upper Miocene, culminating in the Middle Pleistocene, when the crystalline rocks were brought to a brittle field of deformation from the earlier ductile and brittle–ductile fields.

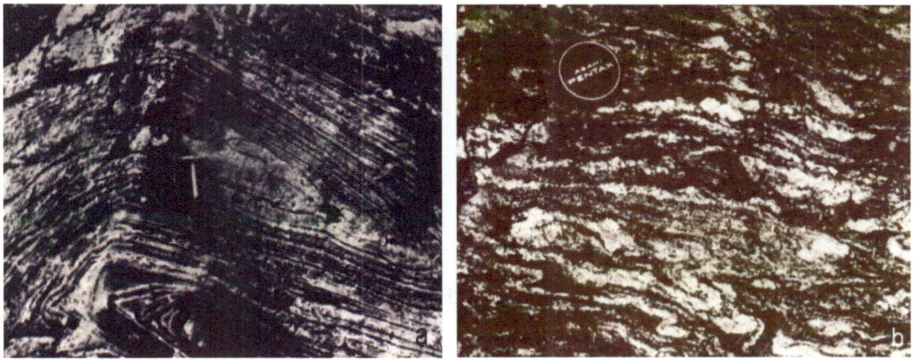

FIG. 17.9

(a) Two generations of isoclinal fold in calc-silicate rock in the Higher Himalayan Crystallines from the Badarinath Temple area. Elliptical structure in the lower part is due to oblique intersection of an early isocline. (b) Small-scale rootless isocline folds, some of which are hook-shaped, in thin layers of psammites and granite enclosed in roughly isoclinal zones of transposed foliation. From Joshimath area in the Bhagirathi Valley.

Reproduced from Roy and Valdiya (1988) with permission.

17.5.4 INDUS-TSANGPO SUTURE ZONE

Tectonically, the most significant event in the evolutionary history of the Himalayas is the development of the Indus-Tsangpo Suture Zone (popularly known by the abbreviated name ITSZ), which occurs south of Kohistan-Ladakh volcanic Arc in the western part of the Himalayas. The Suture Zone extends over about 3000 km from Afghanistan in the west to Myanmar in the east and marks the major zone of collision between the Eurasian and Indian Plates (Gansser, 1991). It includes remnants of oceanic Crust, ophiolitic melanges, flysch units, and continental rocks. The ophiolite zone comprises dunites, peridotites, pyroxenites, and chromitites. These are overlain by the layered anorthosites, gabbros of variable compositions, and basalts. The occurrences of red-coloured chert containing radiolarian fossils are noted along the Indus River section near Leh in Ladakh (Fig. 17.10a).The ophiolite sequence, known as the Nidar ophiolite sequence is well exposed in the Sunda-Nala section in the Ladak region (Fig. 17.10b).

Well-exposed ophilite zone is also observed at other places like Spongtang, and Zaldat Valleys in Ladakh, where also limestone occurs as ‘caught-up’ patches within ‘deep ocean floor’ deposits. The ophiolites in these sections comprise serpentinites, blue schists, gabbros, and pillow lavas.

Chains of felsic plutonic bodies of different dimensions and ages (between 60 and 45 Ma) occur at different places along the Indus-Tsangpo Suture Zone. These are designated as the ‘Andean Type’ collisional granitoids (Le Fort, 1988). These granitoid bodies are thought to have been welled up in great volumes forming stocks, bosses, and large batholithic bodies in successive spurts (Gansser, 1991). A number of felsic and mafic intrusives accompanied the emplacement of the granitoid batholiths (Fig. 17.11).

The structural and kinematic analyses of the Indus-Tsangpo Suture Zone indicate occurrence of superposed deformation fabrics, which first developed during the process of ophiolite emplacement. The continued continental collision between the ophiolite-laden northernmost margin of the Himalayas and the southern part of the Tibetan Plateau of the Eurasian Plate resulted in the back-thrusting and

FIG. 17.10

(a) The Indus River, east of Leh, flowing along the Indus-Tsangpo Suture Zone. The *red colour* in hills is due to the presence of red chert. (b) Nidar ophiolitic melange at the Sumdo-Nala section, Ladakh, showing an exotic caught-up block of limestone (*light-coloured* patches in the middle) within the ultramafic rocks *(green-coloured)*.

Image courtesy: Rajneesh Bhutani.

FIG. 17.11

Intermingling of felsic and mafic intrusive phases in the southern margin of the Ladakh batholith, near Chumathang, Eastern Ladakh.

Image courtesy: Rajneesh Bhutani.

inversion of the structural order along the Indus-Tsangpo Suture Zone during early Paleogene (Figs. 17.12 and 17.13). The ophiolite massifs, ophiolite melanges, and flysch units were thrust northward as nappe sheets during the Eocene. The Indus-Tsangpo Suture Zone, therefore, shows a complex structural anatomy characterized by both south- and north-directed thrust faults and ductile and brittle deformation fabrics (Xu et al., 2015).

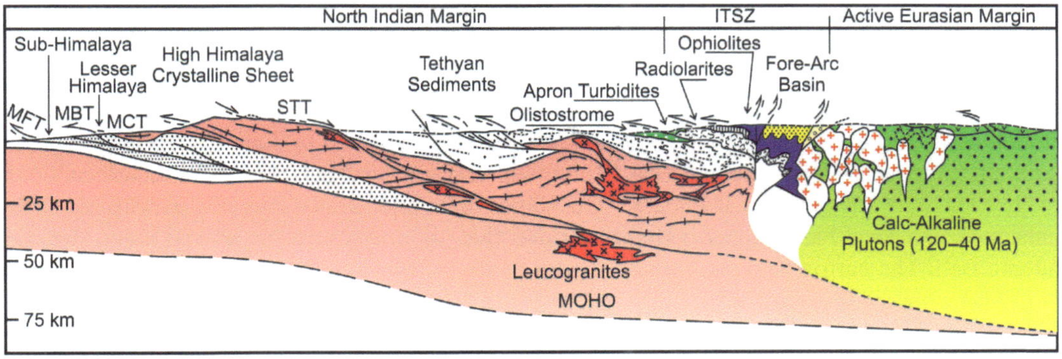

FIG. 17.12

A tectonic section across the Himalayas between the Main Frontal Thrust (also called as the Himalayan Frontal Thrust) in the south and the 'active Eurasian margin' in the north. Compiled and collated from different sources.

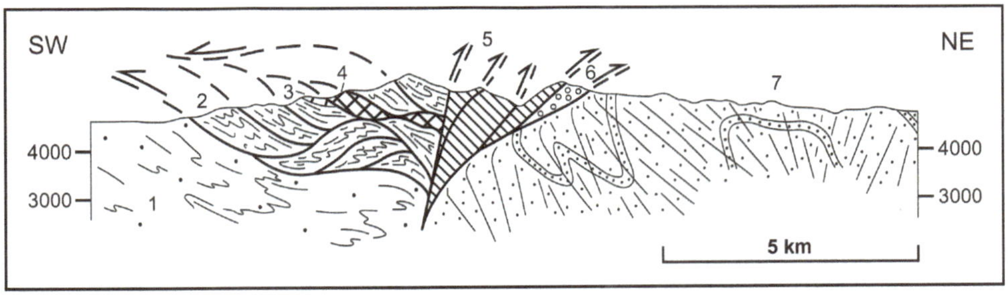

FIG. 17.13

A cross-section of the thrust slices across the Indus-Tsangpo Suture Zone at Saka region in southern Tibet Plateau showing development of a flowerlike structure. (1) Isoclinally folded recumbent folding in north Tethyan sediments showing subhorizontal axial plane cleavages. Thrust slices including radiolarites (2) imbricating with recrystallized limestone (3) and basalt and tuffs (4) and ophiolites (5) backthrusted onto Oligo-Miocene conglomerates (6) and Xizage Formations (7).

Reproduced from Burg and Chen (1984) with permission.

17.5.5 TRANS-HIMALAYAN MAGMATIC ARC

The Trans-Himalaya is a mountain chain just north of Indus-Tsangpo Suture Zone that is a long, linear and complex Trans-Himalaya Plutonic Belt (also known as Gangdese Plutonic Belt) that represents the Cretaceous to early Tertiary Andean-type magmatic arc formed on the southern portion of the Tibet Plateau. The magmatic structures define imbricate plutons making up the calc-alkaline batholith of dominantly gabbroic to granodioritic compositions with subordinate granites. Radiometric ages of the composite rocks range between 175 and 30 Ma. The growing number of geochronological results suggest two major stages separated by a period of relative quiescence: (1) The main magmatic stage is Late Cretaceous (ca. 113–80 Ma), ending with adakitic intrusions, and (2) the second important emplacement episode is the early Paleogene (ca. 65–46 Ma) with a culmination at around 50 Ma. The plutonic belt is similarly divided into two belts: (1) The northern belt essentially consists in Cretaceous, peraluminous, and S-type granites; (2) the southern belt is dominated by Late Cretaceous-Oligocene granitoids with mostly I-type compositions. The southern (Gangdese) belt is overlain by volcanic rocks of equivalent ages (69–43 Ma). There are a few bodies of conglomerates intercalated with red shales, sandstones, and local basaltic and ignimbritic flows that lay unconformably on the Trans-Himalaya Plutonic Belt. The narrow and discontinuous occurrences of few Late Oligocene-Early Miocene (ca. 26–15 Ma) molassic sequence occur at Mt. Kailas (Figs. 17.14 and 17.15).

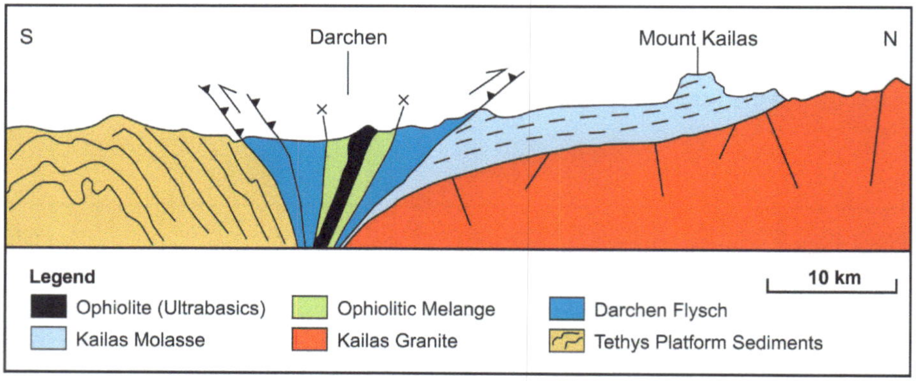

FIG. 17.14

Geological profile section across the Indus-Tsangpo Suture Zone between Darchen and Mt. Kailash.

Based on Gansser (1964).

FIG. 17.15

A majestic view of Mount Kailas in the western part of Tibet showing excellent subhorizontal bedding in molasse deposits, about 15 km north of the Indus-Tsangpo Suture Zone.

17.5.5.1 Chronology of the convergence history

Chronology of the convergence between India and Eurasian Plates based on the Paleomagnetic data from southern Tibet region of Eurasian Plate compared with apparent Polar Wander Paths for the Indian Plate indicates that the intraoceanic Subduction and collision between the two crustal blocks occurred at equatorial latitudes, with progressive suturing during the Maastrichtian to the Danian (72–66 Ma) in the northwestern Himalaya until Ypresian (56–48 Ma) in the eastern Himalaya. This has reportedly produced some Neogene anticlockwise rotational underthrusting of India along the Main Central Thrust. A review of the geological history of the Himalayan-Tibetan Orogen suggests that at least 1400 km of north–south shortening has been absorbed by the zone represented by the Himalayan Mountains since the onset of the Indo-Eurasian collision.

Convergence between Indian and Eurasian Plates continued after their initial contact, up to the present day causing displacement of about 5 cm/year. Several hundred kilometres of shortening and possibly as much as 1500 km took place between the Tethyan Himalaya and the Indian Plate since Paleocene. Large-scale thrust imbrication with propagation of shortening towards the foreland absorbed a major part of this intracontinental shortening. The resulting Himalaya represents a fold-and-thrust wedge formed within the Indian Plate but encompasses a part of the ocean floor deposits of the southern Tibetan marine basins.

The southernmost thrust-bound block is known as the Sub-Himalayas representing the Indian Foreland, bounded in the south by the Main Frontal Thrust that separates the Himalayan domain from the adjacent Gangetic Foredeep in the south.

Seismic studies and the rate of southward advance of the foreland basin indicate that 10–25 mm/year of shortening is currently taken up within the Himalaya. The Himalayan topographic front, that is, the zone where the 'averaged' elevations abruptly change from about 2 km to about 5 km, closely coincides with the Main Central Thrust. This occurrence suggests a genetic relationship between the bending and underthrusting of the Indian Subcontinent beneath the Himalayan blocks and the formation of the Higher Himalaya just north of the Main Central Thrust. The Himalayas consists of a north-dipping pile of tectonostratigraphic units overlain by marine sediments in the north and thrust upon younger foreland basin sediments in the south. Isotopic data suggest that the Higher Himalaya gneisses and schists are composed of late Proterozoic to early Paleozoic rocks representing the upper part of the continental Crust of subducted India. Mesozoic sediments of Tethyan origin, which were originally situated on the northern margin of India, locally occur on the older (possibly Archaean and Palaeoproterozoic) gneiss and schists. The contact that juxtaposes very low-grade to unmetamorphosed hanging wall sediments of the Tethys Himalaya on higher-grade schists and Miocene leucogranites of the Higher Himalayas (also called Main Central Crystalline Sheet) is a north-dipping normal fault. It is responsible for northward shearing of the leucogranites and superposition of unmetamorphosed sediments onto the staurolite-garnet schists. Ductile normal faulting is dated between 24 and 11 Ma. To the east, steeply dipping, brittle normal faults separate Jurassic schists from gneisses indicating several tens of kilometres of down-to-the-north vertical movement. Normal faulting is Miocene to perhaps Pliocene in age and contemporaneous, for at least part of their history, with the structurally lower, southward thrusting on the Main Central Thrust.

REFERENCES

Argand, E., 1924. La tectonique de l'Asie. Congrès Géologique International, Belgique, Comptes Rendues de la 13ème Session, en Belgique 1922. H. Vaillant-Carmanne, Liège, pp. 171–372.

Bhutani, R., Pande, K., Venkatesan, T.R., 2009. Tectono-thermal evolution of the India-Asia collision zone based on the 40Ar-39Ar thermochronology in Ladakh, India. Proc. Indian Acad. Sci. (Earth Planet. Sci.) 113 (4), 737–754.

Borneman, N.L., Hodges, K.V., Van Soest, M.C., Bohon, W., Wartho, J.A., Cronk, S.S., Ahmad, T., 2015. Age and structure of the Shok suture in the Ladak region of northwestern India: implications for slip on the Karakoram fault. Syst. Tectonics 34, 2011–2033.

Brookefield, M.E., Andrews-Speed, C.P., 1984. Sedimentology, petrography and tectonic significance of the shelf flysch and molasse clastic deposits across the Indus Suture zone Ladakh, NW, India. Sediment. Geol. 40, 249–286.

Burg, J.-P., Chen, G.M., 1984. Tectonics and structural zonation of southern Tibet, China. Nature 311, 219–223.

Gansser, A., 1964. Geology of the Himalayas. Interscience Publishers, London. 289p.

Gansser, A., 1991. Facts and theories on the Himalayas. Eclogae Geol. Helv. 84, 33–59.

Gibbons, A.D., Zahirovic, S., Muller, R.D., Whittaker, J.M., Yatheesh, V., 2015. A tectonic model reconciling evidences for collision between India, Eurasia and intra-oceanic arcs of the central-eastern Tethys. Gondw. Res. 28, 451–492.

Johnson, G.D., Opdyke, N.M., Tandon, S.K., Nanda, A.C., 1983. The magnetic polarity stratigraphy of the Siwalik Groups at Haritalyanagar, District Bilaspur, H.P. Himal. Geol. 2, 118–144.

Kumar, R., Nanda, A.C., 1989. Multistoreyed sandbody complexes of the middle Siwalik subgroup of Mohand, Dehradun Valley. J. Geol. Soc. India 34, 597–616.

Le Fort, P., 1988. Granites in tectonic evolution of the Himalaya—a model for its genesis and emplacement. J. Geophys. Res. 86, 10545–10568.

Mahoney, J.J., Duncan, R.A., Khan, W., Gnos, E., McCormick, G.R., 2002. Cretaceous volcanic rocks of the South Tethyan suture zone, Pakistan: implications for reunion hotspot and Deccan Traps. Earth Planet Sci. Lett. 203, 295–310.

Mathur, N.S., 1983. Indus formation of Ladakh Himalaya: its biozonation, correlation and faunal provincialism. In: Thakur, V.C., Sharma, K.K. (Eds.), Geology of the Indus Suture Zonein Ladakh. Wadia Institute of Himalayan Geology, Dehradun, pp. 127–144.

Qureshy, M.N., 1969. The thickening of a basalt layer as a possible cause for the uplift of the Himalayas—a suggestion based on gravity data. Tectonophysics 7, 137–157.

Qureshy, M.N., Kumar, S., 1992. Isostasy and neotectonics of the northwest Himalaya and Foredeep. Mem. Geol. Soc. India 29, 201–222.

Ranga Rao, A., Agarwal, R.P., Sharma, U.N., Bhalla, M.S., Nanda, A.C., 1988. Magnetic-polarity stratigraphy and vertebrate palaeontology of the upper Siwalik subgroup of Jammu hills, India. J. Geol. Soc. India 31, 361–385.

Robertson, A.H.F., Collins, A.S., 2002. Shyok Suture Zone, N. Pakistan: Late Mesozoic-Tertiary evolution of a critical suture separating oceanic Ladakh Arc from the Asian continental margin. J. Asian Earth Sci. 20, 309–351.

Rolland, Y., Percher, A., Picard, C., 2000. Middle Cretaceous back-arc formation and arc evolution along the Asian Margin: the Shyok Suture Zone in the Chogo Lungama-Turmik area (northern Pakistan). Tectonophysics 325, 145–173.

Roy, A.B., 2003. Geological and geophysical manifestations of the reunion plume—Indian lithosphere interactions—evidence from Northwest India. Gondwana Res. 6, 487–500.

Roy, A.B., 2004. The Phanerozoic reconstitution of Indian Shield as the aftermath of break-up of the Gondwanaland. Gondwana Res. 7, 387–406.

Roy, A.B., Valdiya, K.S., 1988. Tectonometamorphic evolution of the Great Himalayan Thrust Sheets in Garhwal region, Kumaun Himalaya. J. Geol. Soc. India 32, 106–114.

Sahni, A., 1984. Cretaceous-Palaeocene terrestrial faunas of India lack of endemism during drifting of the Indian plate. Science 226 (4673), 441–443.

Sharma, K.K., Kumar, S., 1978. Contributions to the geology of Ladakh, northwestern Himalaya. Himal. Geol. 8, 252–287.

Tandon, S.K., 1991. The Himalayan foreland: focus on Siwalik Basin. In: Tandon, S.K., et al. (Eds.), Sedimentary Basins of India. Gyanodaya Prakashan, Nainital, pp. 171–201.

Treloar, P.J., Coward, M.P., William, M.P., Khan, A.M., 1989. Basement-Cover Imbrications South of the Main Mantle Thrust, North Pakistan. Geological Society of America. (Special Paper 232), pp. 137–153.

Upadhyay, R., Sinha, A.K., Chandra, R., Rai, H., 1999. Tectonic and magmatic evolution of the Eastern Karakoram. India. Geodynam. Acta 12 (6), 341–358.

Valdiya, K.S., 1987. Trans-Himadri fault and domal upwarps immediately south of the collision zone: tectonic implications. Curr. Sci. 56, 200–209.

Valdiya, K.S., 1998. Dynamic Himalaya. Universities Press, Hyderabad, pp. 1–178.

Valdiya, K.S., 2010. The Making of India: Geodyanamic Evolution. Macmillan Publishers India Ltd. 816p.

Windley, B.F., 1983. Metamorphism and tectonics of the Himalayas. Geol. Soc. Lond. 140, 849–866.

Xu, Z.-Q., Dilek, Y., Yang, J.-S., Liang, F.-H., Ba, D.-Z., Cai, Z.-H., Li, G.-W., Ji, S.-C., 2015. Crustal structure of the Indus-Tsangpo suture zone and its ophiolites in southern Tibet. Gondwana Res. 27, 507–524.

CHAPTER

HIMALAYAS: POSTCOLLISION EVOLUTIONARY TECTONICS

18

18.1 INTRODUCTION

The Himalayas, the highest, youngest, and the most active continental mountain chain in the world today, includes many fascinating geological and tectonic traits, which began to form about 50 million years ago and continue to evolve even today! In this chapter, we propose to discuss some of those little known aspects that are described as 'controversial' and/or 'debatable'. Regionally, we may have to go beyond the terrain of Indian Subcontinent into Afghanistan, Tajikistan, China, and Myanmar for understanding and comprehending some of these aspects.

18.2 OROCLINAL BENDING OF THE HIMALAYAN ARC

The most characteristic feature about geomorphotectonics of the Himalayas is its arc-shaped pattern bracing the northern margin of Indian Subcontinent (Fig. 18.1). Following up the palaeomagnetic studies from the Himalayan Arc and southern Tibet Plateau, and comparing that with the simulated apparent polar wander paths for the Indian Plate, Klootwijk and Powell (1985) found a consistent pattern of rotations of the Himalayan Arc relative to the Indian Shield. The rotation that varies gradually from 45 degree clockwise in the northwestern Himalayas to slightly anticlockwise in the Lhasa region in southern Tibet is found to be consistent with the model of continental underthrusting of the 'Greater India' beneath the Tibetan Plateau since Early Miocene (Fig. 18.1).

Based on the palaeomagnetic observations, Klootwijk and Powell (1985) proposed a steady-state model for the formation of the Himalayan Arc as follows: (1) collision between the Greater India's northern boundary and the southern Asia that occurred at equatorial latitudes, with progressive suturing from Palaeocene in the northwestern Himalayas until the Early Eocene in the eastern Himalaya; (2) continued convergence and indentation of Greater India into the southern Asia up to the Early Miocene resulted in the southeastward extrusion of Indochina; and (3) the anticlockwise rotational underthrusting of Greater India along the Main Central Thrust during the Neogene, with the Pliocene/Quaternary oroclinal bending of the Himalayan Arc.

BOX 18.1

The term 'orocline' is commonly used for any mountain belt showing a significant curvature.

Indian Shield. https://doi.org/10.1016/B978-0-12-809839-4.00019-9

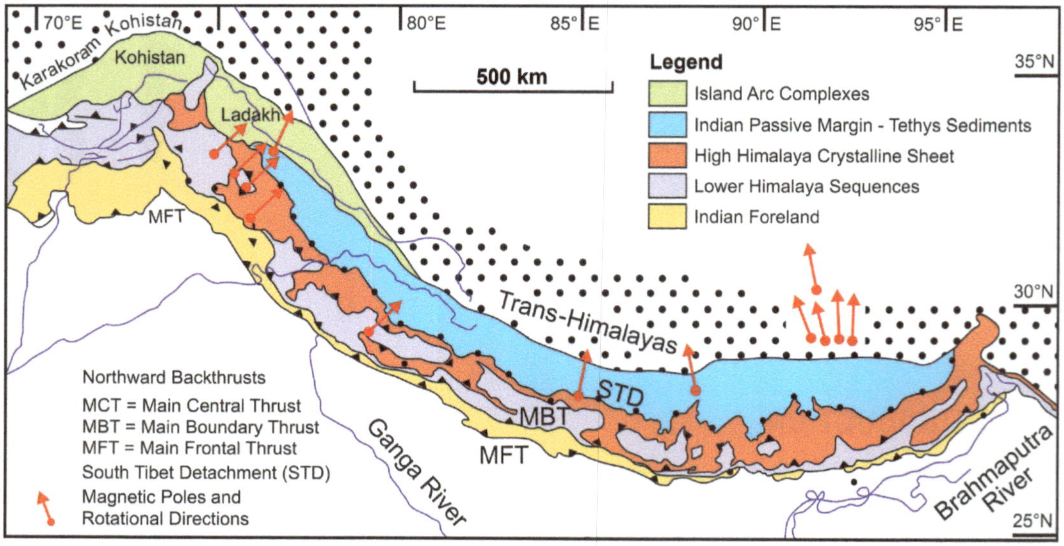

FIG. 18.1

Palaeomagnetic map of the Himalayan Arc and southern Tibet indicating sense of rotation of the segments of the Himalayan Arc and the Lhasa block with respect to the Indian Shield.

Reproduced from Klootwijk and Powell, (1985) with permission.

BOX 18.2

'Steady state' is a state or condition of a system or process that does not change in time or undergo negligible changes over a specified time.

The shape of the Tibetan block, whose southern margin matches excellently with the oroclinal bending of the Himalayas, need also be considered especially for the reason that all the tectonic division of the mountain chains faithfully follow the similar arcuate bending without showing development of any discordant feature like 'syntaxial bending' or any such feature like that in the major part of the eastern and central part of the mountain belt. We cannot rule out the possibility that the Himalayas simulated the preexisting southward curvature of the Tibetan block at least partially.

18.3 HIMALAYAN SYNTAXES AND INDENTER CORNERS

The two Himalayan syntaxes, one formed over the Nanga Parbat in the west and the other around Namcha-Barwa in the east (Fig. 18.2a and b), represent the two most active orogenic pivotal points around which the long, arcuate mountain chain of the Himalayas made sweeping changes in the trend line. The two syntaxes are somewhat different in their regional tectonic frameworks. The eastern syntaxis spans a well-defined 'indenter corner' generated at the eastern edge of the Indian Plate. In contrast to the eastern syntaxis, the tectonics, geomorphic expression, and kinematics of the western Himalayan syntaxis appears more diffuse with regional strain patterns exhibiting negligible vorticity

(Bernard et al., 2000). The western syntaxis shows less influence of the lateral Plate edge than does the eastern syntaxis, where large compressive and shear strains are localized. Despite significant differences in their tectonics, both the eastern and the western syntaxes entrain the Himalaya's two major orogen-traversing rivers: the Indus in the west and the Tsangpo (known as the Brahmaputra as the river drains into the Indian Plains) in the east. Both the rivers cut cross strike gorges through the Himalayas as they turn sharply from the Tibetan Plateau to head southward into the foreland, establishing extreme local relief and facilitating efficient removal of detritus. The cutting of these gorges may be quite recent in age, perhaps due to river capture within the syntaxes that diverted these rivers within the past 10 Ma (Royden et al., 2000).

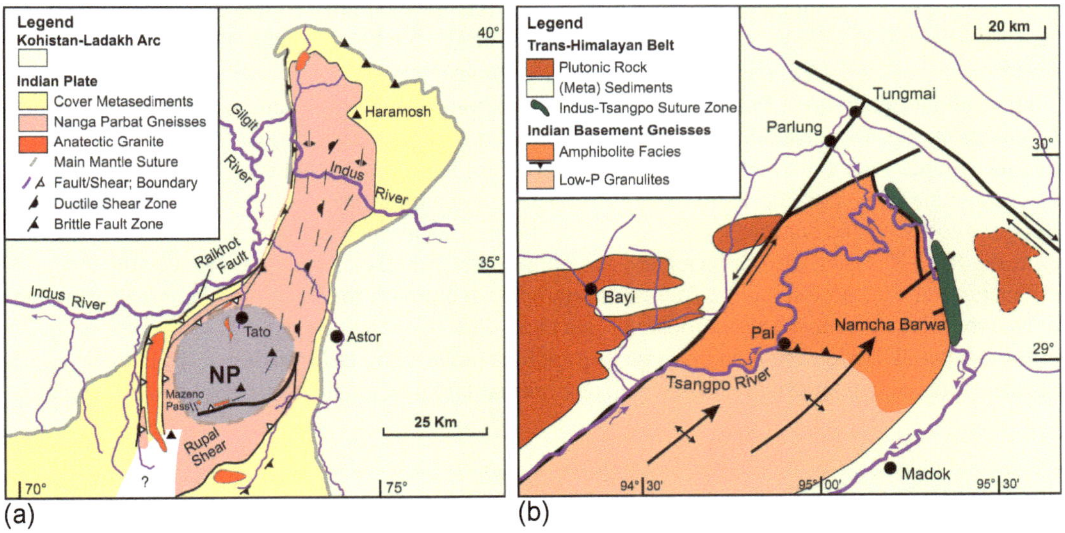

FIG. 18.2

(a) Geological sketch map of Nanga Parbat massif. Grey-coloured region in the core of the massif marks an area characterized by the presence of young (>3 Ma) granites and cordierite-bearing granulites.
(b) Geological sketch map of Namcha-Barwa metamorphic massif. The metamorphic zonation within the massif is approximate.

(a) Reproduced from Schneider et al. (1999) with permission. (b) Reproduced from Burg et al. (1998) with permission.

One additional feature common to both the Himalayan syntaxes is the presence of active metamorphic massif that developed in basement rocks of the Indian Crust. Structural mapping and cooling-age patterns show that the Nanga Parbat massif is a crustal-scale pop-up structure delineated by the active, brittle faults and older shear zones along which the granitoids were emplaced.

The active Namcha-Barwa metamorphic massif in the eastern syntaxis shares several features with the Nanga Parbat. These include rapid exhumation of the antiformal massif, which exhibits metamorphic and structural overprinting during Pleistocene on the Precambrian Indian basement. A spectacular knickpoint has developed where the Tsangpo River crosses the Namcha-Barwa antiform just at a point where local relief becomes the greatest, suggesting that the high rates of differential rock uplift around Namcha-Barwa. Immediate upstream of the knickpoint, the accumulation of a huge amount of sediment is observed along some 30 km stretch of the Tsangpo River. In the downstream part of the knickpoint, the Tsangpo forms a great fall, dropping some 30 m across a bedrock lip. Regional drainage patterns suggest that an ancestral Tsangpo-Irrawaddy river was captured by the Brahmaputra network due to efficient headward cutting.

18.4 OBLIQUE CONVERGENCE, ARC-PARALLEL EXTENSION, AND STRIKE-SLIP FAULTING IN HIGHER HIMALAYA

Arc-parallel extension is an important component of the active deformation pattern of the Himalaya. This extension is accommodated via the arc-perpendicular normal faults. Based on the analysis of ~130 global positioning system geodetic velocities, Styron et al. (2011) suggested about 3 cm/year of arc-parallel extension in the Himalayas during the recent times.

Throughout much of the Phanerozoic, the southern margin of Eurasia remained tectonically active. The collision and the subsequent accretion of continental lithospheric fragments against the Eurasia since the middle Paleozoic have produced the highly deformed Crust that now makes up the orogens of Central Asia (Yin and Harrison, 2000). This is an on-going process. India's late Cretaceous-early Paleogene collision and continued convergence with Eurasia have produced an active deformation zone extending for >2000 km. This deformation has uplifted the Himalayan mountain range and portions of the Tibetan Plateau, producing the highest topographic feature on the Earth, where most of the Indo-Eurasian relative motion is accommodated (Gan et al., 2007).

The arc-parallel extension is an important component of the active deformation of the Himalayas. This extension is accommodated via arc-perpendicular normal faults linked to arc-parallel strike-slip faults.

18.5 RAPID UPLIFT IN HIMALAYAS

As the Himalayan orogenic cycle progressed into the Quaternary, the process of uplift of the Himalayas became quite significant and faster. Cocurrently, there was an increase in the influx rate and volume of sediments, reflected in the later part of the deposition of the Siwalik sediments. This rise was of the magnitude of 3000–4000 m within a very short span of time. The evidence of rapid uplift was also manifested in the fast rate of sedimentation in the Arabian Sea and the Bay of Bengal causing increased heavy mineral distribution in the undersea alluvial fans, showing increased presence of fossils, and signature of early-man settlement in the sources provenance. The Himalayan topography was elevated to much higher heights than the previous levels leading to widespread glaciations. Some movements of younger age have also been reported from the zone of Main Boundary Thrust, which had undergone elevation varying between 200 and 3100 m. Three pulses of uplift have been also identified during Pleistocene in many valleys and gorges of the Himalayas (Valdiya, 1998). Similarly, the Main Central Thrust has also witnessed abrupt rapid uplift of the magnitude of about 3000 m on its southern face (Nakata, 1989). Rapid uplift has also been noticed in the Pir-Panjal (Burbank and Johnson, 1983), in Haramosh Massif in Nanga Parbat (Burbank et al., 1996), and in the Tibetan Plateau (Shackelton and Chang, 1988). The rate of uplift had increased from 3–7 to ~12 mm/year from the Pliocene to the last 60 thousand years (Arita, 1992). The author recorded five episodes of uplifts in different parts of the Himalayas, which are (from the oldest to youngest) (i) at ~45–40 Ma, (ii) at ~40–35 Ma, (iii) at 24–17 Ma, (iv) at 11–7 Ma, and (v) at 3.5 Ma.

18.6 TECTONIC REJUVENATION OF THE HIMALAYAS

Tectonic resurrection and geomorphic rejuvenation led to emergence of the Himalayas as the most prodigeous orographic feature on Earth. Due to the uninterrupted uplift and erosion, there was continuous exhumation of the Greater Himalayas that resulted in the development of extremely rugged topographic features. On the contrary, the Lesser Himalayas was still at low altitudes at the onset of the Quaternary. The tranquillity that persisted during this period of Himalayan tectonics is consistent with the development of features like meandering of Siwalik rivers with the formation of braided flood plains. The Tibetan terrain during that time remained as a low rising peneplained feature with an average elevation of 900 m (Shackelton and Chang, 1988).

The later reactivation along the Main Boundary Thrust and the uplift in the Lesser Himalayas dramatically changed the terrain into 'rugged mountain rampart' topography. This upheaval continued further along with the rise of the Himalayas to higher altitudes by a magnitude of 3–4 km (Gansser, 1991). The tectonic resurrection of the Himalayas along the reactivated faults due to tectonic rejuvenation caused dramatic changes in topography through swinging of the rivers, which ultimately led to the development of highly abraded terrains. The rapid exhumation of the metamorphic massifs caused tectonic rejuvenation especially in the syntaxial zones. The exhumation or uncovering of the massifs is a joint creation of the metamorphic as well as the surface erosional processes which were most pronounced in the Nanga Parbat and the Namcha-Barwa Syntaxes (Zietler, et al., 2001). Such an erosion-related metamorphic exhumation also explains how the massifs developed atop a weak Crust that is hot, dry, and thin (Park and Mackie, 2000). It has been stated that in various kinds of tectonics-related aneurysm like emplacement of vapour-absent anatectic melts during erosional exhumation, exposure of young low-pressure granulites, development of structural relief via growth of antiformal folds and thrusting, formation of a vigorous metamorphic-meteoric hydrothermal system, upward advection of isotherms as evidenced by a shallow brittle to ductile transition, and the development of generally hot resistive Crust are all consistent with the advection of deep crustal material into a relatively weak crustal zone (Zietler et al., 2001). Wobus et al. (2003) speculated that the continuous exhumation observed in the hinterland of Himalayas (such as Nepal) is correlatable with the intense and sustained erosion driven by focussed orographic precipitation at the foot of the Higher Himalayas.

18.7 THE CONUNDRUM OF INVERTED METAMORPHISM

The juxtaposition of the Higher Himalayan Crystallines and Lesser Himalayan Formations across the Main Central Thrust is associated at most locations in the Himalayas with an increase in metamorphic grade with higher structural position (i.e. shallower depth). The Higher Himalayan Crystallines vary substantially in thickness across the entire Himalayas. This has been explained by different authors as: (1) the presence of variable initial thickness, (2) the Main Central Thrust cutting up section at certain locations, and (3) the imbrication within the hanging wall of the Main Central Thrust. In the central Himalaya, where the structural thickness of the Higher Himalayan Crystalline Complex is between 8 and 12 km, the inverted metamorphic section reaches sillimanite grade (Fig. 18.3).

Thermobarometric studies of the Greater Himalayan Crystalline Complex indicate a general decrease in pressure with increasing distance above the Main Central Thrust. Typically, pressures of 7–8 kbar were achieved adjacent to the Main Central Thrust (kyanite grade), whereas peak pressures at the structurally highest levels were only about 3–4 kbar (corresponding to the sillimanite grade).

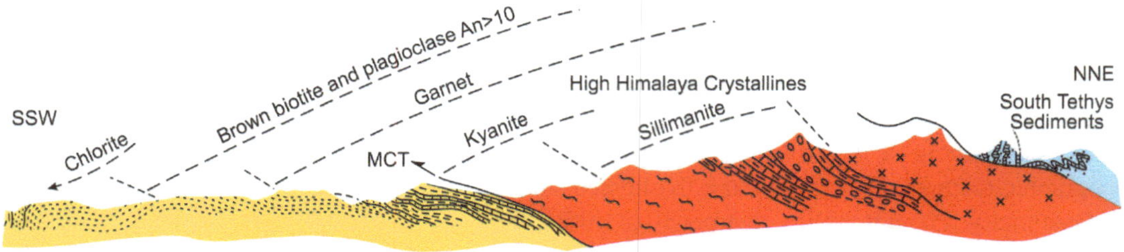

FIG. 18.3

Simplified geologic and metamorphic cross section through the Himalayas in central Nepal.

Reproduced from Le Fort (1986) with permission.

Metamorphism within the Lesser Himalayan Formations increases from zeolite to kyanite grade over a (north-south) distance of 20 km. The region approximately bounded by the garnet isograd in the Lesser Himalayan Formations and the hanging wall gneisses of the Greater Himalayan Crystalline Complex is typically characterized by a highly sheared, 4–8 km thick zone of distributed deformation with a top-to-the-south shear sense, referred to as the Main Central Thrust Zone.

Based on detailed structural and metamorphic studies, Gupta et al. (2010) proposed that the Main Central Thrust is almost certainly responsible for the formation of the inverted metamorphic sequence in the Sikkim Himalayas. According to Dasgupta et al. (2004), the metapelitic rocks of the Sikkim Himalayas show an inverted metamorphic sequence of the complete Barrovian Zones from chlorite to sillimanite + K-feldspar, with the higher grade rocks appearing progressively at the higher structural levels (Fig. 18.4).

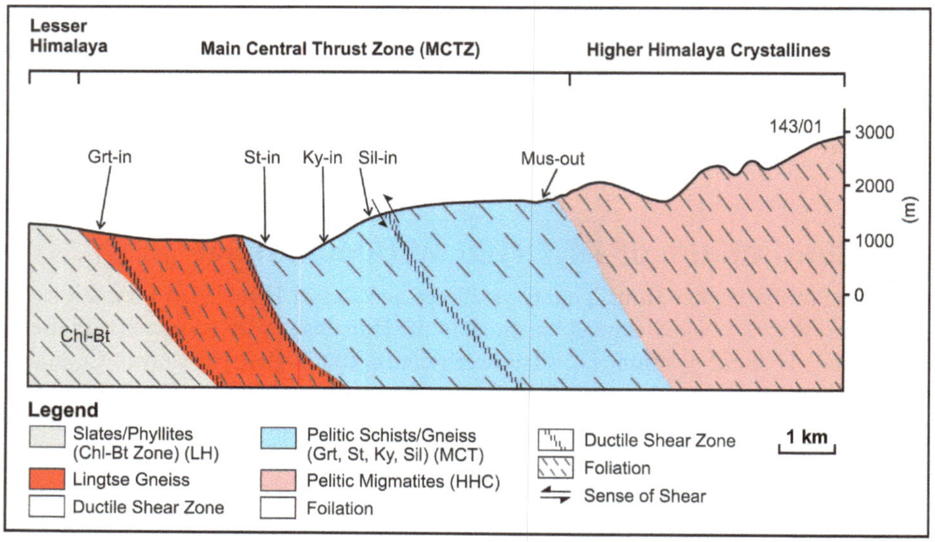

FIG. 18.4

Schematic cross section along approximately east-west transect in eastern Sikkim, illustrating disposition of the Barrovian isograds.

Reproduced from Dasgupta et al. (2004) with permission.

Dasgupta et al. (2004), however, insisted on the careful determination of peak metamorphic P-T conditions on the basis of 'judicious thermobarometric and phase-equilibrium analyses' coupled with detailed structural studies for developing acceptable model for the origin of the 'inverted metamorphic sequence' in the Himalayas.

Harrison et al. (1999), who have analysed a large number of publications on the aspect of inverted metamorphism in the Himalayas, suggest that the key to comprehending the tectonic evolution is to understand the relationships between large-scale faulting (thrusting), anatexis, and inverted metamorphism. After analysing numerous models that have been advanced by different authors to explain the relationship of inverted metamorphism, Harrison et al. (1999) preferred to group all these propositions into four different types:

1. Inverted metamorphism developed within the footwall of the Main Central Thrust and anatexis in the hanging wall are spatially and temporally related by thrusting (Fig. 18.5a).
2. Thrusting results from anatexis (Fig. 18.5b).
3. Anatexis results from normal faulting.
4. Apparent metamorphic inversion in the footwall of Main Central Thrust produced by deformation of two right-way-up metamorphic sequences (Fig. 18.5c).

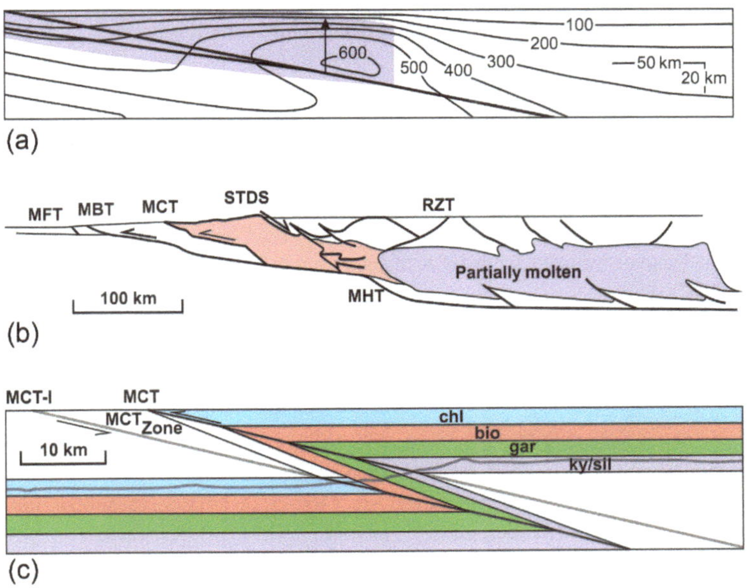

FIG. 18.5

Schematic illustrations of different models proposed to explain the relationship of inverted metamorphism and/or anatexis to large-scale faulting within the Himalaya. (a) Type 1: Inverted metamorphism developed within the footwall of the MCT and anatexis in the hanging wall are spatially and temporally related by thrusting. The direction of material transport is shown by the line with the arrowhead, and the shaded region corresponds to Crust accreted to the hanging wall. (b) Type 2: Thrusting results from anatexis. The illustration shows a partially molten Tibetan middle Crust that is being actively extruded southward. (c) Type 4: Apparent inverted metamorphism in the footwall of the MCT is produced by deformation of two right-way-up metamorphic sequences.

Reproduced from Harrison et al. (1999) with permission.

Based on the critical examination of the various models, Harrison et al. (1999) found that almost all the models referred by them suffer from serious constraints, especially in those that seem to hinge on the faulty assumption that inverted metamorphism and anatexis are synchronous in time. The best model, according to the authors, is a variant of the type 4 model (Fig. 18.5c) that suggests that the inverted metamorphism in the footwall of the Main Central Thrust is 'apparent' produced by deformation of two right-way-up metamorphic sequences in which continuous convergence is manifested by episodic magmatism and out-of-sequence thrusting. This conclusion has a reflection on the thought enshrined in the model presented by Dasgupta et al. (2004) from the Sikkim Himalaya, who mentioned that the available models with some modification are adept to explain the observations on the inverted metamorphism in the Himalayas.

REFERENCES

Arita, K., 1992. Crustal movements of the Nepal Himalaya. In: Proc. Symp. Him. Geo. Shimane University, Matsue, pp. 3–5 (Abstract).

Bernard, M., Shen-Tu, B., Holt, W.E., Davis, D., 2000. Kinematics of active deformation in the Sulaiman Lobe and Range, Pakistan. J. Geophys. Res. 105, 13,253–13,279.

Burbank, D.W., Johnson, G.D., 1983. The late Cenozoic chronologic and stratigraphic development of the Kashmir intermontane basin, northwestern Himalaya. Palaeogeogr. Palaeoclimatol. Palaeoecol. 43, 205–235.

Burbank, D.W., Leland, J., Fielding, E., Anderson, R.S., 1996. Bedrock incision, uplift and threshold hillslopes in the northwestern Himalaya. Nature 379, 505–510.

Burg, J.-P., Nievergelt, P., Oberli, F., Seward, D., Davy, P., 1998. The Namche Barwa syntaxis: evidence for exhumation related to compressional crustal folding. J. Asian Earth Sci. 16, 239–252.

Dasgupta, S., Ganguly, J., Neogi, J., 2004. Inverted metamorphic sequence in the Sikkim Himalayas: crystallization history, P–T gradient and implications. J. Metam. Geol. 2004 (22), 395–412.

Gan, W., Zhang, P., Shen, Z., Niu, Z., Wang, M., Wan, Y., Zhou, D., Cheng, J., 2007. Present-day crustal motion within the Tibetan Plateau inferred from GPS measurements. J. Geophys. Res. 112, B08416. https://doi.org/10.1029/2005JB004120.

Gansser, A., 1991. Facts and theories of Himalayas. Eclogae Geol. Helv. 84, 33–59.

Gupta, S., Dasi, A., Goswami, S., Modak, A., Mondal, S., 2010. Evidence for structural discordance in the inverted metamorphic sequence of Sikkim Himalaya: towards resolving the Main Central Thrust controversy. J. Geol. Soc. India 75, 313–322.

Harrison, M., Grove, M., Lovera, O.M., D'andrea, 1999. The origin of Himalayan anatexis and inverted metamorphism: models and constraints. J. Asian Earth Sci. 17, 755–772.

Klootwijk, C.T., Powell, C.M.A., 1985. The Himalayan Arc: large-scale continental subduction, oroclinal bending and back-arc spreading. Earth Planet. Sci. Lett. 75, 167–183.

Le Fort, P., 1986. Metamorphism and Magmatism During the Himalayan Collision. Geol. Soc. London 19, 159–172.

Nakata, T., 1989. Active faults of the Himalaya of India and Nepal. In: Malinconico, L.L., Lillie, R.J. (Eds.), Tectonics of Western Himalaya. Geol. Soc. Am., Boulder, pp. 243–264.

Park, S., Mackie, R., 2000. Resistive (dry?) lower crust in an active orogen, Nanga Parbat, northern Pakistan. Tectonophysics 316, 359–380.

Royden, L.H., Clark, M.K., Whipple, K.X., Burchfiel, B.C., 2000. River incision and capture related to tectonics of the eastern Himalayan Syntaxis. Eos. Trans. AGU 81, S413.

Schneider, D.A., Edwards, M.A., Kidd, W.S.F., Khan, M.A., Seeber, L., Zeitler, P.K., 1999. Tectonics of Nanga Parbat, western Himalaya: Synkinematic plutonism within the doubly vergent shear zones of a crustal-scale pop-up structure. Geology 27, 999–1002.

Styron, R.H., Taylor, M.H., Murphy, M.A., 2011. Oblique convergence, arc-parallel extension, and the role of strike-slip faulting in the High Himalaya. Geosphere 7, 1–15. https://doi.org/10.1130/GES00606.

Shackelton, R.M., Chang, C., 1988. Cenozoic uplift and deformation of the Tibetian Plateau: the geomorphological evidence. Philos. Trans. R. Soc. Lond. A 327, 365–377.

Valdiya, K.S., 1998. Dynamic Himalaya. Universities Press, Hyderabad, 1–178.

Wobus, C.W., Hodges, K.V., Whipple, K.S., 2003. Has focussed denudation sustained active thrusting at the Himalayan topographic front? Geology 31 (10), 861–864.

Yin, A., Harrison, T.M., 2000. Geologic evolution of the Himalayan-Tibetan orogen. Annu. Rev. Earth Planet. Sci. 28, 211–280. https://doi.org/10.1146/annurev.earth.28.1.211.

Zietler, P.K., Meltzer, A.S., Koons, P.O., Craw, D., Hallet, B., Chamberlain, C.P., Kidd, W.S.F., Park, S.K., Seeber, L., Bishop, M., Shroder, J., 2001. Erosion, Himalayan geodynamics and the geomorphology of metamorphism. Geol. Soc. Am. Today 4–9.

PART

6

QUATERNARY GEOLOGY AND SEISMICITY

CHAPTER

QUATERNARY: A PERIOD OF DRAMATIC MORPHOTECTONIC CHANGE IN INDIAN SHIELD

19

19.1 INTRODUCTION

Quaternary represents the youngest period in the evolutionary history of the Earth. It is divided into an older Pleistocene and a younger Holocene Era. Out of these, the Pleistocene covers the major time span, about 1.81 million years. The youngest Holocene that began about 10,000 years ago has an open-ended upper limit. Dramatic morphotectonic features are recorded during the Quaternary in the Indian Shield that includes deposition of thick layers of sediments, reshaping of landmasses, gross alteration in subsurface physical character, upliftment, and reactivation of faults. Scattered occurrences of Quaternary deposits are known from different parts of the Indian Subcontinent, fringing the entire Peninsular India, all along the western and eastern coastlines. Beyond the coastal regions, thick Quaternary deposits occur in the central part of India along the Narmada and Tapti Valleys, in the Bengal Basin, and in parts of Rajasthan and Gujarat. In the Himalayas, the Quaternary deposits form the younger part of the youngest Siwalik Group. The outcrops of the Siwalik Group are continuously traceable south of the Main Boundary Thrust, almost all along the east-west spread of the Himalayas.

The Quaternary geology is well accounted in various literatures. The present discussion is intended to highlight the stunning geomorphological changes observed in different parts of the Indian Subcontinent, which grossly altered the surface physical character of the Indian landmass, unknown in any other 'Shield' areas of the world. Significantly, the entire reshaping process was operative during the last phase of the Earth's history, more specifically during Holocene, which began about 10,000 years ago, and much of these changes were witnessed by 'man' (Valdiya, 2010). According to Kale (2014), the Indian Subcontinent encompasses all the different geomorphic landscape features that are found on the present day Earth's surface. The geomorphic diversity shown in two topographic sections of the Indian Subcontinent is quite revealing (Fig. 19.1).

BOX 19.1

The Neolithic was the period in the development of human expertise beginning at around 15,000 BC.

Indian Shield. https://doi.org/10.1016/B978-0-12-809839-4.00032-1

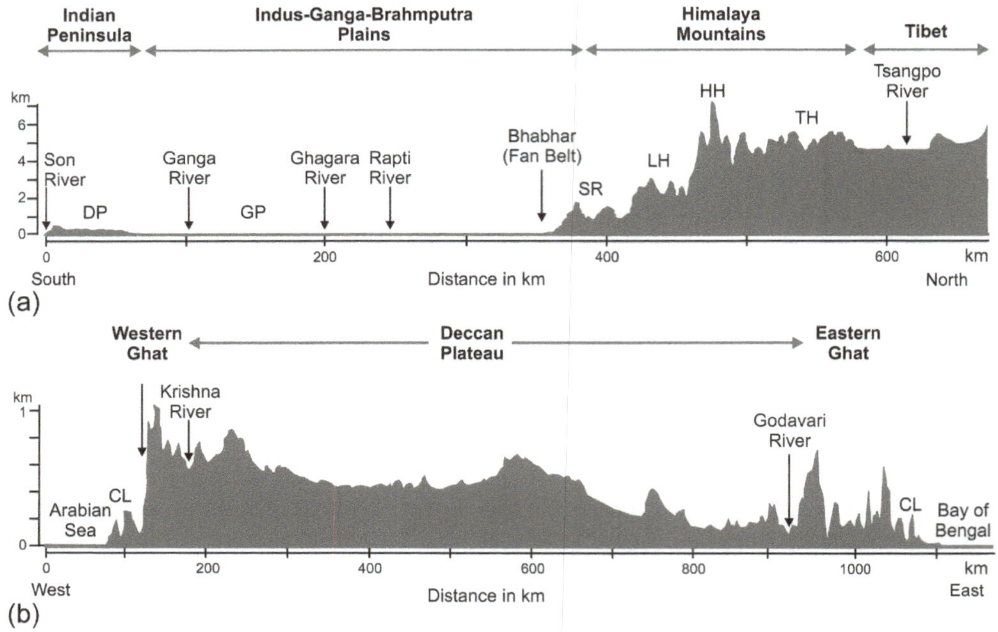

FIG. 19.1

The fault-controlled topographic section of the Indian Subcontinent as shown in two profiles (a) the north-south profile approximately along 83°03′ longitude and (b) the east-west profile approximately along 17°03′ latitude.

Reproduced from Kale (2014) with permission.

Tectonically, the deposition of almost all the Quaternary formations took place in fault-bounded basins excluding, however, the most recent soils and alluvial deposits. Apart from the wide-open valleys and flat plains, uplifted residual hills, inselbergs (isolated mountains and hills), and hill ranges, the geomorphology of much of the Peninsular India is dominated by the presence of a number of plateaus and high-level surfaces at different elevations, at places, occupied by lateritic duricrust. The differential uplift and reactivation of different faults took place in the late Neogene and Quaternary periods (Kale and Vaidyanadhan, 2014).

19.2 LATE QUATERNARY TECTONICS IN HIMALAYAS

The Quaternary tectonics in the Himalayas represents the major morphogenic phase in the evolution of the mountain (Gansser, 1964). During the early Holocene, the Lesser Himalayas was a gentle terrain of low relief, less than a thousand metres above sea level. Evidences indicated that the Stone-Age people across the Himalayan domain freely migrated across both, north and south of this terrain of subdued topography (Valdiya, 2010). It was during this period that the Himalayan ranges started to rise to attain its present spectacular height and form (Valdiya, 1993). The final uplift in the geological history was observed by our early ancestors during the pre-Holocene Neolithic time (Valdiya, 1998).

The lofty mountain barriers of the flat and matured topographic belts developed after the Neolithic times (between 15,000 and 4500–2000 BC) on the pre-existing faults of the Main Boundary Thrust. The uplifting of the 'matured topographic terrains' took place episodically, leading to geomorphic rejuvenation of the modern Himalayas (Valdiya, 1998). Intermittent movement on these fault/thrust surfaces has resulted in the ponding of rivers with the formation of lakes upstream, along active dislocation zones.

The geomorphic rejuvenation of the Lesser Himalaya had left markings in different parts of the rising Himalayan ranges, in the form of tilting and upliftment of the younger sediments adjacent to the reactivated Himalayan Frontal Thrust (marked as HFF in Fig. 19.2a). In the valley of Kosi River in the Kumaun Himalaya, tilted gravel bed is overlain by fluvial terraces at three successive higher levels implying that after tilting of Dun Gravel, at least three phases of uplifting occurred during late Pleistocene and early Holocene time (Fig. 19.2b). In another instance as observed in the Kosi River valley, the oldest bed is perched at 67 m above the bed of Ramganga at Kalagarh in southeastern Garwal (Fig. 19.2c).

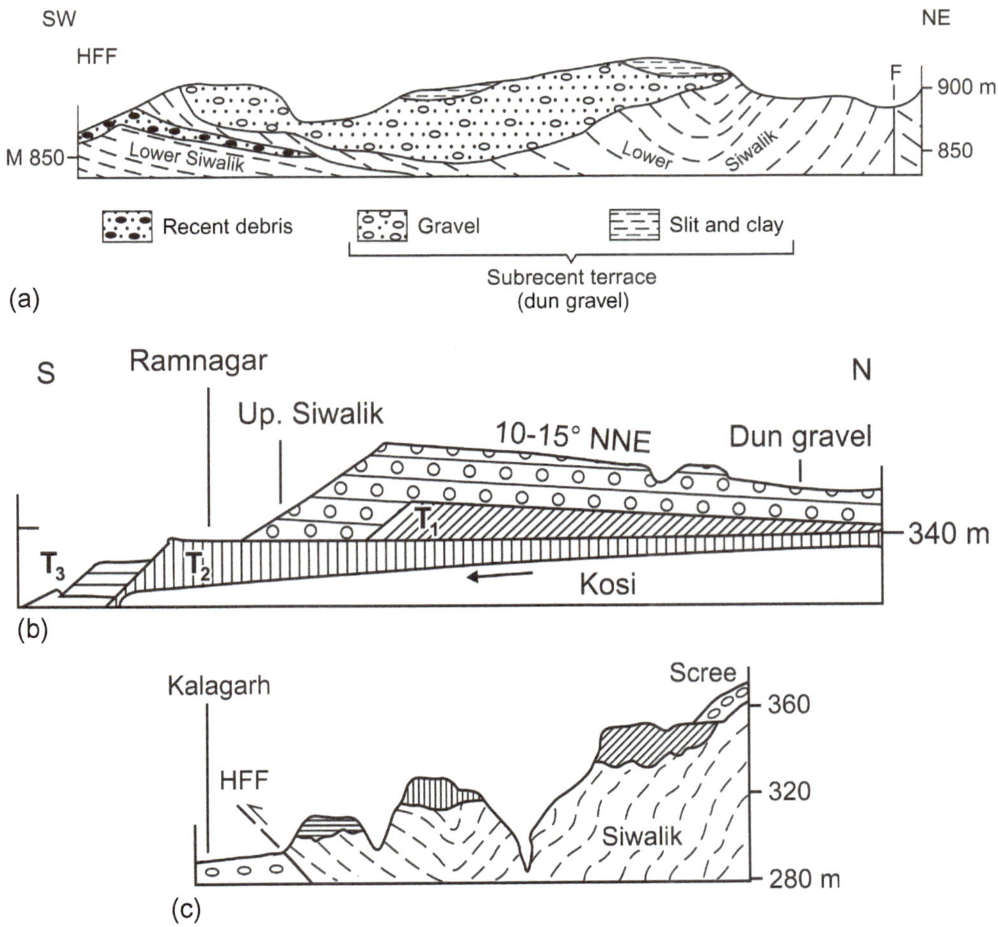

FIG. 19.2

(a–c) Evidence of deformation and uplift on the Himalayan Frontal Thrust (HFF in figures) during the late Quaternary uplift of recent sediments. T1, T2, and T3 marks three levels of uplift during the late Quaternary.

Reproduced from Valdiya (1993) with permission.

There are several examples from the Himalayas that indicate extensive geomorphic rejuvenation in the Quaternary. This is indicated by the presence of three levels of terraces formed corresponding to the three spurts during Pleistocene (Fig. 19.3). At places where a fault crosses a river section, more than three levels are also observed. The terrace has yielded pieces of pottery of the historical Kushan period, dating back to the second to fourth century AD (Valdiya, 1998).

FIG. 19.3

Picture showing development of river terrace near the bending of meandering of a Himalayan river.

19.3 ACTIVE TECTONICS IN PENINSULAR INDIA

The gross physiographic character of the Indian Peninsula is considered a 'single-plateau' geomorphic entity, though there are significant diversities not only in its geological constitution but also in the physiographic types and character. There are several mountain ranges that criss-cross the entire belt (Fig. 19.4), which did not evolve simply as fold mountains. The only exception is the Aravalli Mountain that had a long Precambrian evolutionary history and has undergone upliftment related rejuvenation during late Quaternary (Roy and Jakhar, 2002).

The topography of individual geomorphic entities is generally undulating with prominent granite hills having dome-like appearances in outlines. However, there are quite a few flat-topped hills having mountainous height that cannot be described as orogenic mountain. On the other hand, these mountainous hills appear to have retained the pristine horizontality of bedding or other depositional features even in those that show considerable vertical uplifts (Fig. 19.5). Added to these is the invariable presence of scarp faces and development of water falls especially when such faces cut across the river valleys and stream channels.

The features described above provide unmistakable evidence that the major landforms (valleys, high-level surfaces, and lateritic and duricrusted landforms) could not have developed merely as the legacy of long history of prolonged weathering and erosion. On the other hand, much of the mega- and the microscale geomorphic traits can be attributed to the more-recent block uplift-type tectonic activities that affected the entire Indian Peninsula. The evidence of ancient gently rolling, almost featureless 'peneplain' surfaces marking the top of uplands areas like plateaus and mountains provides proofs of the prolonged

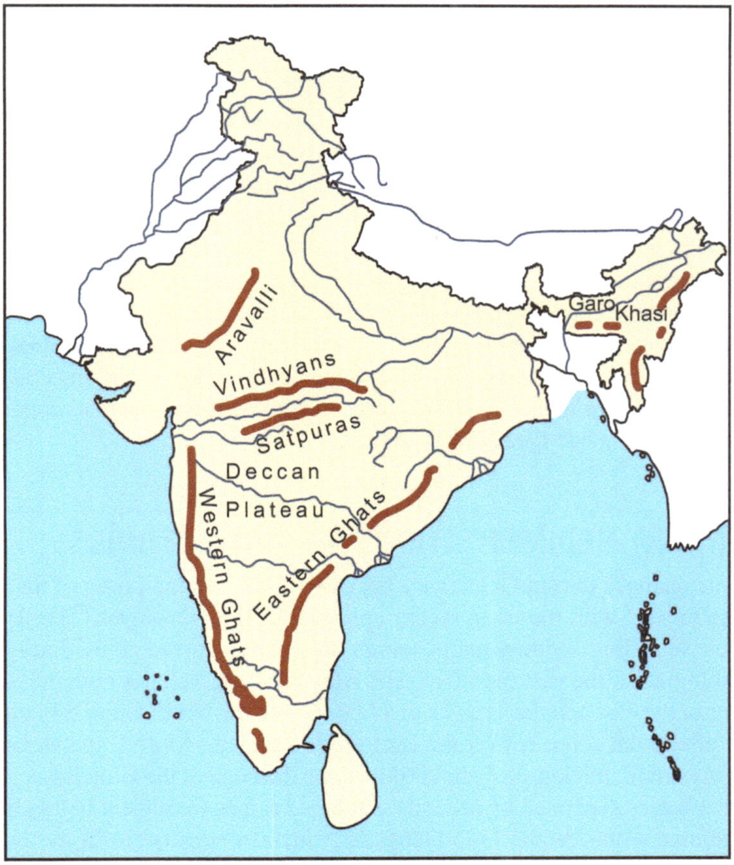

FIG. 19.4

Geomorphotectonic map of India showing occurrences of mountains and plateaus.

FIG. 19.5

Flat-topped hills in the Deccan Trap countries showing more than one erosional surface. Most notable feature is the distinct subhorizontal layering that is parallel to the cap of the uplifted mountain block.

period of denudation reaching the base level of erosion much before its elevation to the mountainous height. Heron (1953) while discussing the physiography of the Aravalli Mountains mentioned about the uplifted 'Jurassic' erosion surfaces. Presumably, the combination of two features resulted because of vertical uplift of much denuded peneplained surfaces. The landscape around Munnar Hills (1450m amsl) lying on the western foot hill of the Anamudi Mountain (Fig. 2.4) might have formed in that way. There are reasons to consider that the present physiographic traits of the Indian Peninsula are examples of active tectonics. According to Kale and Shejwalkar (2008), the flat-topped hills in the Deccan Plateau evolved due to uplift of fault-bounded blocks during the late Quaternary period. Earlier, Radhakrishna (1993) had suggested that the most noticeable feature in the development of spectacular physiographic diversity in the landscape of the Peninsular India was the outcome of sculpturing of the landscape mainly due to the 'neotectonic' activities during late Quaternary period.

19.4 QUATERNARY SEDIMENTATION AND NEOTECTONICS

The morphotectonic changes during Quaternary began with the waning phase of the Siwalik sedimentation and came to an end with the most recent upheaval of the Himalayas. The depocentres had by then shifted to their subsiding southern parts, which ultimately evolved as the Indo-Gangetic Alluvial Plains. Geomorphologically, the vast Indo-Gangetic Alluvial Plain includes both the narrow basin of the Brahmaputra River in the east and the Thar Desert (along with the North Gujarat Plain) in the west. The average thickness of alluvial deposits over the entire Indo-Gangetic Alluvial Plains ranges between 400 and 800 m, with a maximum thickness of about 6 km along the edge of the Himalayas. The belt is divided into a number of subbasins separated by several submerged ridges (basement highs) lying across it.

Quaternary sediments outside the Indo-Gangetic Alluvial Plains occur in the Narmada and Tapti Basins in Peninsular India and along the coastlines. Thick laterite formations (some of which contain rich bauxite deposits) were produced at this time in parts of the central India, Eastern Ghats, and the Konkan coasts in the Western Ghats.

The Thar Desert in the east of the Indus Basin had a fluvial prehistory. This is linked with the establishment of monsoon system over the Subcontinent during the mid-Pleistocene simultaneously with the rise of Aravalli Mountains producing the rain-shadow zone to its west. The series of saline lakes that dot the entire desert land were formed by the segmentation and blocking of river channels due to neotectonic (Quaternary) movements (Roy and Jakhar, 2001).

The Quaternary neotectonic movements caused spectacular geomorphic changes in the entire Subcontinent, primarily through movements of fault-bounded blocks. The Rann of Kachchh in northern Gujarat is a classic example of regional uplift and down faulting during historical times (Roy et al., 2013, 2014). The development of the Ganges-Brahmaputra-Meghna Delta Complex (also known as the Sundarban Delta) is a very important geological landform feature that evolved in three stages (Roy and Chatterjee, 2015) of tectonically influenced delta sedimentation processes during the late Pleistocene.

19.5 SPECTACULAR QUATERNARY DEFORMATIONAL FEATURES

Before winding up the discussion on aspects of geomorphotectonics of the Peninsular India, it would be worthwhile to discuss some spectacular deformation features that have considerably altered the geomorphotectonic landscape features evolved during the late Quaternary. These include (i) terrain evolution and drainage disorganization in western Rajasthan and the neighbouring region; (ii) the abysmally deep Narmada Graben bounded by two mountainous hill ranges, the Satpura and Vindhyan Hills in central India (Fig. 19.4); (iii) the Western Ghats, an unusually high mountain virtually bordering the west coastline of the Peninsular India; and (iv) the late Quaternary uplift of the Shillong Plateau along Dauki Fault.

19.5.1 QUATERNARY DRAINAGE DISORGANIZATION: DISAPPEARANCE OF RIG VEDIC SARASWATI

Several lines of geological evidence confirm the existence of a high-energy fluvial regime in western Rajasthan during the late Quaternary period. The geomorphic description of the extinct river system matches well with the Saraswati River described so vividly in the Rig Veda (Roy and Jakhar, 2001). The Vedic River that presumably flowed parallel to the Aravalli Mountains during its initial stages (Fig. 19.6) had migrated westward during late Quaternary (Holocene) uplift of the Aravalli Mountains. The Quaternary movements, which brought about down-sagging of the northern part of the Aravalli Mountains, also forced the Yamuna River to swap its original course to flow eastward across the flattened 'Mountain range'. The river presumably pirated the Saraswati waters while it drifted eastward to join the River Ganga (Roy and Jakhar, 2001).

FIG. 19.6

Spread of cobbles and gravels at Jayal, in western Rajasthan, presumably deposited by the proto-Saraswati.

Reproduced from Roy and Jakhar (2001) with permission.

19.5.2 NARMADA GRABEN AND ASSOCIATED GEOMORPHOTECTONIC FEATURES

Geomorphic data combined with stratigraphic studies provided significant information to constrain the timing and the extent of fault movements along the lower parts of the Narmada-Son Tectonic Belt. The varying nature and degree of tectonic movements during the late Pleistocene-Holocene period have produced four geomorphic surfaces in the lower Narmada Valley. Two major phases of tectonic movements in a compressive stress regime are recorded along the Narmada-Son Fault, which caused slow synsedimentary subsidence of the basin during late Pleistocene due to differential movement along zone of tectonic dislocation (Chamyal et al., 2002). This was followed by inversion of the basin during the Holocene, marked by differential uplift along the Narmada-Son Fault (Fig. 19.7). The present landscape of the lower Narmada Valley comprises four geomorphic surfaces and has evolved mainly due to tectonic activity along the Narmada-Son Fault in a compressive stress regime (Chamyal et al., 2002). The Holocene period is also marked by basin-inversion, which had earlier undergone subsidence. The inversion of the basin is due to a significant increase in compressive stresses along the Narmada-Son Fault during the early Holocene, resulting in differential uplift of the lower Narmada Valley (Chamyal et al., 2002).

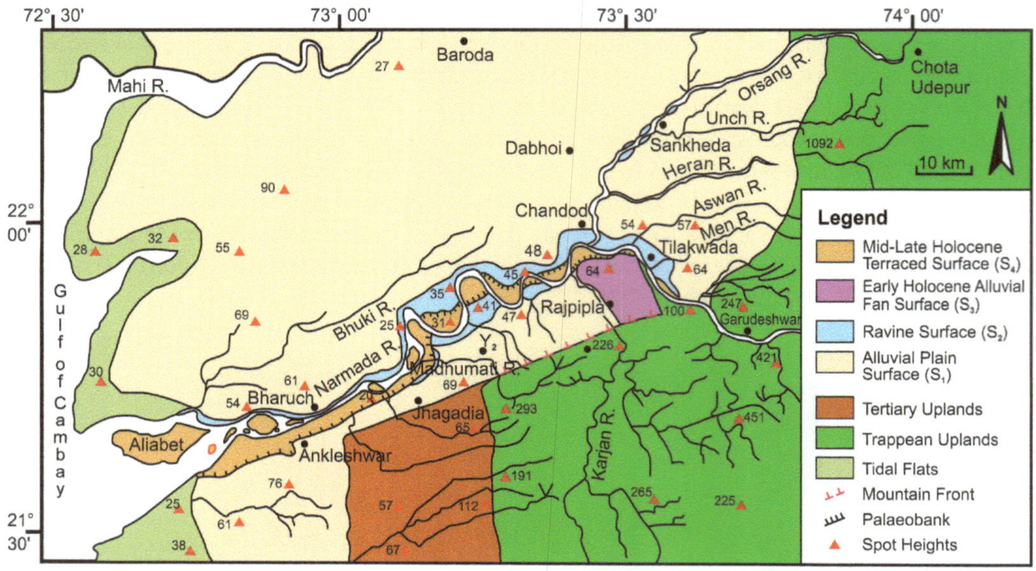

FIG. 19.7

Tectonic map of the lower Narmada Valley. The most notable features are the structural highs and reverse faults to the south of Narmada-Son Fault.

Reproduced from Chamyal et al. (2002) with permission.

Increase of river gradient index in two places in and around Jabalpur indicates vertical uplift due to the north-south compression along the Narmada-Son Fault zone. The geomorphometric database and satellite data interpretation clearly suggest that the Narmada-Son South Fault was active during the Quaternary time. Various tectonically induced landforms were formed all along the Narmada River. These include shifting of river channel (near Amarkantak) and the occurrence of waterfalls, formation of gorge with nearly vertical wall in and around Bhedaghat near Jabalpur. The development of river terraces are common landform features that had formed due to movement along the Narmada-Son Fault.

The morphometric indices and geomorphological observations in the Narmada River suggest the role of active tectonics in the evolution of fluvial landform. The Narmada River is occupied by the rocks belonging to Vindhya and Satpura towards upper reaches and the Deccan Trap towards lower reaches. The river system is controlled by two parallel faults, the Narmada-Son North Fault and the Narmada-Son South Fault (Fig. 19.8). A multitude of roughly north-south trending tear faults offset the Tapti and Narmada Grabens. Recent tectonic activity on these faults is evident from the abrupt eastward swerving for short distances of the northerly flowing tributary rivers and streams joining the Narmada River.

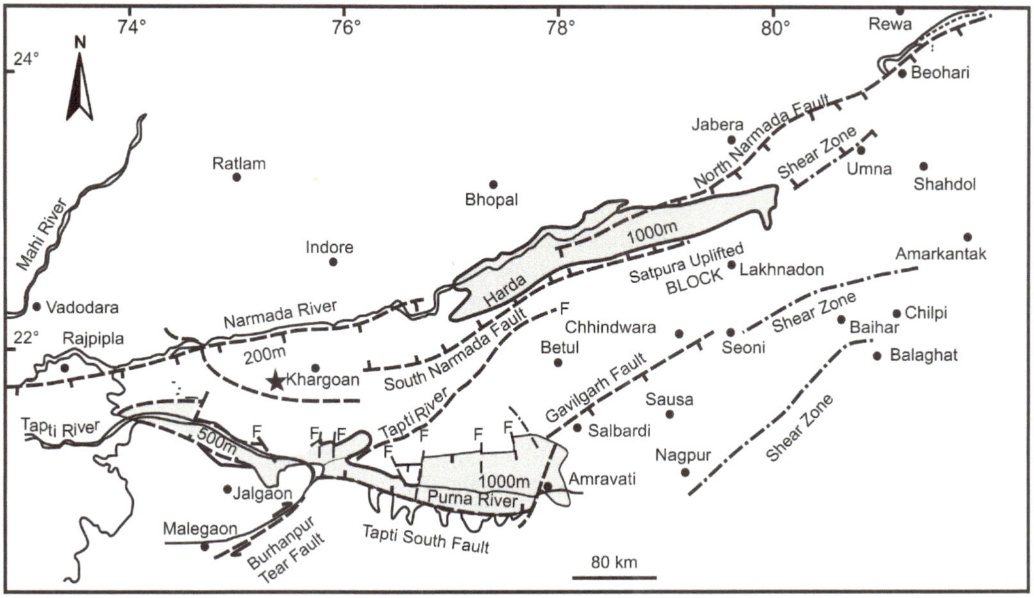

FIG. 19.8

Tectonic framework of the Satputa tectonized belt indicating intimate association between the rift valleys of the Narmada and Tapti Rivers on the two sides. Between the Narmada Graben in the north and the Tapti Graben occurs what has been described as the Central Indian Shear Zone.

Reproduced from Krishnaswamy and Raghunandan (2005) with permission.

The Narmada Rift is define by a pair of parallel normal faults that are dotted with hot springs. Towards the upper reaches, the river shows southwest migrations, while the lower reaches show migration of the river towards north northwest. Migration of river can be assigned to the low resistance bedrock or tectonic uplift along Narmada-Son Fault.

The Satpura Horst shows evidence of uplift to the tune of about 1000 m during Quaternary (Krishnaswamy and Raghunandan, 2005). Matching with the rise, the river valleys are getting deeper and narrower, with simultaneous filling of the river beds by the loads of sediments brought from the eroded river banks. All the different features indicate that the terrain is tectonically active.

19.5.3 WESTERN GHATS

Peninsular India, conventionally thought as a 'Stable Continental Region', has well-preserved records of physiographic diversity and youthful character of the mountain ranges, as significant as those in the Himalayas. This is especially so with the mountain ranges bordering the western margin known as the Western Ghats (also described as the Sahyadri Mountains, Radhakrishna, 1993). The Western Ghats ranges, which run parallel to the West Coast, constitute the most prominent physiographic feature of the Peninsular India. It is a great escarpment that can be traced for more than 1500 km in the form of formidable 'wall' extending from near the Tapti River in the north to Kanyakumari (the southernmost tip of the Peninsula). Lithologically, the Western Ghats include three different rock types: flat-lying Deccan Traps in the north, Dharwar schists and Peninsular Gneisses in the middle, and the granulites and charnockites in the south.

The physiography of the Western Ghats mountain is as 'magnificent and awe-inspiring' as that in the Himalayas, though genetically the two mountains evolved under two different tectonic situation. While the Himalayas evolved entirely during continental collision-related orogeny, the Western Ghats evolved essentially due to the epeirogenic tectonism associated with the vertical movements of the crustal blocks bounded between active faults. The straight line aspect of the West Coast and that of the Western Ghats scarp are very striking. The abrupt termination of the outcrops of the Deccan Trap along the line of the Western Ghats is a clear indication of much younger (post-Deccan Trap) uplift related movements along the fault surfaces (Radhakrishna, 1993). The evidence for this come from the occurrence of long line of thermal springs that occur along the western edge of the Western Ghats scarp. The occurrence of the north-south running linear array of thermal springs is an indication of recent faulting.

19.5.4 DAUKI FAULT AND UPLIFT OF SHILLONG PLATEAU

Shillong Plateau is an important geomorphotectonic feature in the northeast part of Indian Subcontinent. This rectangular-shaped block of the highland is a fault-bounded horst. The east-west trending Dauki Fault borders the Shillong Plateau in the south. The Dauki Fault is a major tectonic feature in the region along which the Shillong Plateau has been lifted up by about 2 km, amsl. In fact, the Dauki Fault forms the east-west running line of the triple-point junction south-east of the Shillong Plateau (Roy and Chatterjee, 2015). The east-west trending subvertical Dauki Fault marks the southern wall-like boundary of the rectangular horst of the Shillong Plateau, thought as a giant 'pop-up' structure (Bilhgam and England, 2001). Studies have indicated that Dauki Fault originated during the Kerguelen Plume outburst at c.117–118 Ma (Fig. 19.9; Roy and Chatterjee, 2015).

Tectonically, a triple-point intersection of lineaments/fractures develops because of local doming up of the Crust during upwelling of the Mantle Plume. The age of the Dauki Fault together with the 3

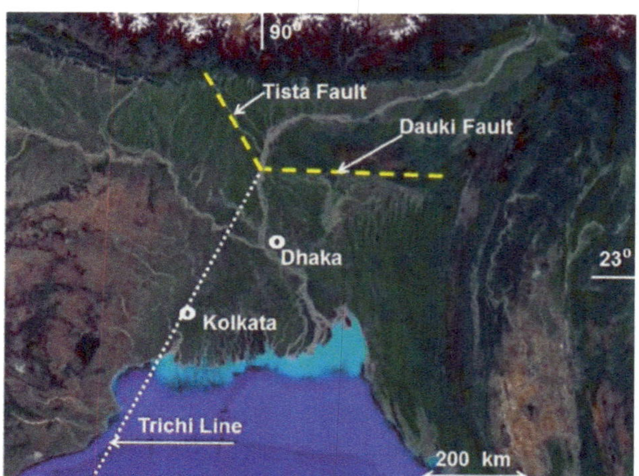

FIG. 19.9

Diagrammatic illustration of Triple-Junction intersection of three lineament features: Dauki Fault, Tista Fault, and the N30°E–S30°W Line (marked by *white*).

Reproduced from Roy and Chatterjee (2015) with permission.

He/4 He ratio (18R/RA) of the flood basalts help correlate these with the upwelling of the Kerguelen Plume (Basu et al., 2001). Besides the evidence of Plume-related magma upwelling, the secondary proof of Plume outburst in the region comes from the coincidence of the north-south trending Jamuna Fault with the Ninety East Ridge, considered the path of the Kerguelen Plume in the Bay of Bengal (Roy et al., 2013, 2014).

REFERENCES

Basu, A. R., Weaver, K. L. and Sengupta, S., 2001 A plume head and tail in the Bengal basin and Bay of Bengal: Rajmahal and Sylhet Traps with surrounding alkali volcanism and the Ninety East Ridge. EOS Trans. Am. Geophys. Union, v.p. VI12A- 0950.

Bilhgam, R. and England, P. 2001 Plateau 'Pop-up' in the great 1897 Assam earthquake. Nature, v.410, pp. 806–809.

Chamyal, L.S., Maurya, S., Bhandari, S., Raj, R., 2002. Late Quaternary geomorphic evolution of the lower Narmada valley, Western India: implications for neotectonic activity along the Narmada-Son Fault. Geomorphology 46, 177–202.

Gansser, A., 1964. Geology of the Himalayas. Interscience Publishers, London. 289 p.

Heron, A.M., 1953. Geology of central Rajputana. Geol. Surv. India Memoir 79. 339 p.

Kale, V.S., 2014. Landscapes and Landforms of India. https://doi.org/10.1007/978-94-017-8029-2_3.

Kale, V.S., Shejwalkar, N., 2008. Uplift along the western margin of the Deccan Basalt Province: is there any geomorphometric evidence? J. Earth Syst. Sci. 117, 959–971.

Kale, V.S., Vaidyanadhan, R., 2014. The Indian peninsular geomorphic landscapes. In: Landscapes and Landforms of India. World Geomorphical Landscapes, Springer Science + Business Media, Dordrecht, pp. 65–87. https://doi.org/10.1007/978-94-017-8029-2_6.

Krishnaswamy, V.S., Raghunandan, K.R., 2005. The Satpura uplift and the paleoclimate of the Holocene and auxiliary evidence from Valmiki Ramayana. J. Geol. Soc. India 66, 161–170.

Radhakrishna, B.P., 1993. Neogene uplift and geomorphic rejuvenation of Indian Peninsula. Curr. Sci. 64, 787–793.

Roy, A.B., Chatterjee, A., 2015. Tectonic framework and evolutionary history of the Bengal Basin in the Indian subcontinent. Curr. Sci. 109, 271–279.

Roy, A.B., Jakhar, S.R., 2001. Late Quaternary drainage disorganization, and migration and extinction of the Vedic Saraswati. Curr. Sci. 81, 1188–1195.

Roy, A.B., Jakhar, S.R., 2002. Geology of Rajasthan: Precambrian to Recent. Scientific Publishers, Jodhpur. 421 p.

Roy, A.B., Chatterjee, A., Chauhan, N.K., 2013. Late quaternary deformation and palaeoseismicity: insight into geomorphotectonic evolution of Kachchh, western Indian Subcontinent. J. Geol. Soc. India 81, 31–40.

Roy, A.B., Chauhan, N.K., Chatterjee, A., 2014. Kachchh Mesozoic domes, western India: study of morphotectono character and evolution. Curr. Sci. 107, 688–693.

Valdiya, K., 1993. Uplift and geomorphic rejuvenation of the Himalaya: in the quaternary period. Curr. Sci. 64, 873–885.

Valdiya, K., 1998. Dynamic Himalaya. Universities Press, Hyderabad. 178 p.

Valdiya, K., 2010. Making of India Geodynamic Evolution. Macmillan. 816 p.

CHAPTER

SEISMICITY IN THE INDIAN SHIELD

20

20.1 CONCEPT OF STABILITY

The Indian Shield is traditionally considered a stable continental block remaining rigid even during the last ~550 million years of the Earth's history, the time period formally described as Phanerozoic. The concept of stability receives support from the fact that there is virtually no record of any orogeny-related post-Precambrian tectonothermal reconstitution of the Indian Crust except in the belt of the Himalayas in the north (Balakrishnan, 1997; Roy, 2004). The idea of stability of the Indian Crust is so deeply entrenched in our thoughts that we tend to characterize the incidence of earthquakes in the region as 'Stable Continental Region' (SCR) seismicity (Rajendran and Rajendran, 2003; Rastogi, 2016; Raval and Veeraswamy, 2014). The basis of such a concept could only be the assumption of the tectonic similarity between the Indian Shield and the Canadian Shield of North America. On the other hand, the following facts tell us a different story.

1. Not only the Himalayas forms the northern margin of the Indian Shield, a considerable part of the Indian Crust now forms parts of the Himalayan edifice. In addition to that, the huge Indo-Gangetic Alluvium Plain resulted due to bending of the 'pre-existing' India in the frontal part of the Himalayas.
2. There are records of high-intensity earthquakes (>7.8 in the Richter scale), which usually occur only in collision-type Suture Zones. Presently, the most seismically susceptible area is Kachchh in northern Gujarat falling in Seismic Zone V (Bureau of Indian Standards, IS 1893:2002) outside the Himalayan terrane.
3. The localized, unusually high heat-flow values (reaching up to $107\,\mathrm{mWm^{-2}}$) over parts of the Peninsular Indian Shield speak of an abnormally hot Crust, unlike the heat-flow pattern in different Shield areas of the world.
4. Bouguer gravity anomaly maps published by the National Geophysical Research Institute, Hyderabad show high positive values (>40 mGal) in several parts of India, which according to Negi et al. (1986) is a reflection of 'mobility' rather than stability (Fig. 2.14).

All the above features indicate that the Indian Shield is tectonically quite unstable (Johnston, 1992) unlike the Canadian and other similar Shield areas of the world. An understanding of the unique character of the Indian Shield in geophysical terms needs a critical analysis of the Phanerozoic geological history (discussed in earlier chapters), which has significantly altered its geological character.

Indian Shield. https://doi.org/10.1016/B978-0-12-809839-4.00020-5

20.2 ASPECTS OF SEISMICITY IN INDIAN SHIELD

The two important aspects that have dramatic impact on the Indian Shield in changing its character as a Stable Continental Region are (i) the development of Lineaments and (ii) the underplating of the lithospheric Crust due to the impingement of number of Plumes. Both these features are discussed in detail in Chapter 16. The Lineaments formed either as new set of fractures or through reactivation of the ancient Precambrian grains during the Jurassic break-up of the Gondwanaland introducing a type of structural heterogeneity in the Crust that is commonly described as block tectonics, in which the movements are transmitted only along narrow zones bordering rigid parts. A classic example of rigid-body movement in a smaller scale is observed in the Sanchor–Nal Sarovar quadrangle, where two major Lineament-bound blocks have undergone rigid-body movement during Pleistocene (Sridhar et al., 1997) (Fig. 20.1a and b). The relative movement between the rigid blocks triggers seismicity, while the individual block remains free from any internal deformation.

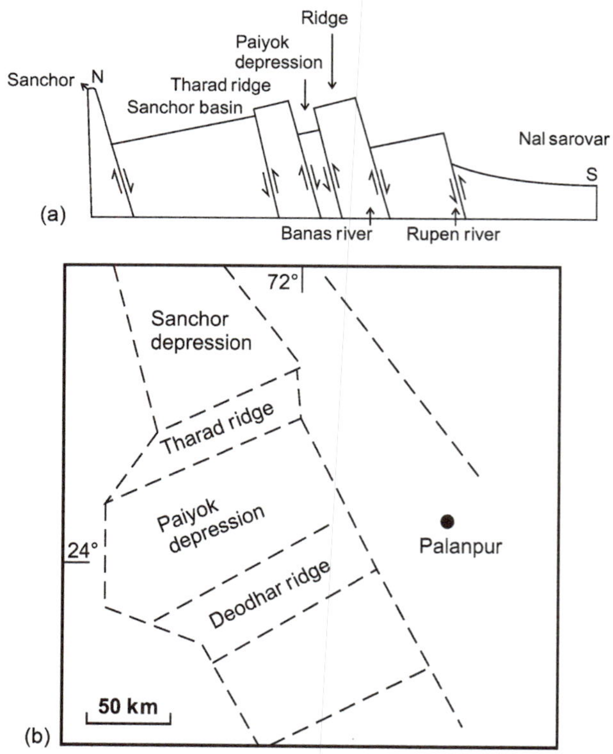

FIG. 20.1

Schematic diagrams illustrating a classic example of block tectonic pattern in the Sanchor–Nal Sarovar quadrangle, northern Gujarat. (a) Cross-sectional view, (b) Plan view.

Reproduced from Sridhar (1997) with permission.

According to Molnar (1988), the study of this aspect of continental deformation has a considerable bearing on the understanding of seismic behaviour of the Peninsular Indian Shield. Murthy (1999) has been able to delineate one such earthquake-free zone in the free-air gravity shaded relief image map of India.

20.2.1 NEOTECTONICS AND ACTIVE FAULT

The epicentres of all the known major present day earthquakes in the Indian Shield occur either on the major Lineaments or reasonably close to that (Murthy and Raval, 2000). This can be taken as a proof of importance of these features in causing earthquakes. The data available with us are, however, pitiably few. To overcome this unfortunate constraint, we will have to get information from the records of neotectonic movements, a considerable part of which took place during the Quaternary period, the last phase of the Earth's history or a little earlier. The term neotectonic is generally used for describing tectonic movements that occurred in the upper part of Cenozoic (Neogene). However, the use of the term 'neotectonic' is better be avoided for the two reasons: (i) the tectonics that describes 'building and reconstitution' of the Earth's Crust is a process of continuity in the Crust-building history, and (ii) the descriptions and interpretations on the initiation of any particular deformation vary from author to author and also from place to place (van Hinsbergen, 2011).

Another commonly used term in areas of recent seismicity is 'active faulting' presumably indicating the tectonic processes occurring during the 'present day'. The same term is being widely used for describing faults in certain parts of India (Verma and Bansal 2016; Rajendran and Rajendran, 2003). It may be interesting to know that outside the Indian Subcontinent, the active faults are defined as those which occur in the vicinity of the Plate boundaries ('Active fault'. *Earthquake Glossary. USGS Earthquake Hazards Program. November 3, 2009)*. The use of the term in the Indian Subcontinent outside the Himalayan terrain may thus appear arbitrary. Accordingly, not much importance is attached now to the active fault studies in India because most of the major faults/fault zones are considered active and have potential to generate large earthquakes in the future.

The study by Copley et al. (2014) on the Tapti Fault in northern Maharashtra is quite revealing. These authors have discussed about the offset of alluvial fan surfaces that they suggest to have resulted due to series of high-magnitude thrust-fault-related earthquakes along the Tapti River basin during the Holocene. The high ratio of fault displacement to length on the alluvial fan offsets implies high stress-drop faulting, as has been observed elsewhere in the Peninsular India.

Considering all these and understanding the fact that the maximum geomorphic changes in the Indian Subcontinent took place only during the Quaternary, more precisely during the late Quaternary (Kothyari et al., 2010), it is advisable to describe all the earthquake producing deformation and seismicity as a Quaternary process.

Along with the records of late Quaternary uplift in the Himalayas, geomorphic changes have been noted in different parts of the Subcontinent. These include drainage disorganization correlated with migration and disappearance of the rivers (Roy and Jakhar, 2001), evolution of saline lakes (Roy, 1999) in western Rajasthan, the formation of abysmally deep valleys bordering the Satpura horst mountain (claimed to be a contemporary event of the Ramayana days) (Krishnaswamy and Raghu Nandan, 2005) or the possible evolution of about 1300 km-long 'great wall' of the Western Ghats standing as a huge scarp facing the Arabian Sea during the Neogene–Quaternary (Radhakrishna, 1993). Besides, there are also reports of block uplifts of the Nilgiri Mountain, Mount Abu, Chota Nagpur Plateau, Shillong Plateau, and many other features from different parts of Peninsular India, all of which must have evolved during the phase of Quaternary deformation. An important point that emerges from the above description is that in all instances of block uplift, downfaulting, and horizontal displacements, the actual movements must have been transmitted along the Lineaments that bound the affected blocks. These earth movements that brought about all geomorphic changes must also have caused countless high-intensity earthquakes, the most of which are without any studied record. The understanding of the past earthquakes is an emerging field of paleoseismic study (Roy et al., 2013, 2017).

20.2.2 PALAEOSEIMICITY

Looking at the poor records of the past earthquakes, there can hardly be any doubt that the information from paleoseismic studies (constrained with proper isotope dates) would be useful in explaining the seismic behaviour of the 'not so stable' Indian Shield.

Scanning through the records of the prehistoric/archaeologic periods, we find some place names that provide indirect hints about calamities that cannot be due to anything else other than earthquakes. Hints come from the names of Harappan Sites like Mohenjo-daro that means 'Mound of Death' or Lothal that means 'Place of Death'. Such massive calamities (described as the Archaeological Catastrophes) could only have caused by incidence of earthquake. At Kalibangan on the Ghaggar River, some palaeoseismic evidence is recorded in a pre-Harappan site hit by an earthquake of over 8 in Richter scale. Some earthquake-related destruction of structures is visible at the archaeological site at Mohenjo-daro. At the Harappan 'metropolis' at Dholavira, in an island in Great Rann of Kachchh, some evidence of earthquake-related destruction is visible at different spots (Fig. 20.2a and b).

(a) (b)

FIG. 20.2

Evidence of ancient earthquake at Harappan metropolis, Dholavira in the Rann of Kachchh: (a) Accumulation of assorted rock debris and soil (a type of seismite) formed during an earthquake, possibly during the last stage of the Dholavira habitation; (b) a tilted wall providing evidence of past earthquake.

Reproduced from Roy, et al. (2013) with permission.

There are a few studies in the Peninsular India that have adduced historical and archaeological evidence of earthquake-induced destruction of temples and other habitats due to the movements along fault surfaces. One such spectacular evidence comes from the sand dune-covered mound of destroyed temples on the bank of the Cauvery River near Mysore, Karnataka (Fig. 20.3a, b) (Roy and Sethumadhav, 2014). The confirmation for the earth movements in the region comes from the studies of Valdiya (1998, 2001) who has highlighted on the tectonic aspects especially referring to the importance of Late Quaternary Lineament-controlled deformation that changed the landform pattern in the region. Roy and Setumadhav (2014) have also emphasized on the shifting of river courses due to uplift-related earth movements that must have ensued repeated earthquakes in the region.

FIG. 20.3

(a) The mysterious spread of sand forming a mound at Talakadu, near Mysore, Karnataka. (b) Evidence of earthquake-related destruction of an ancient temple on the north bank of the meandering Cauvery River at Talakad, near Mysore.

Reproduced from Roy and Sethumadhav (2014) with permission.

Beyond historical and archaeological records, the evidence of earthquake-related deformation comes from the geological confirmation of past earth movement resulting from faulting in recent geological past. Some such deformation in the Kachchh region in the recent past is illustrated below (Fig. 20.4a-c).

Finally, it may be said that the faults are the best evidence of paleoseismicity. The direct geological evidence of past earthquake in sedimentary rocks is provided by seismite, a term used to describe a variety of effects of seismic shocks on unconsolidated sediments during faulting. The scope of the term (seismite) has since been enlarged to include several other earthquake-related depositional features that are dubbed as 'paleoseismogram'.

There is a growing awareness amongst the earth scientists about the studies of faults assuming its importance in understanding the different aspects of paleoseismicity. Tectonic geomorphology is defined as the application of geomorphology to tectonic problems and includes the study of landform assemblages and landscape evolution and development of process-response models for areas and regions affected by fault-related tectonic activity. It is a relatively new branch of applied geomorphology, the study of which requires understanding of a number of disciplines other than geomorphology like structural geology and Quaternary geology. Another emerging subject is Archaeoseismology that is the study of past *earthquakes* deriving from the analysis of *archaeological sites*. There cannot be any better terrane than the Peninsular Indian Shield for initiating such studies.

FIG. 20.4

Illustrations of recent and not so recent deformation from Kachchh in northern Gujarat: (a) A fault scarp in all probability formed during the most recent Bhuj earthquake, Kachchh. Notable feature is the development of inclined alluvial fans facing the fault-related uplift. (b) Trace of Allah Bund Fault marked by low–raised ridge amidst the dried-up Rann west of the Pacham Island, formed during the 16 June 1819 earthquake in the Rann of Kutch; (c) A 3D false-colour Google image of Khadir Island showing prominent scarp face along the northern face.

Reproduced from Roy et al. (2013) with permission.

REFERENCES

Balakrishnan, T.S., 1997. Major tectonic elements of the Indian subcontinent and contiguous areas: a geophysical view. Geol. Soc. India Mem. 38, 155.

Copley, A., Mitra, S., Alastair Sloan, R., Gaonkar, S., Reynolds, K., 2014. Active faulting in apparently stable peninsular India: Rift inversion and a Holocene-age great earthquake on the Tapti Fault. J. Geophys. Res. Solid Earth 119, 6650–6666.

Johnston, A.C., 1992. Interplate not always stable. Nature 355, 213–214.

Kothyari, G., Pant, P., Joshi, M., Khayingshing, L., Malik, J., 2010. Active faulting and deformation of quaternary landform sub-Himalaya, India. Geochronometria 37, 63–71.

Krishnaswamy, V.S., Raghu Nandan, K.R., 2005. The Satpura uplift and the Palaeoclimate of the Holocene and auxiliary evidence from the Valmiki Ramayana. J. Geol. Soc. India 2005 (66), 161–170.

Molnar, P., 1988. Continental tectonics in the aftermath of plate tectonics. Nature 1988 (335), 131–137.

Murthy, Y.S., 1999. Images of the gravity field of India: their salient features. J. Geol. Soc. India 1999 (54), 221–235.

Murthy, Y.S., Raval, U., 2000. Bouguer gravity image of India and the neighbouring ESCAP region: a study of tectonics and natural resources. J. Geophys. 21, 59–70.

Negi, J.G., Pandey, O.P., Agarwal, P.K., 1986. Super mobility of hot Indian Lithosphere. Tectonophysics 11, 135–147.

Radhakrishna, B.P., 1993. Neogene uplift and geomorphic rejuvenation of the India Peninsula. Curr. Sci. 1993 (64), 787–793.50.

Rajendran, K., Rajendran, C.P., 2003. Seismogenesis in the stable continental regions and implications for hazard assessment: two recent examples from India. Curr. Sci. 85, 896–903.

Rastogi, B.K., 2016. Seismicity of Indian Stable continental region. J. Earth Sci. Eng. 3, 57–93.

Raval, U., Veeraswamy, K., 2014. Energy resources and mid-continental (stable continental region) earthquakes over India: association with mantle plume-affected regions. Curr. Sci. 107, 1154–1160.

Roy, A.B., 1999. Evolution of saline lakes in Rajasthan. Curr. Sci. 76, 290–295.

Roy, A.B., 2004. The Phanerozoic reconstitution of Indian Shield as the aftermath of break-up of the Gondwanaland. Gondwana Res. 7, 387–406.

Roy, A.B., Jakhar, S.R., 2001. Late Quaternary drainage disorganization, and migration and extinction of the Vedic Saraswati. Curr. Sci. 81, 1188–1195.

Roy, A.B., Sethumadhav, M.S., 2014. The 'mystic' sand dune-covered temples of Talakad, Mysore district, Karnakata: evidence of earthquake-related destruction. Curr. Sci. 107, 246–254.

Roy, A.B., Chatterjee, A., Chauhan, N.K., 2013. Late quaternary deformation and palaeoseismicity: insight into geomorphotectonic evolution of Kachchh, western Indian Subcontinent. J. Geol. Soc. India 81, 31–40.

Roy, A.B., Bhu, H., Sharma, P., Vaishnav, K., 2017. Deserted nineteenth century Paliwal villages around Jaisalmer, western Rajasthan, India: historical evidence of palaeoseismicity. Curr. Sci. 112, 2017.

Sridhar, V., Chamyal, L.S., Merh, S.S., 1997. Drainage of North Gujarat: present and past. J. Geol. Soc. India 48, 417–426.

Valdiya, K.S., 1998. Late quaternary movements and landscape rejuvenation in southeastern Karnataka and adjoining Tamil Nadu in Southern Indian Shield. J. Geol. Soc. India 1998 (51), 139–166.

Valdiya, K.S., 2001. Tectonic resurgence of the Mysore plateau and surrounding regions in cratonic southern India. Curr. Sci. 81, 1068–1089.

van Hinsbergen, D.J.J., 2011. Short note on the use of Neotectonic and Palaeotectonic nomenclature. Turk. J. Earth Sci. 20, 161–165.

Verma, M., Bansal, B.K., 2016. Active fault research in India: achievements and future perspective. Geomat. Nat. Haz. Risk 7, 65–84.

PART

7

EPILOGUE

CHAPTER

CORRELATION AND EVOLUTIONARY HISTORY OF INDIAN SHIELD

21

21.1 PRECAMBRIAN EVOLUTION OF INDIAN SHIELD

The term 'Indian Shield' very often finds place in literature although several aspects of this Precambrian crustal block continue to be ill-defined or almost unknown. The triangular-shaped Peninsular India that lies south of the Indo-Gangetic Alluvial Plain is traditionally described as the Indian Shield or as the Peninsular Indian Shield, though neither its shape nor the geophysical characters match with the best known Shield areas of the world (Mahadevan, 1994). Available information, however, suggests that the Peninsular India is only a remnant of the pristine Indian Shield, which had much wider geographical extent prior to its decimation during the post-Precambrian Phanerozoic Eon (Roy, 2004).

The limits of the pristine Indian lithospheric Crust are yet to be properly constrained. It is known to extend laterally well beyond the western and the eastern coasts (Kaila and Harinarayan, 1976). The correlation of the granulite facies rocks of the Eastern Ghats Granulite Belt with those of the Antarctica suggests that the two belts were continuous prior to the separation of the Antarctica along with the opening of Bengal basin followed by the opening of the Indian Ocean. Like the granulite belt of the Eastern Ghats Granulite Belt, the Southern Granulite Belt along with Sri Lanka and Madagascar once formed a continuous Precambrian terrane in the south and southwest of Peninsular India, which were separated subsequent due to the break-up of the continents.

The northern limit of the Precambrian Indian Shield is generally placed along the Indus-Tsangpo Ophiolite Belt, which is commonly taken as the zone of collision between the Indian and Eurasian Plates. However, according to Kaila and Harinarayan (1976) and Kaila (1981), the northern limit of India coincides with the southern margin of the Tien Shan 'Mobile Fold belt' that passes south of Ordos and Shanxi blocks, finally turning northeastwards reaching Tianjin, southeast of Peking (now Beijing) (Mahadevan, 1994). The suggestion finds support in the study of Tarim Craton, China, inferring connection of northern India along the northwestern margin of South China Craton during Neoproterozoic (Wu et al., 2017).

Westwards, the boundary of the Precambrian Indian Shield is placed along the southward continuation of Pamir-Alay Fracture Zone passing along the western boundary of the Badakhshan Mountains joining further south with the Chaman Fault system, west of the Sulaiman and Kirthar ranges. The western continental margin is generally placed along a line extending from west of Laccadives to where the Owen Fracture Zone meets the collision boundary on the land.

The concept of 'Greater India' emerged when it was understood that much of the Indian Crust has been accommodated in the growth of the Himalayas. Hosts of models have been proposed suggesting the possible northern extension of the Indian Subcontinent. One of the models came from Argand's suggestion way back in 1924, much before the emergence of the Plate Tectonic model (Fig. 21.1).

Indian Shield. https://doi.org/10.1016/B978-0-12-809839-4.00021-7

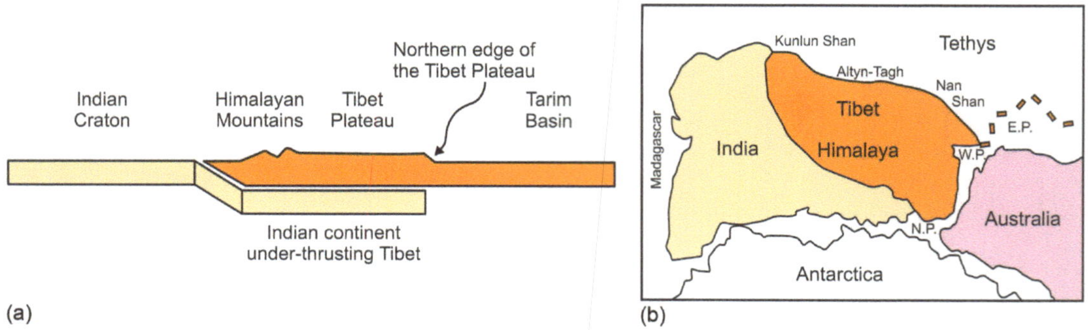

FIG. 21.1

Greater India reconstructions (a) based on underthrusting model of Argand (1924). (b) Redrafting of 'under thrusting model' in which the Himalayas and Tibet are considered to be underlain by the Indian Continental Crust. *NP*, Naturaliste Plateau; *WP*, Wallaby Plateau; *EP*, Exmouth Plateau.

Reproduced from Ali and Aitchison (2008) with permission.

Understandably, the Indian Shield that evolved as a stabilized Precambrian crustal block covered much wider spatial extent than that of the Peninsular India. However, the actual size and shape of this Precambrian crustal block still remains little known or grossly misunderstood because of the successive events of break-up and reconstitution that took place during the Phanerozoic (Roy, 2003, 2004). The Peninsular India constitutes only a part of the 'Greater India' (Argand, 1924) that evolved during the Precambrian and remained virtually undisturbed (by any orogenic event or by thermal perturbation caused by Plume impingement) till the initiation of its break-up in the Phanerozoic.

The earliest proposal on the broad tectonic division of the Indian Shield (should read Peninsular Indian Shield) comes from Fermor (1936) who divided it into two broad tectonometamorphic domains: Charnockitic and Non-Charnockitic belts. A feature of great significance is the cross-cutting relationship between the Charnockitic and the Non-Charnockitic terranes indicated in the *Geological Map of India* published by Geological Survey of India (1993), especially in the case of the Eastern Ghats Granulite Belt that lies astride the structural grains of the three Precambrian crustal blocks (Dharwar, Bastar, and Singhbhum) occurring on its western side.

The combined terranes are known to have their counterparts outside the domain of the Peninsular India. Apart from the possibility of their being 'exotic' or 'suspect terranes' (Radhakrishna, 1989), the granulite and other high-grade metamorphic rocks comprising the Eastern Ghats Granulite Belt and the Southern Granulite Belt (even considering the revised boundaries) appear distinctly different from the components of the Non-Charnockite belt in terms of their lithostratigraphic character, evolutionary history, the age of cratonization, post-cratonization basin formation, and metallogenic history. The Shillong Plateau that occurs in the northeastern part of the Indian Shield is an isolated block surrounded by alluvial covers. Based on the lithological character of the belt, we would prefer to include this as another 'exotic' block, similar to those of the Eastern Ghats Granulite Belt and the Southern Granulite Belts.

Characterization of the Charnockite belts as exotic or suspect terranes implies that the Non-Charnockitic domain constituted the primary crustal block that constituted the 'Proto-India' (Dobmeir and Raith, 2003) or 'Core Central Region' (Mahadevan, 1994) to which the granulite belts were later accreted. In the present description, however, the term Proto-India is preferred as it appears a little more evocative in describing the framework of the Indian Shield.

According to Naqvi et al. (1974), the Proto-India is constituted of a collage of 'Protocontinents' separated by rift valleys (Fig. 21.2). The Protocontinents (numbered 1–6) include Dharwar, Bastar, Singhbhum, Rajmahal, Bundelkhand, and Aravalli. The accreted/exotic terranes, numbered 7, 8, and 9 in Fig. 21.2, include Eastern Ghats Granulite Belt, Southern Granulite Belt, and the truncated Shillong Plateau. The Protocontinents and the accreted terranes together constitute the Indian Shield.

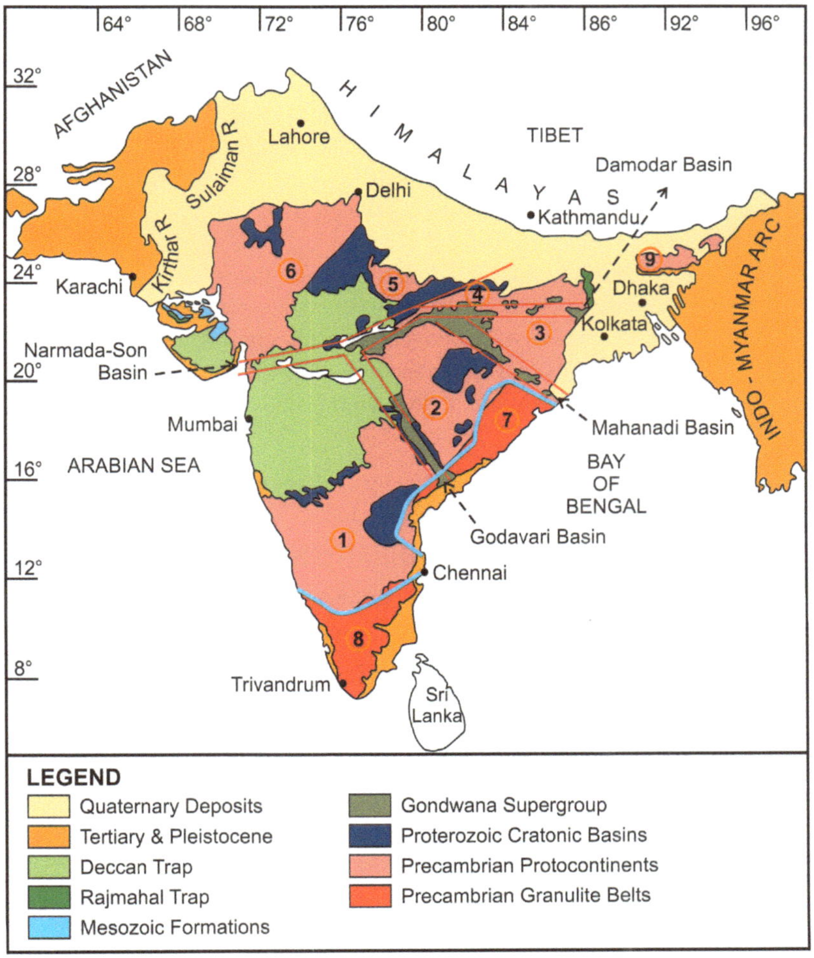

FIG. 21.2

Tectonic divisions of Indian Subcontinent superimposed on the Geological Survey of India Map (1993). *Blue line* marks the boundary between Charnockite and Non-Charnockite terranes. Numbers 1–6 are the different Protocontinents, while 7, 8, and 9 are the three exotic blocks or suspect terranes that were accreted to the Protocontinents.

Detailed studies on the growth history of all the Protocontinents clearly indicate that their evolution was entirely on the sialic Crust that evolved during 3400/3500 Ma or earlier. The geochemical character of the oldest dated granitic gneisses also indicated that the pristine sialic basement evolved much earlier than the suggested age. Based on the available isotope data, an older petrogenetic history is hinted for the primeval Crust from which the present-day granite gneisses had evolved.

Developing on such an old Archaean basement, the lithotectonic components making up the supracrustal components show distinct Precambrian growth history summarized in the following comparative table (Table 21.1).

Even a casual glance at Table 21.1 would indicate a very distinctive history of evolution including the period of cratonization, post-cratonic basin development, and metallogenic traits of the different Protocontinents. The most significant point that emerges in the summary is the fact that none of the features described in individual Protocontinent domain is repeated or followed in any of the adjacent domains. All the described features are restricted entirely and exclusively within the domain of individual Protocontinent, and there is not a single evidence of any feature that cross-cuts the Protocontinent boundary. In view of this, the Protocontinents are also described as the fundamental nuclei. The collage of these Protocontinents constitutes the Proto-India, which with the addition of the 'accretionary terranes' formed the Precambrian Indian Shield.

As a postscript, it may be worth mentioning that the source energy for the crustal build-up and tectonostratigraphic development were essentially intracontinental in nature and in no way responsible for any crustal build-up outside the domain of the concerned Protocontinent. Such being the scale of evolutionary module, it is unthinkable that any global-scale tectonic force had been responsible or was operative in building up the 'Proto-India' or the 'Core Central Region' of the Precambrian Indian Shield.

21.2 PHANEROZOIC RECONSTITUTION

21.2.1 THE CONCEPT OF STABILITY OF INDIAN SHIELD

The Indian Shield, which evolved as a completely cratonized Precambrian Crust, has presumably remained stable even during the Phanerozoic. The concept of stability receives support from the fact that there is virtually no record of any orogeny-related post-Precambrian tectonothermal reconstitution of the Indian Crust, except in the region that now constitutes the Himalayas. In fact, the intrusion of the youngest 'orogenic' granite-forming event recorded from different parts of the Shield dates back to about 850 million years before present. A totally different picture about the crustal stability of the Indian Shield emerges when we consider that a part of the Indian Shield area in the present-day Kachchh (nee Kutch) region of northern Gujarat is stated to occur within the Seismic Zone 5 (cf. Bureau of Indian Standards, IS: 1893:2002). The high seismic susceptibility of the Kachchh region is quite striking considering that the belt does not form a part of any Suture Zone. Irrespective of the cause, the present-day seismicity of the Kachchh region, therefore, poses a serious question on the concept of stability of the Indian Crust as a whole.

The significant information that questions the stability of the Indian Shield during the Phanerozoic comes from the data on the gravity anomaly pattern and the heat-flow values over the entire region of the Indian Shield. The Bouguer anomaly gravity map (prepared by National Geophysical Research Institute, Hyderabad, 1974) shows the dominance of negative Bouguer anomalies over the major part of the Indian Shield. The strongest negative values occur along the arcuate belt of the Sub-Himalayas

Table 21.1 Comparative Study of Different Protocontinents

	Dharwar	Bastar	Singhbhum	Rajmahal	Bundelkhand	Aravalli
History of early evolution	Archaean greenstone belts over basement Peninsular gneiss	Proterozoic fold belts on Archaean basement including greenstone belts	Proterozoic fold belts on Archaean basement including greenstone belts	Proterozoic fold belts on Archaean gneissic terrane	Dominantly granite-gneiss basement	Proterozoic fold belts evolved on granite-gneiss basement
Cratonization	Cratonization at ~2.5 Ga	Cratonization at ~1.8 Ga	Cratonization at ~1.8 Ga	No definite age data	Cratonization at ~ 2.5 Ga	Cratonization at ~ 0.85 Ga
Post-cratonization history	Formation of Proterozoic cratonic basin	Younger Proterozoic cratonic basin	–	–	Younger Proterozoic cratonic basins	Younger Proterozoic cratonic basins
Metallogeny	Gold, banded iron formation, copper, diamond	Copper, banded iron formation, Mn ore, diamond	Copper, banded iron formation, uranium, nickel, platinum	Mica pegmatite	Diamond	Rock-phosphate, lead-zinc, uranium

and the adjacent Indo-Gangetic Alluvial Plains in the south (Fig. 16.5). The belts of positive Bouguer anomalies are recorded in parts of the west coast, over a considerable part in the Saurashtra-Kachchh region of the northern Gujarat and in the southwestern Rajasthan. Similar positive Bouguer anomalies are seen along parts of the east coast and over the belt of Rajmahal-Sylhet Trap in the northeastern part of the Indian Shield. According to Negi et al. (1986), the gravity anomaly pattern observed in the Indian Shield is a reflection of its 'mobility'.

The thermal structure of the Indian Shield like the gravity picture is also considered characteristic of stable continental areas. The average heat-flow value of the present-day India is $-56\,mW/m^2$ that compares well with the global average of continental heat flow of $50\,mW/m^2$ (Gupta, 1982; Mahadevan, 1994). In strong contrast to this view, workers like Negi et al. (1986), Rao et al. (1976), and Singh (1985) suggest that the Indian Shield represents an abnormally hot and considerably thin lithospheric Crust. The higher heat-flow values are recorded sporadically over a considerable part of the Indian Shield. The important high heat-flow regions include the Mesozoic-Cenozoic basins of Assam-Bengal region in the northeast India, parts of the east and west coasts, the Cambay Basin in northern Gujarat, and over the linear belt that runs parallel to the Narmada-Son Lineament zone in the central part of the Indian Subcontinent. The geophysical features, especially the gravity anomaly and the heat-flow values as indicated above, seem to counter the commonly held notion of overall stability of the Indian Shield (not considering the part that now constitutes the Himalayas). In fact, there are a number of features, which make the Indian Shield quite unique unlike any known stable crustal blocks like the Canadian, Ukrainian, Baltic, and some other Shield areas of the world (Roy, 2004).

21.2.2 THE FACTORS CAUSING PHANEROZOIC RECONSTITUTION OF INDIAN SHIELD

The important Phanerozoic geological events, which have grossly altered the shape, size, and lithostratigraphic character of the Indian Shield subsequent to the Jurassic break-up, are the following:

(i) Plume outburst-related features which affected the Indian Crust during Cretaceous-Eocene
(ii) Continental collision during the Himalayan Orogeny
(iii) Post-collision tectonism and seismicity

These are the important global events, which have not only affected the Indian Crust in pruning its shape and size but also have considerably altered its geophysical and tectonic character.

The most significant consequence of the Plume impingements was the chipping of the Gondwanic Indian Crust with simultaneous opening of a new ocean. The earliest stripping event was during the impingement of Kerguelen Plume that triggered separation of Antarctica from the Indian crustal block during the early Cretaceous. Close to the separation of Antarctica, a new hotspot activity was initiated at ~80 Ma, south of Bhubaneswar in the eastern part of the Indian Subcontinent, which is thought to be continuous with the northern extension of 85° East Ridge (Michael and Krishna, 2011). The Bengal Basin in eastern part of the Indian Subcontinent, according to Roy and Chatterjee (2015), evolved coevally with the 85° East Ridge in the Bay of Bengal that is linked with the Afanasi-Nikitin Ridge and is considered a late-Cretaceous product of the Crozet hotspot (Mahoney et al., 1996). The linking of Afanasi-Nikitin Ridge with the Crozet Plume may, however, appear a debatable issue.

On the western side, the separation of the Madagascar from the Indian continental block took place during the Marion Plume activity at 88–90 Ma (Storey et al., 1995; Torsvik et al., 1998). Evidence for this comes from the occurrence of acid and mafic volcanic rocks of similar age in different parts of the central and northern Kerala, St Mary's Island off the Karnataka coast, and also from Madagascar (Kumar et al., 2001; Storey et al., 1995; Torsvik et al., 1998). Further evidence of Marion Plume-related activity is marked by the event of basin closing in the Kachchh region in western India much to the north of dismembered zone in the south, about 84 million years ago, during Upper Cretaceous (Santonian, 83–86 Ma) (Roy et al., 2017). The time of basin inversion coincides with the event of impingement of the Marion Plume under the Indian Lithosphere as it moved over the Plume head around that time. Significantly, this was also the time of opening of a new basin in the Kachchh region in northwestern Gujarat and also in the Bagh Basin in Central India, much to the east of Kachchh (Roy et al., 2017).

Further trimming of the Indian Lithosphere took place during the youngest Reunion Plume during the period of transition from Cretaceous to Paleogene. The main effect of the Reunion Plume outburst was the further modifications of the west coastline with the consequent vertical rise of the Western Ghats. The Reunion Plume activity has also caused widespread basaltic lava spread over a significant part of the west Central India.

With the trimming of the Indian Shield by the Plume outburst-related break-up, evolved the crustal block that continued its northward journey and ultimately collided with the Eurasian Plate. The Himalayas emerged from the grand continental collision event. The evolving Himalayas incorporated a considerable part of the Indian Shield into its domain. The rise of the Himalayas was followed by the formation of a huge depression in the front of the mountain because of the ducking of the leading edge of the truncated 'Shield' underneath the rising mountains. This depression, which is filled up by the two important river systems, the Indus and the Ganga, constituted the great basins of the Indo-Gangetic Alluvial Plain.

Transformation of the Indian Shield did not end with the rise of the Himalayas and the formation of the Indo-Gangetic Alluvial Plain in its south. During the Quaternary or more precisely during the late Quaternary, the triangular southern part of the remnant of the Indian Shield witnessed dramatic geomorphic transformation in different parts. These include drainage disorganization causing migration and disappearance of major rivers (often described as the 'extinct Saraswati River and its tributaries', Roy and Jakhar, 2002), the evolution of saline lakes in western Rajasthan (Roy, 1999), the formation of abysmally deep valleys bordering the Satpura horst mountain (Krishnaswamy and Raghunandan, 2005), and the evolution of the about 1300 km-long 'great wall' of the Western Ghats, standing as a huge scarp facing the Arabian Sea (Radhakrishna, 1993). Besides, there are also reports of block uplifts of the Nilgiri Mountains in the southern part of the Western Ghats, Mount Abu, Chota Nagpur Plateau, Shillong Plateau, and many other mountains and plateau-like features in different parts of Peninsular India. An important point that emerges from the above description is that in all instances of block uplift, down-faulting, and horizontal displacements, the actual movements must have been transmitted along the lineaments that bounded the affected blocks. The Earth movements that brought about all geomorphic changes must also have caused countless high-intensity earthquakes; the most of which are without any studied record. In other words, the Precambrian crustal block that evolved as a stable Indian Shield during the Precambrian underwent spectacular changes in Phanerozoic in terms of its shape and size and in the geophysical traits while turning the crustal block into an unstable, seismically susceptible crustal block, the Indian Subcontinent.

REFERENCES

Ali, J.R., Aitchison, J.C., 2008. Gondwana to Asia: Plate Tectonics, paleogeography, and biological connectivity of the Indian Subcontinent from the idle Jurassic through latest Eocene (166–35 Ma). Earth-Sci. Rev. 88, 145–166.

Argand, E., 1924. In: La technique de l'Asie. Proc. 13th Int. Geol. Congr. 7, pp. 171–372.

Dobmeir, C.J., Raith, M.M., 2003. Crustal architecture and evolution of the Eastern Ghats Belt and adjacent regions of India. In: Yoshida, M., Windley, B.F., Dasgupta, S. (Eds.), Proterozoic East Gondwana Supercontinent Assembly and Breakup. vol. 206. Geol. Soc., London, pp. 145–168. Spl. Publ.

Fermor, L.L., 1936. An attempt at the correlation of the ancient schistose formations of Peninsular India. Mem. Geol. Surv. India 70, 51.

Gupta, M.L., 1982. Heat flow in the Indian peninsula—its geological and geophysical implications. Tectonophysics 8, 71–90.

Kaila, K.L., 1981. Structure and seismotectonics of the Himalaya-Pamir-Hindukush region and the Indian Plate boundary. Geodynamics Series. Am. Geophys. Union 3, 272–293.

Kaila, K.L., Harinarayan, 1976. In: Evolution of the Himalaya based on seismotectonics and deep seismic sounding. Proc. Himalayan Geology Seminar. Section-II. Structure, Tectonics, Seismicity and Evolution, pp. 1–30.

Krishnaswamy, V.S., Raghunandan, K.R., 2005. The Satpura uplift and the Palaeoclimate of the Holocene and auxiliary evidence from the Valmiki Ramayana. J. Geol. Soc. India 66, 161–170.

Kumar, A., Pande, K., Venkatesan, T.R., 2001. The Karnataka Late Cretaceous dykes as products of the Marion hot spot at the Madagascar-India breakup event—evidence from ^{40}Ar-^{39}Ar geochronology and geochemistry. Geophys. Res. Lett. 28, 2715–2718.

Mahadevan, T.M., 1994. Deep Continental Structure of India: A Review. Geological Society of India, Bangalore, p. 569. Mem. 28.

Mahoney, J.J., White, W.M., Upton, B.G.J., Neal, C.R., Scrutton, R.A., 1996. Beyond EM-1: lavas from Afanasy-Nikitin Rise and the Crozet Archipelago, Indian Ocean. Geology 24, 615–618.

Michael, L., Krishna, K.S., 2011. Dating of the 85°E Ridge (northeastern Indian Ocean) using marine magnetic anomalies. Curr. Sci. 100, 1314–1322.

Naqvi, S.M., Rao, D., Narayan, H., 1974. Archaean Protocontinental Growth of the Indian Shield and antiquity of rift valleys. Precambrian Res. 1, 345–398.

Negi, J.G., Pandey, O.P., Agrawal, P.K., 1986. Supermobility of hot Indian Lithosphere. Tectonophysics 11, 135–147.

Radhakrishna, B.P., 1989. Suspect tectono-stratigraphic terrane elements in the Indian subcontinent. J. Geol. Soc. India 34, 1–24.

Radhakrishna, B.P., 1993. Neogene uplift and geomorphic evolution of the Indian Peninsula. Curr. Sci. 64, 787–792.

Rao, R.U.M., Rao, G.V., Narayan, H., 1976. Radioactive heat generation and heat flow in the Indian shield. Earth Planet. Sci. Lett. 30, 57–64.

Roy, A., Chatterjee, B., Alokesh, C., Chauhan, N.K., 2017. Geological evolution of Kachchh: an epitome of successive Phanerozoic events. Curr. Sci. 112, 1051–1056.

Roy, A.B., 1999. Evolution of saline lakes in Rajasthan. Curr. Sci. 76, 290–295.

Roy, A.B., 2003. Geological and geophysical manifestations of the Reunion Plume-Indian Lithosphere interactions—evidence from NW India. Gondwana Res. 6, 487–500.

Roy, A.B., 2004. The Phanerozoic reconstitution of Indian Shield as the aftermath of break-up of the Gondwanaland. Gondwana Res. 7, 387–406.

Roy, A.B., Chatterjee, A., 2015. Tectonic framework and evolutionary history of the Bengal Basin in the Indian subcontinent. Curr. Sci. 109, 271–279.

Roy, A.B., Jakhar, S.R., 2002. Late Quaternary drainage disorganization, and migration and extinction of the Vedic Saraswati. Curr. Sci. 81, 1188–1195.

Singh, R.N., 1985. Thermal structure of the Indian Shield. Indian J. Earth Sci. 12, 155–158.

Storey, M., Mahoney, J.J., Sauders, A.D., Duncan, R.A., Kelley, S.P., Coffin, M.F., 1995. Timing hot spot-related volcanism and the breakup of Madagascar and India. Science 267, 852–855.

Torsvik, T.H., Tucker, R.D., Ashwald, L.D., Eidea, E.A., Rakotosolofod, N.A., de Wite, M.J., 1998. Late Cretaceous magmatism in Madagascar: palaeomagnetic evidence for a stationary Marion hotspot. Earth Planet. Sci. Lett. 164, 221–232.

Wu, G., Xiaoa, Y., Bernard, B., Mac, D., Lic, X., Zhu, G., 2017. Ca. 850 Ma magmatic events in the Tarim Craton: age, geochemistry and implications for assembly of Rodinia supercontinent. Precambrian Res. https://doi.org/10.1016/j.precamres.2017.10.020.

Index

Note: Page numbers followed by *f* indicate figures, *t* indicate tables, and *b* indicate boxes.

H

N

O

P

Q

R

S

T

Kolibri 360 GmbH
Ferdinand-Jühlke-Straße 7
99095 Erfurt, Deutschland
produktsicherheit@kolibri360.de